Protein Kinases

Frontiers in Molecular Biology

SERIES EDITORS

B. D. Hames

Department of Biochemistry and Molecular Biology
University of Leeds, Leeds LS2 9JT, UK

AND

D. M. Glover

Department of Biochemistry
University of Dundee, Dundee DD1 4HN, UK

OTHER TITLES IN THE SERIES

Protein Kinases

EDITED BY

James Robert Woodgett

*Ontario Cancer Institute, Division of
Cellular and Molecular Biology,
Toronto, Ontario, Canada*

OXFORD UNIVERSITY PRESS
Oxford New York Tokyo

Oxford University Press, Walton Street, Oxford OX2 6DP

Oxford   New York
Athens   Auckland   Bangkok   Bombay
Calcutta   Cape Town   Dar es Salaam   Delhi
Florence   Hong Kong   Istanbul   Karachi
Kuala Lumpur   Madras   Madrid   Melbourne
Mexico City   Nairobi   Paris   Singapore
Taipei   Tokyo   Toronto
and associated companies in
Berlin   Ibadan

Oxford is a trade mark of Oxford University Press

Published in the United States
by Oxford University Press Inc., New York

A catalogue record for this book is available from the British Library

Library of Congress Cataloging in Publication Data
Protein kinases / edited by James Robert Woodgett. — 1st ed.
p. cm. — (Frontiers in molecular biology)
Includes bibliographical references.
1. Protein kinase.   I. Woodgett, James Robert.   II. Series.
QP606.P76P378   1994   574.19'25 — dc20   94–25411
ISBN 0 19 963409 2 (Hbk)
ISBN 0 19 963408 4 (Pbk)

Typeset by
Footnote Graphics, Warminster, Wilts.
Printed in Great Britain by
The Bath Press, Avon

Preface

A statistical fisherman plumbing the depths of genomic sequences is more likely to land a protein kinase gene than anything else. Since the first of these enzymes was discovered over 35 years ago, estimates of the total number of variants per organism have been steadily upgraded. Current 'guesstimates' predict over a thousand, in which case this catalytic unit will represent around 1% of our entire genetic repertoire. What is so unique about this structure to justify such a huge investment in the gene pool? Why are there so many? What do they do and how are they controlled?

The answers to such basic questions are best found by consideration of the family of enzymes as a whole. By definition, all protein kinases are capable of transferring the terminal phosphate group from ATP to an amino acid residue within a protein. However, beyond this shared catalytic process, protein kinases are as diverse in function as they are in phylogenetic distribution. They play a role in virtually all regulatable biological processes, from ion transport to metabolic pathways to DNA replication to differentiation. Their versatility begets their ubiquity and they pervade every area of biology. To counter the sense of overwhelming complexity associated with the protein kinase field, one of the aims of this book was to create a manageable resource, a compilation of protein kinase information in an accessible form.

By convention, the book is divided into chapters. However, this segregation serves primarily to identify authors with their contributions. Protein kinases are not easily pigeon-holed. By their very nature they affect and are affected by other processes. Therefore, the book's organization is arbitrary, playing second fiddle to the collection of principles discerned from the study of many different protein kinases that are often equally applicable to others. For example, the structural aspects of this family entered a new era with the elucidation of the first protein kinase crystal structure. Owing to the common thread of structural conservation in the super-family, this milestone provided molecular insights into a variety of areas including the mechanisms by which protein kinase activities are modulated. At the time of writing the structures of three distinct protein kinases have been determined. This will likely increase at a rate comparable to the discovery of new enzymes.

The book is not meant to be, nor is, encyclopaedic. Indeed, since perhaps only 15% of all protein kinases expressed in a given organism have been isolated to date, such an opus would be rather premature. However, the discovery of the first few protein kinases had a greater relative impact than today's new recruits. The enzymes thus far characterized have provided paradigms likely to be applicable to tomorrow's kinases. The idea was that now is good a time to take stock of this extended family before we are all overwhelmed by sheer complexity. It is hoped that the drawing together of seemingly disparate pieces of information will lead to better assimilation and understanding of these remarkable proteins and stimulate further study into their physiological roles.

Toronto
January 1994

J.W.

Contents

3 Protein kinase C and its relatives   68

PETER J. PARKER

6 Receptor protein-tyrosine kinases   177

TONY HUNTER and RICHARD A. LINDBERG

7 Non-receptor protein-tyrosine kinases   212

SARA A. COURTNEIDGE

8 Genetic approaches to protein kinase function in lower eukaryotes   243

SIMON E. PLYTE

Contributors

JOSEPH AVRUCH
Diabetes Research Labs, MGH East, Building 149, 13th Street, Charlestown, MA 02129, USA.

SARA A. COURTNEIDGE
Differentiation Programme, European Molecular Biology Laboratory, Meyerhofstrasse 1, 69012 Heidelberg, Germany.

MAREE C. FAUX
St Vincent's Institute of Medical Research, 41 Victoria Parade, Fitzroy, Victoria 3065, Australia.

KATHLEEN L. GOULD
Department of Cell Biology, School of Medicine, Vanderbilt University, Nashville, TN 37232, USA.

COLIN HOUSE
St Vincent's Institute of Medical Research, 41 Victoria Parade, Fitzroy, Victoria 3065, Australia.

SHU-HONG HU
St Vincent's Institute of Medical Research, 41 Victoria Parade, Fitzroy, Victoria 3065, Australia.

TONY HUNTER
Molecular Biology and Virology Laboratory, The Salk Institute, PO Box 85800, San Diego, CA 92186, USA.

BRUCE E. KEMP
St Vincent's Institute of Medical Research, 41 Victoria Parade, Fitzroy, Victoria 3065, Australia.

JOHN M. KYRIAKIS
Diabetes Research Labs, MGH East, Building 149, 13th Street, Charlestown, MA 02129, USA.

RICHARD A. LINDBERG
Department of Immunology, AMGEN, 1840 Dehavilland Drive, Thousand Oaks, CA 91320, USA.

ANTHONY R. MEANS
Department of Pharmacology, Duke University Medical Center, 043 CARL Building, Research Drive, Durham, North Carolina 27710, USA.

KEN I. MITCHELHILL
St Vincent's Institute of Medical Research, 41 Victoria Parade, Fitzroy, Victoria 3065, Australia.

PETER J. PARKER
Protein Phosphorylation Laboratory, ICRF Laboratories, 44 Lincolns Inn Fields, London WC2A 3PX, UK.

SIMON E. PLYTE
Division of Cell and Molecular Biology, Ontario Cancer Institute, 500 Sherbourne Street, Toronto, Ontario M4X 1K9, Canada.

ELZBIETA RADZIO-ANDZELM
Department of Chemistry, 0654 University of California at San Diego, 9500 Gilman Drive, La Jolla, CA 92093-0654, USA.

SUSAN S. TAYLOR
Department of Chemistry, 0654 University of California at San Diego, 9500 Gilman Drive, La Jolla, CA 92093-0654, USA.

TONY TIGANIS
St Vincent's Institute of Medical Research, 41 Victoria Parade, Fitzroy, Victoria 3065, Australia.

Abbreviations

AP-1	activator protein-1
ATP	adenosine triphosphate
BDNF	brain-derived neurotrophic factor
CAK	Cdk-activating kinase
CAP	catabolite gene activation protein
CAT	chloramphenicol acetyl transferase
CDC	cell division cycle
Cdk	cyclin-dependent protein kinase
CNBr	cyanogen bromide
CRE	cAMP response element
CREB	cAMP response element binding protein
CSF	colony stimulating factor
DAG	diacylglycerol
DCCD	dicyclohexyl carbodiimide
EGF	epidermal growth factor
EGFR	epidermal growth factor receptor
ERK	extracellular signal regulated kinase
FGF	fibroblast growth factor
FN	fibronectin
FPLC	fast protein liquid chromatography
FSBA	5-p-fluorosulphonylbenzoyladenosine
GAP	GTPase activating protein
GGF	glial growth factor (heregulin, NDF)
GS	glycogen synthase
GSK-3	glycogen synthase kinase-3
HGF	hepatocyte growth factor
I-2	inhibitor-2
Ig	immunoglobulin
IGF	insulin-like growth factor
ISPK	insulin stimulated protein kinase
MAP	mitogen activated protein (microtubule associated protein)
MARCKS	myristoylated alanine-rich C kinase substrate
MBP	myelin basic protein
MEK	MAP, Erk kinase (MAP kinase kinase)
MLC	myosin light chain
MPF	maturation (M-phase) promoting factor
NDF	neu-differentiation factor
NGF	nerve growth factor

NMT	N-myristoyl transferase
PCNA	proliferating cell nuclear antigen
PCR	polymerase chain reaction
PDBu	phorbol dibutyrate
PDGF	platelet-derived growth factor
PI3-kinase	phosphatidylinositol 3' kinase
PKA	cAMP-dependent protein kinase
PKC	protein kinase C
PKI	protein kinase inhibitor (of PKA)
PLC	phospholipase C
PMA	phorbol 12-myristate 13-acetate
PP-1	protein phosphatase-1
PP-2A	protein phosphatase-2A
PRE	pheromone response element
PtdIns	phosphatidylinositol
PTK	protein-tyrosine kinase
Rsk	ribosomal S6 protein kinase
RTPK	receptor protein-tyrosine kinase
S6K	S6 protein kinase
SAPK	stress-activated protein kinase
SCF	stem cell factor
SDS-PAGE	sodium dodecyl sulphate polyacrylamide gel electrophoresis
SH2	Src homology domain 2
SH3	Src homology domain 3
SKAIPS	S6 kinase autoinhibitory pseudosubstrate peptide
TGF	transforming growth factor
TPA	tetradecanoylphorbol acetate
TRE	tetradecanoylphorbol acetate response element
VEGF	vascular endothelial growth factor
VPC	vulval precursor cell

1 | Cyclic AMP-dependent protein kinase

SUSAN S. TAYLOR and ELZBIETA RADZIO-ANDZELM

1. Introduction

The protein kinase family contains hundreds of diverse but related enzymes that regulate nearly all aspects of growth, differentiation, and division in the eukaryotic cell (1, 2). Phosphorylase kinase was the first specific protein to be discovered and purified in the 1950s (3, 4), and it was followed a decade later by cAMP-dependent protein kinase (PKA) (5). Both phosphorylase kinase and PKA contain multiple subunits. Phosphorylase kinase is a stable $\alpha_4\beta_4\gamma_4\delta_4$ complex, and its catalytic γ-subunit shows extensive sequence similarities with PKA (6, 7). Phosphorylase kinase can be activated either allosterically by calcium binding to its δ-subunit (calmodulin) or by phosphorylation (for review see 8). Like phosphorylase kinase, PKA is composed of different subunits. The inactive holoenzyme of PKA is a hetero-tetramer containing both regulatory (R) and catalytic (C) subunits. However, unlike phosphorylase kinase, activation of PKA involves dissociation of the subunits (for reviews see 9, 10). This mechanism of activation, in fact, appears to be rather distinct within the protein kinase family. Specifically, activation in response to PKA is mediated by cAMP binding with high affinity to the R-subunit, and this promotes dissociation of the complex into an $R_2(cAMP)_4$ dimer and two free and active catalytic subunits (11–13). The affinity of regulatory and catalytic subunits for one another is decreased by 4–5 orders of magnitude in response to cAMP binding (14).

Localization is also important for kinase function. PKA is primarily a cytoplasmic enzyme. When the holoenzyme is activated, however, the free catalytic subunit not only phosphorylates cytoplasmic substrates, it can also migrate into the nucleus (15–18) where it can presumably phosphorylate proteins that are important for the regulation of gene transcription such as the cAMP response element binding protein (CREB) (19). Thus, the R-subunit not only serves as an inhibitor of catalytic activity, it is also a cytoplasmic anchor that prevents the catalytic subunit from entering the nucleus (17, 18).

The specificity of PKA was first elucidated by mapping *in vivo* phosphorylation sites and then studying peptide analogues of those sites (20–22, see also Chapter 2). In this way, it was deduced that PKA has a preference for basic residues preceding the phosphorylation site. The two general consensus sequences are Arg–Arg–

Xaa–Ser/Thr–Yaa and Arg–Xaa–Arg–Xaa–Xaa–Ser/Thr–Yaa, where Xaa is any residue and Yaa is a hydrophobic amino acid (23). By convention, the site of phosphotransfer is designated as the P-site with the P−2, P−3, and P−6 sites having a preference for Arg. The P+1 site is typically a hydrophobic residue.

2. Inhibitors of the catalytic subunit

Like most protein kinases, it is essential that the enzyme be sequestered in an inactive state in the absence of the activation signal. Two different classes of molecules are known to inhibit the catalytic subunit. The first are the R-subunits, discussed above, that mediate reversible inhibition in the absence of cAMP. The other class of inhibitors are the heat-stable protein kinase inhibitors (PKIs), small proteins that bind tightly to the free catalytic subunit (24). As yet, no physiological mechanism has been elucidated for reversing the inhibition by PKI nor has the physiological role of the PKIs been clearly deduced.

At least four unique genes code for the R-subunits, and these constitute two general families. The R_I-subunits elute first from anion exchange residues, are not autophosphorylated, and require the high-affinity binding of ATP for the formation of stable holoenzyme (10, 25–27). Two genes, $R_I\alpha$ and $R_I\beta$, are known (28, 29). The R_{II}-subunits are autophosphorylated and are slightly larger (30, 31). $R_{II}\alpha$ and $R_{II}\beta$ genes have been identified (32, 33). The R-subunits all have a well-defined domain structure as summarized in Fig. 1 and contain both variable and conserved segments. In all cases, the carboxy-terminus is composed of two tandem cAMP binding domains. These arose by gene duplication, and because these domains are related to the cAMP binding domain of the catabolite gene activation protein (CAP) (34), they can be modelled based on the crystallographic coordinates of CAP (35). Each R-subunit also contains a segment that resembles a protein substrate, and this segment is essential for inhibition of the catalytic subunit. Mutations in this region of both the R_I and R_{II} subunits interfere with holoenzyme formation (36–38). The inhibitor regions of various R-subunits are compared in Table 1. As seen in this

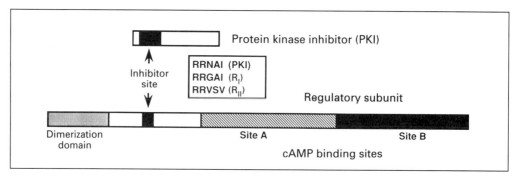

Fig. 1 Comparison of the domain structures of the R-subunits and PKI. The consensus sequence located at the P−3 to P+1 inhibitor site in each protein is also indicated in the box. In the R_I-subunit and PKI, this is a pseudophosphorylation site while this region constitutes an autophosphorylation site in the R_{II}-subunit.

Table 1 Comparison of sequences in the inhibitor regions of type I and II regulatory subunits and PKI. The consensus site flanking the P site is blocked. Also blocked are the P−11 Phe and the P−6 Arg in PKI alpha that are known to be important for the high affinity binding of PKI.

Pseudophosphorylation site

```
                10                    20
T D V E S V I S S F A S S A R A G  R R N A L  P D I Q S S L A T G G S P   PKI alpha: Ki = 0.14 nM (48)
T D V E T T Y A D F I A S G R T G  R R N A I  H D I L V S S A S G N S N   PKI beta: Ki = 0.2 nM (51, 52)
        T T Y A D F A S G R T G    R R N A I  H D                         PKI alpha(5–24): Ki = 2.3 nM (50)
        T T Y A D A I A S G R T G  R R N A I  H D                                          Ki = 270 nM (54)
        S V I S S F A S S A R A G  R R N A I                              PKI beta(5–22): Ki = 90 nM (48)

                                   L R R A S L G                          Km = 4.7 µM (21)
                                   L R R A A L G                          Kd = 270 nM (21)
                                     -3 -2  P+1
         -11        -6
```

```
         80              90
E D E I S P P P P N P V V K G R R  R R G A I  S A E V Y T E E D A A S Y   Bovine alpha^a (27)
D E E I S P T P P N P V V K A R R  R R G G V  S A E V Y T E E D A V S Y   Mouse beta (29)
K A T E K V E A Q N N N N I T R R  R R G A I  S S E P L G D K P A T P L   Dictyostelium (139)
```

R_I subunits (app. K_dS = 0.2 nM) (138)

```
S E D E E – D L D V F I P G R F D  R R V S V  C A E T Y N P D E E E E D   Bovine alpha (31)
E A A E A G A F N A F V I N R F T  R R A S V  C A E A Y N P D E E E D D   Rat beta (32)
T R E K T S T P P L P M H F N A Q  R R T S V  S G E T L Q P P N N F D D D Yeast BCY 1 (140)
```

R_II subunits (app. K_dS = 0.3 nM) (138)

Phosphorylation site

^a Mouse and bovine sequences in this region are identical.

comparison, the R_{II}-subunits contain an autophosphorylation site in this auto-inhibitor segment while the R_I-subunits have a pseudophosphorylation site. This segment occupies the peptide binding site of the catalytic subunit in the holoenzyme complex and prevents other substrates from binding. The R-subunits thus function as competitive inhibitors with respect to peptide substrates (14, 39, 40). In addition to this small consensus sequence flanking the phosphorylation site, P−3 through P+1, the high affinity binding of the R-subunits requires additional sites of interaction, and these most likely lie carboxy-terminal to the consensus site (41, 42). The amino-terminal portion of the R-subunits constitutes a dimerization domain that is important for docking of the holoenzyme to specific cytoplasmic locations (43). In the case of the R_I-subunits, the two protomers in the dimer are linked covalently by two interchain disulphide bonds (44). Disulphide bonded hetero-dimers between $R_I\alpha$ and $R_I\beta$ were also identified recently (45).

Several different isoforms of PKI have also been identified. PKIα was first purified in 1973 (46) and later sequenced (47). Its physical properties and inhibitory properties have been characterized extensively using a variety of techniques and by carefully analysing many peptide analogues (24). More recently a PKIβ isoform was identified (48), and there are at least two variants of PKIβ due to alternate splicing (49). The PKIs, like the R-subunits, contain a substrate-like sequence and inhibit the catalytic subunit by a similar mechanism (50, 51). Like the R_I-subunits, they have a pseudophosphorylation site and formation of a stable complex requires the high affinity binding of MgATP (52, 53). The high affinity binding of PKI, in addition to the consensus site, requires residues on the amino-terminal side of the consensus site. Of particular importance are the P−11 Phe and the P−6 Arg (24, 54).

3. The catalytic subunit structure

For a variety of reasons, the catalytic subunit of PKA is probably the best under-stood biochemically of all the protein kinases. It is one of the smallest protein kinases (7) and can be purified readily (55). Its unique dissociative mechanism of activation makes it one of the simplest protein kinases. It can be expressed readily in *E. coli* (56). For all of these reasons a great deal of information has accrued about this protein kinase since its discovery (for reviews see 9, 57, 58). The recent solution of a crystal structure of the catalytic subunit for the first time puts all of this biochemical information into a structural context (59, 60). This relatively simple protein kinase thus provides not only a detailed description of one protein kinase, but also establishes many general rules that will almost certainly apply to the entire protein kinase family (61, 62).

3.1 Isoforms

Three genes are known that code for catalytic subunits in mammalian tissues. The Cα-subunit is expressed in most tissues and is the major form of the catalytic subunit (7, 63). Expression of the Cβ-subunit is more tissue-specific with levels being particularly high in neuronal tissues (64–66). Cγ was found in testis, and its expression appears to be the most limited (67). The three isoforms are highly conserved with most of the variations located in the first 40 residues.

The catalytic subunit is highly conserved in all vertebrates, as well as in *Aplysia* (68), *Drosophila* (69), and *C. elegans* (70). In more primitive phyla, such as yeast, the three catalytic subunits, TPK1, 2, and 3, are larger than their mammalian counterpart because of additional sequences at the amino-terminus (71). The catalytic subunit was recently isolated from *Dictyostelium discoideum*, and this enzyme contains over 300 additional residues at its amino-terminus (72, 73).

3.2 General domain structure

The general organization of the catalytic subunit is summarized in Fig. 2, where the overall domain structure is correlated also with the intron/exon boundaries of the mouse catalytic subunit (74). The catalytic core, conserved throughout the protein kinase family, extends from residues 40 to 300. It includes an ATP binding domain (residues 40–127) and a peptide binding/catalytic domain. The amino-terminal 40 residues consist of two segments. The amino-terminal Gly is myristoylated (75), and exon I encodes a myristoylation motif (residues 1–14). Exon II encodes a helix motif that precedes the ATP binding domain. The ATP binding domain is encoded by exons III–V, while the peptide binding/catalytic domain is encoded by exons VI–IX. The final 40 residues are encoded by the last exon. The two forms of the catalytic subunit in *Aplysia* result from alternate splicing within the ATP binding domain (68); however, alternate splicing does not account for variations in the Cα, Cβ, and Cγ isoforms in vertebrates.

3.3 General architecture

The overall folding of the polypeptide chain of the mouse recombinant catalytic subunit and the location of various conserved residues are shown in Fig. 3. The

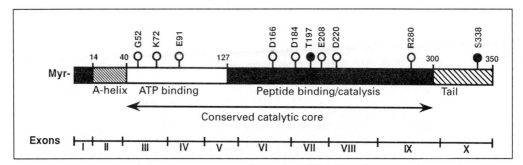

Fig. 2 General domain structure of the catalytic subunit. The catalytic core is conserved throughout the protein kinase family, and the invariant or highly conserved residues that extend throughout this core are indicated by closed circles or open circles; phosphorylation sites are closed circles. The location of the intron/exon borders are indicated below with the exons labelled as I–X.

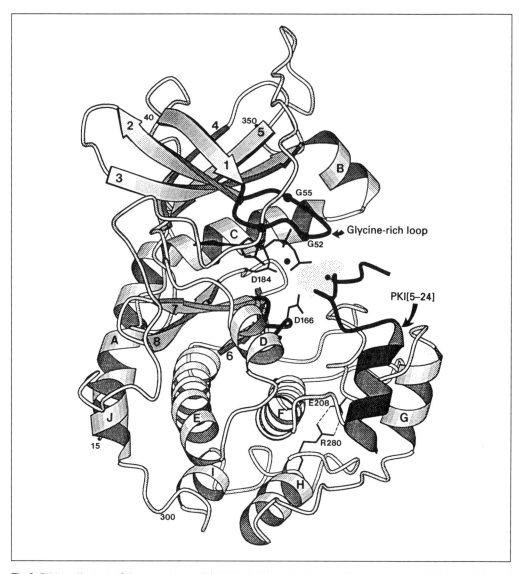

Fig. 3 Ribbon diagram of the mouse recombinant catalytic subunit cocrystallized with an inhibitor peptide and MgATP. The conserved catalytic core extends from residue 40 to 300. The amino-terminal extension contains a myristoylation motif with the amino-terminal glycine being acylated. In this structure of the recombinant enzyme, residues 1–14 are not seen. The long Λ-helix that precedes the conserved core (residues 10–31) may also be a conserved feature of most protein kinases (79). The carboxy-terminal tail wraps as an extended chain around the surface of both lobes. ATP and the activating Mg^{2+} ion that bridges the β- and γ-phosphates are indicated in bold. The inhibitor peptide, PKI(5–24), is shaded dark with the methyl group of Ala at the P position shown as a dot. The site where phosphotransfer takes place is shaded. Several conserved side-chains are also indicated (Asp166, Asp184, Glu208, and Arg280). The conserved glycines in the glycine-rich loop (Gly50, 52, and 55) and Lys72 are indicated by black circles. The glycine-rich loop linking β-strands 1 and 2 and the catalytic loop between β-strands 6 and 7 are highlighted in black.

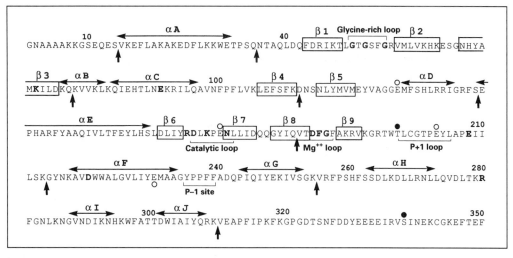

Fig. 4 The sequence of the mouse Cα-subunit correlated with the crystal structure. α-Helices are indicated by arrows; β-strands are boxed; highly conserved residues are in bold; phosphorylation sites are indicated by black dots; intron/exon junctions are indicated by arrows. Charged residues important for recognition of the P−2, P−3, and P−6 arginines are indicated by open circles. The glycine loop, the catalytic loop, the Mg^{2+} binding loop, the P+1 loop, and the P−11 site are all indicated.

general features of secondary structure are correlated with the sequence in Fig. 4. The structure shown in Fig. 3 is a ternary complex where the recombinant catalytic subunit was co-crystallized with MgATP and a 20-residue inhibitor peptide from PKI (76, 77).

The catalytic subunit, in general, consists of two lobes, and catalysis takes place at the cleft between the two lobes. The small lobe in the conserved catalytic core (residues 40–127) is dominated by an antiparallel β-sheet and is associated primarily with ATP binding. In contrast, the large lobe (residues 128–300) is dominated by α helices with only a small segment of β-sheet at the base of the cleft. The large lobe is associated with catalysis and peptide binding.

The first 40 residues preceding the conserved catalytic core interact with the surface of both lobes on the side that is opposite to the active site. The polypeptide chain begins with an amino-terminal myristic acid which is not visualized in the crystals of the non-myristoylated recombinant enzyme (59, 60). The structure of the recombinant enzyme crystallized in the absence of detergents begins with residue 15 that initiates an α-helix that binds initially to the surface of the large lobe (59). When the myristoylated enzyme purified from porcine heart was crystallized, the first 15 residues could be visualized, as well as the fatty acid, as described later (78). In this structure of the mammalian enzyme, the helix is extended by one turn, and the fatty acid binds to a hydrophobic pocket on the surface of the large lobe. The A-helix (residues 10–31) binds to the surface of both lobes, and Trp30 near the carboxy-terminus of the helix binds in a deep hydrophobic pocket that lies precisely between the two lobes. Recent analysis of the *Dictyostelium* catalytic subunit suggests that this A-helix motif may be a conserved feature of all protein kinases (79).

The conserved core begins approximately with β-strand 1 (residue 43). Between β-strands 1 and 2 lies the first highly conserved sequence motif, a glycine-rich loop. β-Strand 3 contains another conserved residue, Lys72, first identified as part of the ATP binding site by affinity labelling with fluorosulfonylbenzoyl adenosine (FSBA) (80). The only two helical segments in the small lobe lie between β strands 3 and 4. The short B-helix is quite exposed to solvent and is one element of secondary structure that is probably not conserved throughout the protein kinase family (54, 81–83). The C-helix contains another conserved residue, Glu91, that forms an ion pair with Lys72. Its involvement in ATP binding was predicted initially on the basis of labelling with a hydrophobic carbodiimide, dicyclohexyl carbodiimide (DCCD) (84). β-Strands 4 and 5 complete the β-sheet. The extended chain (residues 120–127) between β-strand 5 and the D-helix is the only link between the two lobes in the core. This chain contributes to the binding of the adenine ring, with the α-carbonyl of Glu121 hydrogen bonding to the N^6 nitrogen on the adenine ring. This hydrogen bonding was predicted by analogue studies (85, 86).

The large lobe is stabilized by a bundle of four antiparallel helices (Helix D, E, F, and H). Between helix E and F is a small β-sheet that lies at the active site near the cleft interface. β-Strands 6 and 7 are linked by the catalytic loop shown in Fig. 5. Arg165 precedes the loop and is highly conserved in most protein kinases. In the catalytic subunit this guanidinium side chain forms a tight electrostatic contact with Thr197 (60), an essential (87) and stable phosphorylation site (88). This interaction very likely helps to correctly orient the catalytic loop. This catalytic loop (Asp166–Asn171) interacts with many parts of the molecule. Asp166, one of the invariant residues, most likely serves as a catalytic base. Lys168 is conserved in all

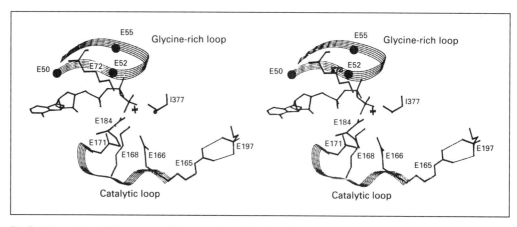

Fig. 5. Stereo view of the conserved loops at the active site. Most of the conserved residues at the active site that contribute to ATP binding and catalysis are shown. Also indicated is Arg165 that precedes the catalytic loop and (P)Thr197 that forms a bivalent ion pair with Arg165. Residues belonging to the protein are designated as E. The P-site Ala in PKI(5–24) is designated as I377 with the dot indicating the position of the methyl side chain. The Mg^{2+} ion bridging the β- and γ-phosphates is designated as a cross.

Ser/Thr-specific protein kinases and binds to the α-phosphate of ATP in the ternary complex containing ATP and PKI(5–24) (62, 76, 89, 90). In a binary complex with a 20-residue phosphorylated peptide, Lys168 binds to the phosphate at the P-site on the peptide (91). Lys168 also hydrogen bonds to the side chain hydroxyl of Thr201 and the α-carbonyl of the P−2 Arg in the peptide (91). Glu170 participates in recognition of the P−2 Arg in the inhibitor peptide. Asn171, another invariant residue, stabilizes the loop by hydrogen bonding to the backbone carbonyl of Asp166. Asn171 also chelates a second inhibitory magnesium ion (76, 77, 89). This loop thus serves as a hub of activity and communicates with many different regions of the protein.

β-Strands 8 and 9 are linked by another critical loop referred to as the DFG loop or the Mg^{2+} binding loop. Asp184 in the DFG loop chelates the essential magnesium ion that bridges the β- and γ-phosphates of ATP. Its role in ATP binding was predicted earlier on the basis of its reactivity with DCCD in the absence of ATP (84) and its subsequent cross-linking by DCCD to Lys72 (92). The backbone carbonyl of Phe187 hydrogen bonds to the ε and N1 nitrogens of Arg165, and this presumably helps to stabilize the interaction of both loops with the essential phosphorylation site. The peptide segment that extends from β-strand 9 to helix G covers much of the surface at the outer edge of the cleft interface. A critical feature of this surface is Thr197. The loop that follows this phospho-Thr, Leu198 through Leu205, participates in substrate recognition. Three additional residues in the large lobe are conserved and appear to stabilize the structure. Asp220 hydrogen bonds to the backbone nitrogens of Arg165 and Tyr164 respectively, and thus may help to stabilize the catalytic loop while Glu208 and Arg280 form a buried ion pair (60).

The conserved residues and their functions are also summarized in Table 2. Most of the conserved charged residues, in particular, were also identified as being functionally important on the basis of charge-to-alanine scanning mutagenesis of the yeast catalytic subunit prior to the structure solution (93). Their predictions are completely consistent with the crystal structure.

3.4 ATP binding site

The small lobe constitutes a unique nucleotide binding motif, and this domain, as well as its interactions with ATP, are summarized in Fig. 6 (59, 76, 90). The adenine ring is buried at the base of the cleft under β-strands 1 and 2. The adenine ring binds in a confined, mostly hydrophobic pocket. The specific interactions of ATP with conserved residues at the active site are summarized in Table 2 (76, 77, 89). With the exception of contacts to the γ-phosphate and hydrophobic interactions with the adenine ring, most of the interactions of the protein with the nucleotide involve the small lobe. Backbone amides of Phe54 and Gly55 in the glycine-rich loop hydrogen bond to the β-phosphate. Lys72 in β-strand 3 ion pairs with the α- and β-phosphates and is stabilized by Glu91 in the C-helix. An activating magnesium ion bridges the β- and γ-phosphates, and it in turn is chelated by Asp184 from the

Table 2 Conserved sequence motifs in the catalytic subunit. Highly conserved residues are shown in bold

Motif	Residue location		Function
		Small lobe	
Glycine-rich loop	**Gly50**		
	Thr51		
	Gly52		
	Ser53	$\beta1-\beta2$	**Phosphate anchor:** α-NH_2 of Phe54 and Gly55
	Phe54		H-bond to β-PO_4 of ATP
	Gly55		
	Arg56		
Additional residues	**Val57**		Side chain of Val57 lies over ribose ring
	Lys72	β-3	Ion-pairs to α-PO_4 and Glu91
	Glu91	C-helix	Ion-pairs with Lys72
		Large lobe	
Catalytic loop	**Arg165**		Ion-pairs with (P)Thr197
	Asp166		predicted catalytic base
	Leu167	$\beta6-\beta7$	
	Lys168		Ion-pairs with γ-PO_4
	Pro169		Ion-pairs with P−2 Arg
	Glu170		ligand to inhibitory Mg^{2+}
	Asn171		H-bonds to α-C=O of Asp166
Mg^{2+} binding loop	**Asp184**		Ligands to 1° and 2° Mg^{2+} ions
	Phe185	$\beta8-\beta9$	
	Gly186		
	Phe187		α-C=O H-bonds to ε N of Arg165
Stabilizing residues	**Asp220**	F-helix	Hydrogen bonds to backbone amides of Arg165 and Tyr164
	Glu208	Follows P+1 loop	Ion-pairs with Arg280
	Arg280	Follows H-helix	Ion-pairs with Glu208

large lobe as seen in Fig. 5. Lys168 in the catalytic loop interacts with the γ-phosphate, and this residue also is part of the large lobe.

3.5 Peptide recognition

As indicated earlier, the peptide specificity of the catalytic subunit was deduced from known phosphorylation sites and from peptide analogues, and recognition of the peptide involves multiple interaction sites (22, 23). Although the requirements for peptide recognition are not absolute, there is a high preference for basic residues preceding the phosphorylation site. In contrast to small substrates with K_ms in the μmolar range, physiological inhibitors such as PKI and the R-subunits

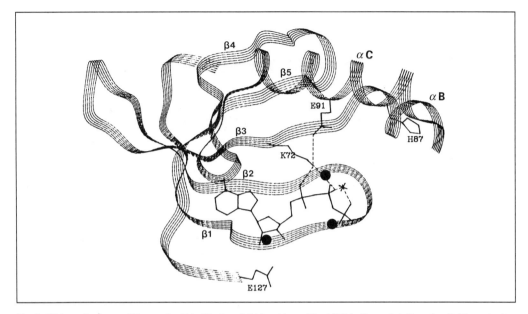

Fig. 6 Ribbon diagram of the nucleotide binding fold (residues 40–127) in the catalytic subunit. The adenine ring of ATP is buried beneath β-strands 1 and 2 and is anchored by hydrogen bonding to the α-carbonyl of Glu121. The α- and β-phosphates are anchored by Lys72 and by hydrogen bonding to backbone amides in the glycine-rich loop. The α-carbons of Gly50, 52, and 55 are indicated by black circles. Lys72 is positioned by Glu91. The activating Mg^{2+} that bridges the β- and γ-phosphates of ATP is shown by a cross. His87 that binds to the phosphate of Thr197 in the closed conformation is also shown, as is Glu127 that binds to the P − 3 Arg in the peptide recognition site.

bind with a high affinity, typically in the sub-nM range. The peptide corresponding to the major inhibitory site in PKI, residues 5–24, has a K_i = 2.3 nM (50) while intact PKI has a K_i of 0.2 nM (52). This peptide, PKI(5–24), was cocrystallized with the catalytic subunit, and this binary complex provides a structural basis for peptide recognition by this enzyme (94).

The general minimum consensus site will be shared in part by all substrates and inhibitors of this complex. This portion of the peptide forms an extended chain that occupies the cleft between the two lobes. An extended conformation of this consensus site region was predicted on the basis of NMR prior to the structure solution (57, 95–97). We have adopted a nomenclature that designates the phosphorylation site as the P-site. Residues that lie carboxy terminal to this are plus while aminoterminal residues are minus. The residues that are important for recognition of the P−3 Arg, the P−2 Arg, and the P+1 hydrophobic residue are seen in Fig. 7 and are also summarized in Table 3. Glu127 and Asp328 bind to the P−3 Arg while Glu170 and Glu230 bind to the P−2 Arg. Leu198, Pro202, and Leu205 line the P+1 binding site and form a hydrophobic pocket.

Most of the carboxylate residues that recognize the various arginines in the peptide were identified prior to the structure solution. Glu170 and Glu127, for example, were identified chemically based on their reactivity with a water soluble

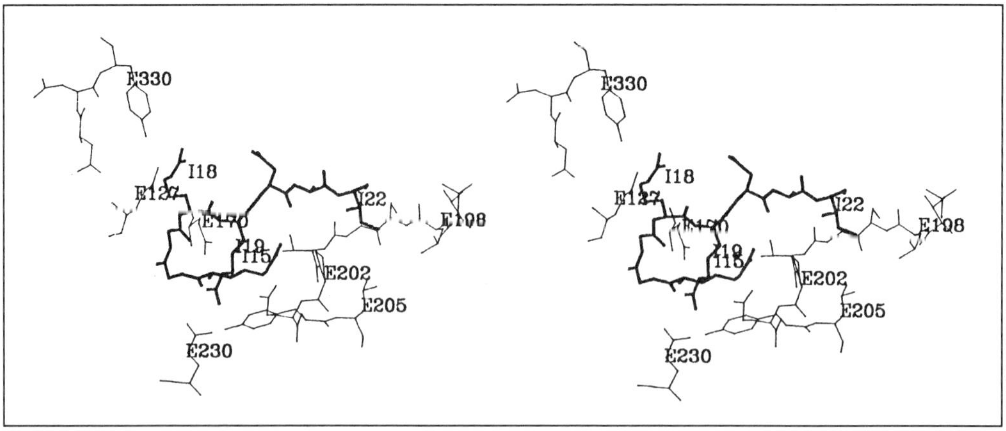

Fig. 7 Stereo view of the consensus peptide recognition site. The P−3, P−2, and P−6 arginines and their interactions with multiple carboxylates are shown. The hydrophobic P+1 loop, and its interaction with the P+1 Ile, is also shown. The peptide is shown in bold.

Table 3 Amino acid side chains that participate in peptide recognition

Recognition site in PKI(5−24)	Amino acids
P−2 Arg	Glu170
	Glu230
P−3 Arg	Asp328
	Glu127
P+1 Ile	Leu198
	Pro202
	Leu205
P−6 Arg	Glu203
	Tyr204
P−11 Phe	Tyr235
	Pro236
	Phe239
	Arg133

carbodiimide, 1-ethyl-3(3-dimethylamino-propyl)-carbodiimide (EDC) (98). Glu203, in contrast, appears to be in a hydrophobic environment based on its reactivity with a hydrophobic carbodiimide, DCCD (84). Asp328 is part of a very negatively charged cluster on the carboxy-terminal tail. The residues comprising this cluster are very reactive with EDC in the absence of peptide but are protected in the ternary complex (98). This region may serve as an initial docking site for peptides as predicted by the circular dichroism studies of Reed *et al.* (99). Since this region also shows conformational flexibility (78), it may also account for the malleability of the catalytic subunit that was predicted on the basis of salt-induced changes in sulphydryl group reactivity (100–102). All of the core carboxylates that contribute to peptide recognition were also identified genetically by charge-to-alanine scanning mutagenesis of the yeast catalytic subunit (TPK1). Gibbs and Zoller not only identified the carboxylates correctly, but also predicted accurately which Glu was important for the recognition of which Arg (93).

The high affinity binding of PKI is due to the region that lies on the amino-terminal side of the consensus site. This segment forms an amphipathic helix that lies on the surface of the large lobe. The P−11 Phe, shown by analogue studies to be important for binding (24, 54), lies in a hydrophobic pocket that consists of Tyr235, Pro236, and Phe239. The P−15 Tyr also comes close to this pocket. This Tyr can be phosphorylated, and phosphorylation causes a 10-fold increase in pKi, although the physiological significance of this modification and whether it occurs *in vivo* have not been determined (103). The P−6 Arg ion pairs with Glu203 as seen in Fig. 7. When the crystal structure of the inhibitor complex is compared with the predictions based on peptide analogues, the correlation is remarkably consistent and accurate (22, 62, 94). The low resolution NMR structure of PKI(5–22) in solution is also consistent with the helical and extended segments seen in the complex where PKI(5–24) is bound to the catalytic subunit (104).

Like PKI, the R-subunits also contain an inhibitor site that closely resembles a peptide substrate. The portion of this site corresponding to the consensus site most likely occupies the peptide binding site at the cleft in a manner that closely resembles PKI. However, unlike PKI, the region of the R-subunit that is responsible for its high affinity binding appears to lie carboxy-terminal to the consensus site. This conclusion is based on several findings. First, a proteolytic fragment of the R-subunit that begins just before the P−3 Arg binds very well to the catalytic subunit (105). Second, using charge-to-alanine scanning mutagenesis, Gibbs *et al.* (41) identified several mutant forms of the yeast catalytic subunit that were catalytically intact using heptapeptide substrates but which could not be regulated by the R-subunit. Most of these mapped to the surface of the catalytic subunit that lies just beyond the consensus site in the crystal structure. Thr197 and the residues that bind to this phosphate also appear to be particularly important for R-subunit recognition (87, 106, 107). Orellana and McKnight independently identified two mutant forms of the catalytic subunit that gave an unregulated phenotype (42). These were His87Gln and Trp196Arg. Both lie on the same surface of the catalytic subunit that flanks Thr197.

3.6 Conformational flexibility

Like hexokinase, the catalytic-subunit is capable of existing in different conformational states. The structure of the ternary complex and of the binary complex with bound PKI(5–24) both correspond to a closed conformation where the MgATP and the consensus portion of the inhibitor peptide are sandwiched tightly between two lobes. In contrast, when the mammalian enzyme was crystallized in the presence of a Tyr7–iodinated analogue of PKI(5–24), the protein crystallized in a more open conformation (78, 108). This open conformation was also found in the crystallized apoenzyme whereas the conformation of the ternary complex of the myristylated mammalian enzyme with PKI(5–24) and MgATP was indistinguishable from the recombinant enzyme ternary complex.

The open conformation results from a concerted rotation of a portion of the small lobe. This segment rotates as a rigid body with the β-sheet rotating approximately 15°. In the closed conformation of the binary complex (60) the single electrostatic interaction linking the small and large lobes is between His87 and the phosphate of Thr197. In the open conformation this interaction is broken leaving these two side chains approximately 6 Å apart. In the absence of ATP and peptide, there is little to hold the structure in a stable closed conformation. In the ternary complexes with the inhibitor peptide, the closed conformation is stabilized by two additional interactions. Lys168 in the catalytic loop interacts electrostatically with the γ-phosphate of ATP, and Asp184 binds to the magnesium ion that bridges the β- and γ-phosphates of ATP. In a substrate:ATP complex, these two interactions would only contribute to stabilization of the closed conformation prior to phosphotransfer. The crystal structure of a phosphopeptide binary complex shows clearly that Lys168 in the large lobe remains associated with the phosphate that is covalently attached to the P-site Ser of the peptide (91). It can thus no longer serve as a bridge between the two lobes.

Conformational flexibility in solution is also apparent from low angle X-ray scattering. Olah *et al.* (109) observed changes in the radius of gyration (R_g) and predicted that the free catalytic subunit has a significant degree of conformational flexibility that most likely corresponds to an opening and closing of the cleft. Low angle neutron scattering also showed a change in the R_g when the free catalytic subunit was compared to the ternary complex with ATP and PKI(5–24) (110). These studies also demonstrated that the inhibitor peptide alone, but not MgATP alone, was sufficient to cause the reduction in the radius of gyration. This is in contrast to hexokinase where ATP alone is sufficient to close the cleft (111).

3.7 Mechanism of catalysis

All kinetic evidence is consistent with a preferred ordered mechanism of catalysis with ATP binding preceding peptide binding (112–114). The catalytic subunit then catalyses the direct transfer of the γ-phosphate of ATP to its peptide substrate, resulting in an inversion of configuration (115). This direct transfer mechanism is quite consistent with the crystal structure and is reinforced in particular by two recent structures where the enzyme was cocrystallized with a substrate peptide and with a phosphorylated substrate peptide (91). While earlier NMR results using a substitution inert cobalt complex of ATP with Co^{2+} could not distinguish between an associative and dissociative process (57, 95, 116), the crystallographic results can clearly accommodate an associative mechanism.

Based on the pH dependency of the reaction, Yoon and Cook (117) first predicted that the phosphotransfer reaction involved a carboxylate functioning as a catalytic base. Subsequent analysis suggests that the observed pH dependency correlates with the P−2 arginine in the peptide, but this does not rule out the involvement of a catalytic base at the site of phosphotransfer. Based on the crystal structure, Asp166 is the obvious candidate since it is only 4 Å from the methyl side chain of the P-site Ala in PKI(5–24). The structure of the substrate peptide shows clearly that the hydroxyl side chain of the P-site Ser is within hydrogen bonding distance of Asp166 (91).

By measuring the rate constants as a function of viscosity, one can distinguish between the chemical and dissociative steps in the reaction pathway. The k_{cat}, for example, is very sensitive to viscosity, indicating that product release and not the chemical step is rate-limiting (118). Since the release of phosphopeptide, like the chemical step, is also very fast, the release of ADP is the rate-limiting step in the overall phosphotransfer reaction.

4. Post-translational modifications

The catalytic subunit is an enzyme that phosphorylates other substrate proteins; however, it is also modified itself by post-translational modifications. It was the first protein shown to be myristoylated at its amino-terminus (75), and the catalytic subunit is also a phosphoprotein. Because these phosphates are resistant to removal by phosphatases, they were first termed 'stable' phosphates (88). The crystal structure provides a molecular basis for understanding how each of these modifications contributes to the overall structure of the catalytic subunit.

The myristic acid is esterified to the amino-terminal Gly, and replacement of Gly with Ala is sufficient to abolish myristoylation, although it has little effect on activity (119). Since *E. coli* has no N-myristoyl transferase (NMT), the recombinant C subunit is not myristoylated. However, when NMT is coexpressed with the catalytic subunit in *E. coli*, the myristoylated enzyme is synthesized (120). The non-myristoylated recombinant enzyme is more labile to thermal denaturation than the mammalian enzyme (56). The addition of the myristoylate restores thermal stability so that the recombinant enzyme now resembles the mammalian enzyme (121). The holoenzyme, in general, is approximately 10° more stable to thermal denaturation than the free subunit, and the stabilizing effect of the myristoylate is seen both in the free catalytic subunit and in the holoenzyme (121).

In the original crystal structure of the recombinant mouse enzyme, the first 14 residues of the catalytic subunit were not visualized, suggesting this region was disordered (59). When the enzyme was crystallized in the presence of detergent, residues 10–14 were seen and a detergent molecule occupied an acyl pocket on the surface of the large lobe near the amino-terminus (60). The recent structure solution of the mammalian enzyme shows the myristoyl moiety binding to this pocket (78). The general location of this acyl binding site is shown in Fig. 8. Several different parts of the molecule converge to form the acyl binding site — the αC–β4 loop from the small lobe, the E-helix from the large lobe, and the beginning of the

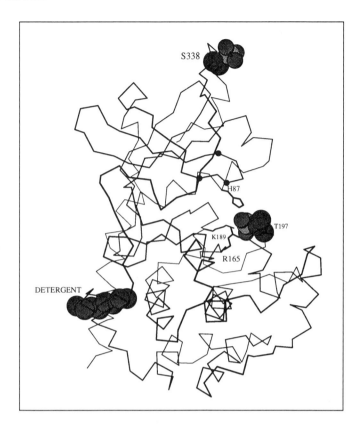

Fig. 8 Post-translational modifications in the catalytic subunit. The α-carbon backbone of the binary recombinant enzyme is shown with the detergent molecule, (P)Thr197, and (P)Ser338 shown as CPK structures. In the myristylated mammalian enzyme the detergent binding site is filled by the myristoyl moiety. Several side-chains that bind to (P)Thr197 (Arg165, Lys189, and His87) are also shown.

carboxyl-terminal tail. The myristoyl moiety appears to anchor the amino-terminus to the surface of the large lobe. There is no evidence that the catalytic subunit associates with membranes as do some other myristoylated proteins such as pp60[src] (122, 123).

Two stable phosphorylation sites, Ser338 and Thr197, were identified during the sequence determination of the catalytic subunit (7, 88). Both are phosphatase-resistant, although, of the two, Thr197 is more stable (12, 88, 124). The recombinant enzyme is also phosphorylated at these sites and, in addition, Ser10 and Ser139 are phosphorylated (125). Three different isoforms of the recombinant enzyme can be resolved by FPLC, and these each differ by a single phosphate (126). The predominant isoform II is phosphorylated at Ser10, Ser338, and Thr197. Phosphorylation of Ser10 can be autophosphorylated in the mammalian enzyme, but it is not clear whether this occurs *in vivo* (124). The two major isoforms of the mammalian catalytic subunit are not due to differences in phosphorylation state (127).

Recent evidence indicates that in mammalian cells the initial transcript for the catalytic subunit is not phosphorylated, suggesting that phosphorylation of Thr197 may be a terminal step in processing the mature and active enzyme (128). Steinberg *et al.* (87) also showed that the dephosphorylated mammalian enzyme is very inactive. The isolation of kinase-negative (kin⁻) mutants from S49 mouse lymphoma cells may eventually shed some light on this processing. These transdominant mutants show negligible kinase activity (129) but have no mutations in the structural gene for the catalytic subunit. Steinberg (128) recently showed that the mutation in these kin⁻ cells is most likely related to phosphorylation since the catalytic subunit transcript is present, but it is insoluble and is not phosphorylated on Thr197. Although the phosphorylation of the recombinant enzyme occurs as an autocatalytic process in *E. coli.*, probably intermolecular (unpublished observations), the results of Steinberg raise the possibility that there could be another kinase that is responsible for the phosphorylation of the catalytic subunit *in vivo* in mammalian cells. Alternatively, a chaperon type of protein could be required to facilitate autophosphorylation.

The crystal structure reveals why these phosphates are so resistant. Ser338 lies at a turn near the carboxy-terminus. It forms an ion pair with Lys342 and is also close to Arg336. In addition, it can hydrogen bond with the backbone amide of Ile339. Thr197 is stabilized even more by interactions with multiple side chains (Fig. 9). Lys189 in β-strand 9, Arg165 preceding catalytic loop, His87 in the C-helix of the small lobe, and Thr195 all interact directly with this phosphate in the closed conformation of the enzyme. This phosphate thus appears to play a major role in assembling the surface on the outer edge of the cleft. Many protein kinases contain

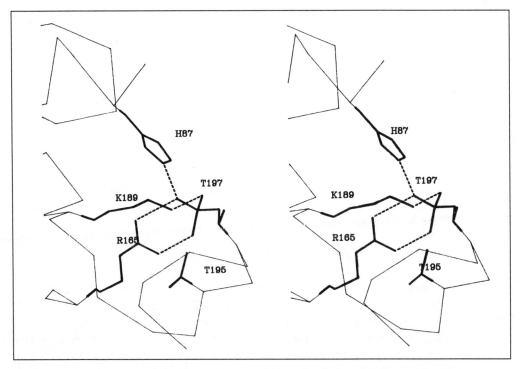

Fig. 9 Stereo view of phosphothreonine 197 and its interactions with neighbouring residues. (P)Thr197 is stabilized by interactions with multiple side-chains from both lobes.

a homologous phosphorylation site that is critical for their activation. Thr161 in cdc2 (130), Thr183 in MAP kinase (131, 132), and Tyr1162/1163 in the insulin receptor (133) are probably functionally analogous to Thr197, although in these other enzymes the phosphorylation of this key residue is dynamic and correlates with activation. There is no evidence so far to indicate that the phosphate on Thr197 in the catalytic subunit turns over once it is phosphorylated.

As discussed previously, the environment around (P)Thr197 changes when the cleft opens (78). In this open conformation, His87 no longer interacts with the phosphate, so the one stable bond between the small lobe and the large lobe is severed.

The general location of all three post-translational modifications relative to the overall structure is shown in Fig. 8.

5. The catalytic subunit as a template for other protein kinases

While the crystal structures of other protein kinases will obviously be essential before a detailed understanding of this family of enzymes can be achieved, in the interim this structure of the catalytic subunit of PKA can serve as a general template for the folding of other protein kinases (59, 61). Several criteria dictate that this enzyme will be a valid template. First is the finding of conserved residues that are scattered throughout the core. These residues, summarized earlier in Fig. 2, are primarily important for ATP binding and catalysis and will almost certainly play a similar role in all protein kinases. They can thus be considered as fixed points. The few remaining conserved residues, such as the buried ion pair Glu208 and Arg280, are presumably important for structural stability. The other observation making it likely that one structure can serve as a template for the entire family is that the inserts associated with different protein kinases all seem to lie at regions on the surface, frequently in connecting loops. Inserts can, therefore, be easily accommodated without perturbing the general folding of the polypeptide chain (59, 134). The validity of these predictions was confirmed with the recent structure solution of cdk2 (83).

The predicted sequence alignment of several other protein kinases with the catalytic core is shown in Fig. 10, and this alignment is correlated with the secondary structure predicted by the catalytic subunit. The catalytic core, extending from residue 40 through to approximately 300, is shared by all eukaryotic members of the protein kinase family (1). The representative protein kinases aligned in Fig. 10 not only emphasize the sequence similarities, but also highlight the highly conserved residues that are scattered throughout the core (81, 135). The alignment of MLCK and pp60src with the elements of secondary structure in the catalytic subunit is based on modelling using the coordinates of the mouse Cα-subunit (79, 81). PKC, cdc2, and the EGF receptor (now shown) have also been modelled in this way (82, 135, 136).

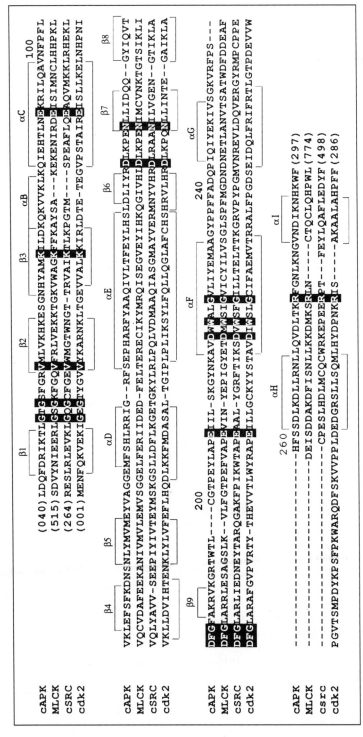

Fig. 10 Sequence alignment of several representative protein kinases. The src sequence is taken from reference 137. The alignment of MLCK (81) and src are based on modelling using the crystallographic coordinates of the catalytic subunit. The alignment of cdk2 is based on its crystal structure (83). Highly conserved residues are highlighted in black. Alpha-helices and β-strands are indicated, and the numbering correlates with Fig. 4. The brackets below the cdk2 sequence indicate the location of helices and strands in cdk2 that correlate directly with similar elements of secondary structure in the catalytic subunit.

The alignment of cdk2 is based on its recently solved crystal structure (83). This structure represents an inactive form of cdk2 that lacks the activating phosphate at Thr160. As seen in Fig. 10, eight of the nine β-strands in PKA are present in cdk2, and these are superimposible in the two structures. All of the α-helices in the core, with the exception of helix B, are also present in cdk2 and, like the β-strands, the helix bundle (D, E, F, and H) is superimposible. Cdk2 contains a 25-residue insert in the large lobe, and this, as predicted, is easily accommodated without perturbation of the folding of the conserved core. The only region that shows significant difference (residues corresponding approximately to 184 through 198 in the catalytic subunit) contains the essential phosphorylation site, Thr160. This residue is equivalent to Thr197 in the catalytic subunit, and Marcote *et al.* (82) predict that phosphorylation of this Thr will lead to a conformational change that will cause cdk2 to resemble the catalytic subunit in this region since most of the residues whose side-chains are predicted to bind to (P)Thr160 are conserved in cdk2.

Acknowledgements

This work was supported by a grant from the Lucille P. Markey Charitable Trust, NIH, and the American Cancer Society (S.S.T.). E.R.-A. is supported by USPHS training grant T32DK07233.

References

1. Hanks, S. K., Quinn, A. M., and Hunter, T. (1988) The protein kinase family: conserved features and deduced phylogeny of the catalytic domains. *Science*, **241**, 42.
2. Krebs, E. G. (1985) The phosphorylation of proteins: a major mechanism for biological regulation. *Biochem. Soc. Trans.*, **13**, 813.
3. Krebs, E. G. and Fischer, E. H. (1956) The phosphorylase b to a converting enzyme of rabbit skeletal muscle. *Biochim. Biophys. Acta*, **20**, 150.
4. Krebs, E. G., Graves, D. J., and Fischer, E. H. (1959) Factors affecting the activity of muscle phosphorylase b kinase. *J. Biol. Chem.*, **234**, 2867.
5. Walsh, D. A., Perkins, J. P., and Krebs, E. G. (1968) An adenosine 3',5'-mono phosphate dependent protein kinase from rabbit skeletal muscle. *J. Biol. Chem.*, **243**, 3763.
6. Reimann, E. M., Titani, K., Ericsson, L. H., Wade, R. D., Fischer, E. H., and Walsh, K. A. (1984) Homology of the γ subunit of phosphorylase b kinase with cAMP-dependent protein kinase. *Biochemistry*, **23**, 4185.
7. Shoji, S., Ericsson, L. H., Walsh, D. A., Fischer, E. H., and Titani, K. (1983) Amino acid sequence of the catalytic subunit of bovine type II adenosine cyclic 3',5'-phosphate dependent protein kinase. *Biochemistry*, **22**, 3702.
8. Pickett-Gies, C. A. and Walsh, D. A. (1986) Phosphorylase kinase. In *The enzymes: control by phosphorylation* (ed. P. D. Boyer and E. G. Krebs), Vol. 17, p. 395. Academic Press, New York.
9. Beebe, S. J. and Corbin, J. D. (1986) Cyclic nucleotide-dependent protein kinases. In *The enzymes: control by phosphorylation part A* (ed. E. G. Krebs and P. D. Boyer), Vol. XVII, p. 43. Academic Press, New York.

10. Taylor, S. S., Buechler, J. A., and Yonemoto, W. (1990) cAMP-dependent protein kinase: framework for a diverse family of regulatory enzymes. *Annu. Rev. Biochem.*, **59**, 971.

11. Gill, G. N. and Garren, L. D. (1969) On the mechanism of action of adrenocorticotropic hormone: the binding of cyclic-3′,5′-adenosine monophosphate to adrenal cortical protein. *Proc. Natl. Acad. Sci. USA*, **63**, 512.

12. Tao, M., Salas, M. L., and Lipmann, F. (1970) Mechanism of activation by adenosine 3′:5′-cyclic monophosphate of a protein phosphokinase from rabbit reticulocytes. *Proc. Natl. Acad. Sci. USA*, **67**, 408.

13. Brostrom, M. A., Reimann, E. M., Walsh, D. A., and Krebs, E. G. (1970) Cyclic 3′,5′-AMP-stimulated protein kinase from cardiac muscle. *Adv. Enzyme Regul.*, **8**, 191.

14. Granot, J., Mildvan, A. S., Hiyama, K., Kondo, H., and Kaiser, E. T. (1980) Magnetic resonance studies of the effect of the regulatory subunit on metal and substrate binding to the catalytic subunit of bovine heart protein kinase. *J. Biol. Chem.*, **255**, 4569.

15. Nigg, E. A., Hilz, H., Eppenberger, H. M., and Dutley, F. (1985) Rapid and reversible translocation of the catalytic subunit of cAMP-dependent protein kinase type II from the Golgi complex to the nucleus. *EMBO J.*, **4**, 2801.

16. Kuettel, M. R., Squinto, S. P., Kwast-Welfeld, J., Schwoch, G., Schweppe, J. S., and Jungmann, R. A. (1985) Localization of nuclear subunits of cyclic AMP-dependent protein kinase by the immunocolloidal gold method. *J. Cell. Biol.*, **101**, 965.

17. Meinkoth, J. L., Ji, Y., Taylor, S. S. and Feramisco, J. R. (1990) Dynamics of the cyclic AMP-dependent protein kinase distribution in living cells. *Proc. Natl. Acad. Sci. USA*, **87**, 9595.

18. Adams, S. R., Harootunian, A. T., Buechler, Y. J., Taylor, S. S., and Tsien, R. Y. (1991) Fluorescence ratio imaging of cyclic AMP in single cells. *Nature*, **349**, 694.

19. Hagiwara, M., Brindle, P., Harootunian, A. T., Armstrong, R., Rivier, J., Vale, W. *et al.* (1993) The coupling of hormonal stimulation and transcription via cyclic AMP-responsive factor CREB is rate limited by nuclear entry of protein kinase A. *Mol. Cell. Biol.*, **13**, 4852.

20. Zetterqvist, O., Ragnarsson, U., Humble, E., Berglund, L., and Engstrom, L. T. (1976) The minimum substrate of cyclic AMP-stimulated protein kinase, as studied by synthetic peptides representing the phosphorylatable site of pyruvate kinase (type L) of rat liver. *Biochem. Biophys. Res. Comm.*, **70**, 696.

21. Kemp, B. F., Graves, D. J., Benjamini, E., and Krebs, E. G. (1977) Role of multiple basic residues in determining the substrate specificity of cyclic AMP-dependent protein kinase. *J. Biol. Chem.*, **252**, 4888.

22. Walsh, D. A., Glass, D. B., and Mitchell, R. D. (1992) Substrate diversity of the cAMP-dependent protein kinase: regulation based upon multiple binding interactions. *Curr. Opin. Cell Biol.*, **4**, 241.

23. Zetterqvist, Ö. Z., Ragnarsson, U., and Engstrom, L. (1990) Substrate specificity of cyclic AMP-dependent protein kinase. In *Peptides and protein phosphorylation* (ed. B. E. Kemp), p. 171. CRC Press, Boca Raton.

24. Walsh, D. A., Angelos, K. L., Van Patten, S. M., Glass, D. B., and Garetto, L. P. (1990) The inhibitor protein of the cAMP-dependent protein kinase. In *Peptides and protein phosphorylation* (ed. B. E. Kemp), p. 43. CRC Press, Boca Raton.

25. Hofmann, F., Beavo, J. A., Bechtel, P. J., and Krebs, E. G. (1975) Comparison of adenosine 3′:5′-monophosphate-dependent protein kinases from rabbit skeletal and bovine heart muscle. *J. Biol. Chem.*, **250**, 7795.

26. Neitzel, J. J., Dostmann, W. R. G., and Taylor, S. S. (1991) The role of MgATP in the activation and reassociation of cAMP-dependent protein kinase I: consequences of replacing the essential arginine in site A. *Biochemistry*, **30,** 733.

27. Titani, K., Sasagawa, T., Ericsson, L. H., Kumar, S., Smith, S. B., Krebs, E. G., and Walsh, K. A. (1984) Amino acid sequence of the regulatory subunit of bovine type I adenosine cyclic 3',5'-phosphate dependent protein kinase. *Biochemistry*, **23,** 4193.

28. Lee, D. C., Carmichael, D. F., Krebs, E. G., and McKnight, G. S. (1983) Isolation of a cDNA clone for the type I regulatory subunit of bovine cAMP-dependent protein kinase. *Proc. Natl. Acad. Sci. USA*, **80,** 3608.

29. Clegg, C. H., Codd, G. G., and McKnight, G. S. (1988) Genetic characterization of a brain-specific form of the type I regulatory subunit of cAMP-dependent protein kinase. *Proc. Natl. Acad. Sci. USA*, **85,** 3703.

30. Rosen, O. M. and Erlichman, J. (1975) Reversible autophosphorylation of a cyclic 3':5'-AMP-dependent protein kinase from bovine cardiac muscle. *J. Biol. Chem.*, **250,** 7788.

31. Takio, K., Smith, S. B., Krebs, E. G., Walsh, K. A., and Titani, K. (1984) Amino acid sequence of the regulatory subunit of bovine type II adenosine cyclic 3',5'-phosphate dependent protein kinase. *Biochemistry*, **23,** 4200.

32. Jahnsen, T., Hedin, L., Kidd, V. J., Beattie, W. G., Lohmann, S. M., Walter, V. *et al.* (1986) Molecular cloning, cDNA structure, and regulation of the regulatory subunit of type II cAMP-dependent protein kinase from rat ovarian granulosa cells. *J. Biol. Chem.*, **261**(26), 12352.

33. Scott, J. D., Glaccum, M. B., Zoller, M. J., Uhler, M. D., Helfman, D. M., McKnight, G. S., and Krebs, E. G. (1987) The molecular cloning of a type II regulatory subunit of the cAMP-dependent protein kinase from rat skeletal muscle and mouse brain. *Proc. Natl. Acad. Sci. USA*, **84,** 5192.

34. Weber, I. T., Takio, K., Titani, K., and Steitz, T. A. (1982) The cAMP-binding domains of the regulatory subunit of cAMP-dependent protein kinase and the catabolite gene activator protein are homolgous. *Proc. Natl. Acad. Sci. USA*, **79,** 7679.

35. Weber, I. T., Steitz, T. A., Bubis, J., and Taylor, S. S. (1987) Predicted structures of the cAMP binding domains of the type I and II regulatory subunits of cAMP-dependent protein kinase. *Biochemistry*, **26,** 343.

36. Ji, Y. and Taylor, S. S. (1990) cAMP-dependent protein kinase I: consequences of mutations in the autoinhibitory site of the regulatory subunit. *FASEB J.*, **4,** A2073.

37. Wang, Y. H., Scott, J. D., McKnight, G. S., and Krebs, E. G. (1991) A constitutively active holoenzyme form of the cAMP-dependent protein kinase. *Proc. Natl. Acad. Sci. USA*, **88,** 2446.

38. Buechler, Y. J., Herberg, F. W., and Taylor, S. S. (1993) Regulation defective mutants of type I cAMP-dependent protein kinase: consequences of replacing Arg94 and Arg95. *J. Biol. Chem.*, **268,** 16495.

39. Rangel-Aldao, R. and Rosen, O. M. (1976) Dissociation and reassociation of phosphorylated and nonphosphorylated forms of cAMP-dependent protein kinase from bovine cardiac muscle. *J. Biol. Chem.*, **251,** 3375.

40. Flockhart, D. A. and Corbin, J. D. (1982) Regulatory mechanisms in the control of protein kinases. *CRC Crit. Rev. Biochem.*, **12,** 133.

41. Gibbs, C. S., Knighton, D. R., Sowadski, J. M., Taylor, S. S., and Zoller, M. J. (1992) Systematic mutational analysis of cAMP-dependent protein kinase identifies unregulated catalytic subunits and defines regions important for the recognition of the regulatory subunit. *J. Biol. Chem.*, **267,** 4806.

42. Orellana, S. A. and McKnight, G. S. (1992) Mutations in the catalytic subunit of cAMP-dependent protein kinase result in unregulated biological activity. *Proc. Natl. Acad. Sci. USA*, **89**, 4726.

43. Scott, J. D., Stofko, R. E., McDonald, J. R., Comer, J. D., Vitalis, E. A., and Mangili, J. A. (1990) Type II regulatory subunit dimerization determines the subcellular localization of the cAMP-dependent protein kinase. *J. Biol. Chem.*, **265**, 21561.

44. Bubis, J., Vedvick, T. S., and Taylor, S. S. (1987) Antiparallel alignment of the two protomers of the regulatory subunits of the regulatory subunit dimer of cAMP-dependent protein kinase I. *J. Biol. Chem.*, **262**, 14961.

45. Taskén, K., Skålhegg, B. S., Solberg, R., Andersson, K. B., Taylor, S. S., Lea, T. *et al.* (1993) Novel isozymes of cAMP-dependent protein kinase exist in human cells due to formation of $R^I\alpha$–$R^I\beta$ heterodimeric complexes. *J. Biol. Chem.*, **268**, 21276.

46. Ashby, C. D. and Walsh, D. A. (1973) Characterization of the interaction of a protein inhibitor with adenosine 3′-5′-monophosphate-dependent protein kinases. *J. Biol. Chem.*, **248**, 1255.

47. Scott, J. D., Fischer, E. H., Takio, K., Demaille, J. G., and Krebs, E. G. (1985) Amino acid sequence of the heat-stable inhibitor of the cAMP-dependent protein kinase from rabbit skeletal muscle. *Proc. Natl. Acad. Sci. USA*, **82**, 5732.

48. Van Patten, S. M., Ng, D. C., Th'ng, J. P., Angelos, K. L., Smith, A. J., and Walsh, D. A. (1991) Molecular cloning of a rat testis form of the inhibitor protein of cAMP-dependent protein kinase. *Proc. Natl. Acad. Sci. USA*, **88**, 5383.

49. Scarpetta, M. A. and Uhler, M. D. (1993) Evidence for two additional isoforms of the endogenous kinase inhibitor of cAMP-dependent protein kinase in mouse. *J. Biol. Chem.*, **268**, 10927.

50. Cheng, H.-C., van Patten, S. M., Smith, A. J., and Walsh, D. A. (1986) An active twenty-amino-acid-residue peptide derived from the inhibitor protein of the cyclic AMP-dependent protein kinase. *Biochem. J.*, **231**, 655.

51. Scott, J. D., Fischer, E. H., Demaille, J. G., and Krebs, E. G. (1985) Identification of an inhibitory region of the heat-stable protein inhibitor of the cAMP-dependent protein kinase. *Proc. Natl. Acad. Sci. USA*, **82**, 4379.

52. Whitehouse, S. and Walsh, D. A. (1983) $MgATP^{2+}$-dependent interaction of the inhibitor protein of the cAMP-dependent protein kinase with the catalytic subunit. *J. Biol. Chem.*, **258**, 3682.

53. Van Patten, S. M., Fletcher, W. H., and Walsh, D. A. (1986) The inhibitor protein of the cAMP-dependent protein kinase–catalytic subunit interaction. *J. Biol. Chem.*, **261**, 5514.

54. Glass, D. B., Lundquist, L. J., Katz, B. M., and Walsh, D. A. (1989) Protein kinase inhibitor-(6–22)-amide peptide analogs with standard and nonstandard amino acid substitutions for phenylalanine 10. Inhibition of cAMP-dependent protein kinase. *J. Biol. Chem.*, **264**, 14579.

55. Kinzel, V. and Kubler, D. (1976) Single step purification of the catalytic subunit(s) of cyclic 3′,5′-adenosine monophosphate-dependent protein kinase(s) from rat muscle. *Biochem. Biophys. Res. Commun.*, **71**, 257.

56. Slice, L. W. and Taylor, S. S. (1989) Expression of the catalytic subunit of cAMP-dependent protein kinase in *Escherichia coli*. *J. Biol. Chem.*, **264**, 20940.

57. Bramson, H. N., Kaiser, E. T., and Mildvan, A. S. (1984) Mechanistic studies of cAMP-dependent protein kinase actions. *CRC Crit. Rev. Biochem.*, **15**, 93.

58. Taylor, S. S., Buechler, J. A., and Knighton, D. R. (1990) cAMP-dependent protein

kinase: mechanism for ATP: protein phosphotransfer. In *Peptides and protein phosphorylation* (ed. B. E. Kemp), p. 1. CRC Press, Boca Raton.

59. Knighton, D. R., Zheng, J., Ten Eyck, L. F., Ashford, V. A., Xuong, N.-h., Taylor, S. S., and Sowadski, J. M. (1991) Crystal structure of the catalytic subunit of cAMP-dependent protein kinase. *Science*, **253**, 407.

60. Knighton, D. R., Bell, S. M., Zheng, J., Ten Eyck, L. F., Xuong, N.-h., Taylor, S. S., and Sowadski, J. M. (1993) 2.0 Å refined crystal structure of the catalytic subunit of cAMP-dependent protein kinase complexed with a peptide inhibitor and detergent. *Acta Cryst.*, **D49**, 357.

61. Taylor, S. S., Knighton, D. R., Zheng, J., Ten Eyck, L. F., and Sowadski, J. M. (1992) Structural framework for the protein kinase family. *Ann. Rev. Cell Biol.*, **8**, 429.

62. Taylor, S. S., Knighton, D. R., Zheng, J., Sowadski, J. M., Gibbs, C. S., and Zoller, M. J. (1993) A template for the protein kinase family. *Trends Biochem. Sci.*, **18**, 84.

63. Uhler, M. D., Carmichael, D. F., Lee, D. C., Chivia, J. C., Krebs, E. G., and McKnight, G. S. (1986) Isolation of cDNA clones for the catalytic subunit of mouse cAMP-dependent protein kinase. *Proc. Natl. Acad. Sci. USA*, **83**, 1300.

64. Cadd, G. G. and McKnight, G. S. (1989) Distinct patterns of cAMP-dependent protein kinase gene expression in mouse brain. *Neuron*, **3**, 71.

65. Showers, M. O. and Maurer, R. A. (1986) A cloned bovine cDNA encodes an alternate form of the catalytic subunit of cAMP-dependent protein kinase. *J. Biol. Chem.*, **261**, 16288.

66. Uhler, M. D., Chrivia, J. C., and McKnight, G. S. (1986) Evidence for a second isoform of the catalytic subunit of cAMP-dependent protein kinase. *J. Biol. Chem.*, **261**, 15360.

67. Beebe, S., Oyen, O., Sandberg, M., Froysa, A., Hansson, V., and Jahnsen, T. (1990) Molecular cloning of a tissue-specific protein kinase (C gamma) from human testis — representing a third isoform for the catalytic subunit of cAMP-dependent protein kinase. *Mol. Endocrinol.*, **4**, 465.

68. Beushausen, S., Bergold, P., Struner, S., Elsete, A., Roytenberg, V., Schwartz, J. H., and Bayley, H. (1988) Two catalytic subunits of cAMP-dependent protein kinase generated by alternative RNA splicing are expressed in *Aplysia* neurons. *Neuron*, **1**, 853.

69. Foster, J. L., Higgins, G. C., and Jackson, F. R. (1988) Cloning, sequence, and expression of the *Drosophila* cAMP-dependent protein kinase catalytic subunit gene. *J. Biol. Chem.*, **263**, 1676.

70. Gross, R. W., Bagchi, S., Lu, X., and Rubin, C. S. (1990) Cloning, characterization, and expression of the gene for the catalytic subunit of cAMP-dependent protein kinase in *Caenorhabditis elegans*. Identification of highly conserved and unique isoforms generated by alternative splicing. *J. Biol. Chem.*, **265**, 6896.

71. Toda, T., Cameron, S., Sass, P., Zoller, M. J., and Wigler, M. (1987) Three different genes in *S. cerevisiae* encode the catalytic subunits of the cAMP-dependent protein kinase. *Cell*, **50**, 277.

72. Mann, S. K. and Firtel, R. A. (1991) A developmentally regulated, putative serine/threonine protein kinase is essential for development in *Dictyostelium*. *Mech. Dev.*, **35**, 89.

73. Bürki, E., Anjard, C., Scholder, J. C., and Reymond, C. D. (1991) Isolation of two genes encoding putative protein kinases regulated during *Dictyostelium discoideum* development. *Gene*, **102**, 57.

74. Chrivia, J. C., Uhler, M. D., and McKnight, G. S. (1988) Characterization of genomic

clones coding for the Cα and Cβ subunits of mouse cAMP-dependent protein kinase. *J. Biol. Chem.*, **263**, 5739.

75. Carr, S. A., Biemann, K., Shoji, S., Parmalee, D. C., and Titani, K. (1982) *n*-Tetradecanoyl in the NH₂ terminal blocking group of the catalytic subunit of the cyclic AMP-dependent protein kinase from bovine cardiac muscle. *Proc. Natl. Acad. Sci. USA*, **79**, 6128.

76. Zheng, J., Knighton, D. R., Ten Eyck, L. F., Karlsson, R., Xuong, N.-h., Taylor, S. S., and Sowadski, J. M. (1993) Crystal structure of the catalytic subunit of cAMP-dependent protein kinase complexed with MgATP and peptide inhibitor. *Biochemistry*, **32**, 2154.

77. Zheng, J., Trafny, E. A., Knighton, D. R., Xuong, N.-h., Taylor, S. S., Ten Eyck, L. F., and Sowadski, J. M. (1993) 2.2 Å refined crystal structure of the catalytic subunit of cAMP-dependent protein kinase complexed with MnATP and a peptide inhibitor. *Acta Cryst.*, **D49**, 362.

78. Zheng, J., Knighton, D. R., Xuong, N.-h., Taylor, S. S., Sowadski, J. M., and Ten Eyck, L. F. (1993) Crystal structures of the myristylated catalytic subunit of cAMP-dependent protein kinase reveal open and closed conformations. *Prot. Sci.*, **2**, 1559.

79. Veron, M., Radzio-Andzelm, E., Tsigelny, I., Ten Eyck, L. F., and Taylor, S. S. (1993) A conserved helix motif complements the protein kinase core. *Proc. Natl. Acad. Sci. USA*, **90**, 10618.

80. Zoller, M. J., Kerlavage, A. R., and Taylor, S. S. (1979) Structural comparisons of cAMP-dependent protein kinases I and II from porcine skeletal muscle. *J. Biol. Chem.*, **254**, 2408.

81. Knighton, D. R., Pearson, R. B., Sowadski, J. M., Means, A. R., Ten Eyck, L. F., Taylor, S. S., and Kemp, B. E. (1992) Structural basis for intrasteric regulation of myosin light chain kinase. *Science*, **258**, 130.

82. Marcote, M. J., Knighton, D. R., Basi, G., Sowadski, J. M., Brambilla, P., Draetta, G., and Taylor, S. S. (1993) A three-dimensional model of the cdc2 protein kinase: identification of cyclin and suc1 binding regions. *Mol. Cell. Biol.*, **13**, 5122.

83. De Bondt, H. L., Rosenblatt, J., Jancarik, J., Jones, H. D., Morgan, D. O., and Kim, S. H. (1993) Crystal structure of cyclin-dependent kinase 2. *Nature*, **363**, 595.

84. Buechler, J. A. and Taylor, S. S. (1988) Identification of Asp 184 as an essential residue in the catalytic subunit of cAMP-dependent protein kinase. *Biochemistry*, **27**, 7356.

85. Hoppe, J., Freist, W., Marutzky, R., and Shaltiel, S. (1978) Mapping the ATP binding site in the catalytic subunit of cAMP-dependent protein kinase; spatial relationship with the ATP binding site of the undissociated enzyme. *Eur. J. Biochem.*, **90**, 427.

86. Flockhart, D. A., Freist, W., Hoppe, J., Lincoln, T. M., and Corbin, J. D. (1984) ATP analog specificity of cAMP-dependent protein kinase, cGMP-dependent protein kinase, and phosphorylase kinase. *Eur. J. Biochem.*, **140**, 289.

87. Steinberg, R. A., Cauthron, R. D., Symcox, M. M., and Shuntoh, H. (1993) Auto-activation of catalytic (C alpha) subunit of cyclic AMP-dependent protein kinase by phosphorylation of threonine 197. *Mol. Cell. Biol.*, **13**, 2332.

88. Shoji, S., Titani, K., Demaille, J. G., and Fischer, E. H. (1979) Sequence of two phosphorylated sites in the catalytic subunit of bovine cardiac muscle adenosine 3′:5′-monophosphate-dependent protein kinase. *J. Biol. Chem.*, **254**, 6211.

89. Bossemeyer, D., Engh, R. A., Kinzel, V., Ponstingl, H., and Huber, R. (1993) Phosphotransferase and substrate binding mechanism of the cAMP-dependent protein kinase catalytic subunit from porcine heart as deduced from the 2.0 Å structure of the

complex with Mn^{2+} adenyl imidodiphosphate and inhibitor peptide PKI(5–24). *EMBO J.*, **12**, 849.

90. Taylor, S. S., Zheng, J., Radzio-Andzelm, E., Knighton, D. R., Ten Eyck, L. F., Sowadski, J. M., *et al.* (1933) cAMP-dependent protein kinase defines a family of enzymes. *Phil. Trans. R. Soc. Lond. B*, **340**, 315.

91. Madhusudan, E. A., Xuong, N.-h., Adams, J. A., Ten Eyck, L. F., Taylor, S. S., and Sowadski, J. M. (1994) cAMP-dependent protein kinase: crystallographic insights into substrate recognition and phosphotransfer. *Prot. Sci.*, **3**, 176.

92. Buechler, J. A. and Taylor, S. S. (1989) Dicyclohexyl carbodiimide crosslinks two essential residues at the active site of the catalytic subunit of cAMP-dependent protein kinase: Asp 184 and Lys 72. *Biochemistry*, **28**, 2065.

93. Gibbs, C. S. and Zoller, M. J. (1991) Rational scanning mutagenesis of a protein kinase identifies functional regions involved in catalysis and substrate interactions. *J. Biol. Chem.*, **266**, 8923.

94. Knighton, D. R., Zheng, J., Ten Eyck, L. F., Xuong, N.-h., Taylor, S. S., and Sowadski, J. M. (1991) Structure of a peptide inhibitor bound to the catalytic subunit of cyclic adenosine monophosphate-dependent protein kinase. *Science*, **253**, 414.

95. Granot, J., Mildvan, A. S., and Kaiser, E. T. (1980) Studies of the mechanism of action and regulation of cAMP-dependent protein kinase. *Arch. Biochem, Biophys.*, **205**, 1.

96. Bramson, H. N., Thomas, N. E., Miller, W. T., Fry, D. C., Mildvan, A. S., and Kaiser, E. T. (1987) Conformation of Leu–Arg–Arg–Ala–Ser–Leu–Gly bound in the active site of adenosine cyclic 3′,5′-phosphate dependent protein kinase. *Biochemistry*, **26**, 4466.

97. Rosevear, P. R., Fry, D., Mildvan, A., Doughty, M., O-Brian, C., and Kaiser, E. T. (1984) NMR studies of the backbone protons and secondary structure of pentapeptide and heptapeptide substrate bound to bovine heart protein kinase. *Biochemistry*, **23**, 3161.

98. Buechler, J. A. and Taylor, S. S. (1990) Differential labeling of the catalytic subunit of cAMP-dependent protein kinase with a water-soluble carbodiimide: identification of carboxyl groups protected by MgATP and inhibitor peptides. *Biochemistry*, **29**, 1937.

99. Reed, J., Kinzel, V., Kemp, B. E., Cheng, H.-C., and Walsh, D. A. (1985) Circular dichroic evidence for an ordered sequence of ligand/binding site interactions in the catalytic reaction of the cAMP-dependent protein kinase. *Biochemistry*, **24**, 2967.

100. Jiménez, J. S., Kupfer, A., Gottlieb, P., and Shaltiel, S. (1981) Substrate-mediated channeling of a chemical reagent to the active site of cAMP-dependent protein kinase. *FEBS Lett.*, **130**, 127.

101. Kupfer, A., Jiménez, J. S., Gottlieb, P., and Shaltiel, S. (1982) On the protein accommodating site of the catalytic subunit of adenosine cyclic 3′,5′-phosphate dependent protein kinase. *Biochemistry*, **21**, 1631.

102. Jiménez, J. S., Kupfer, A., Gani, V., and Shaltiel, S. (1982) Salt-induced conformational changes in the catalytic subunit of adenosine cyclic 3′,5′-phosphate dependent protein kinase. Use for establishing a connection between one sulfhydryl and the γ-P subsite in the ATP site of this subunit. *Biochemistry*, **21**, 1623.

103. Van Patten, S. M., Heisermann, G. J., Cheng, H.-C., and Walsh, D. A. (1987) Tyrosine kinase catalyzed phosphorylation and inactivation of the inhibitor protein of the cAMP-dependent protein kinase. *J. Biol. Chem.*, **262**, 3398.

104. Reed, J., de Ropp, J. S., Trewhella, J., Glass, D. B., Liddle, W. K., Bradbury, E. M. *et al.* (1989) Conformational analysis of PKI(5–22) amide, the active inhibitory fragment of the inhibitor protein of the cyclic AMP-dependent protein kinase. *Biochem. J.*, **264**, 371.

105. Weldon, S. L. and Taylor, S. S. (1985) Monoclonal antibodies as probes for functional domains in cAMP-dependent protein kinase II. *J. Biol. Chem.*, **260**, 4203.

106. Levin, L. R. and Zoller, M. J. (1990) Association of catalytic and regulatory subunits of cyclic AMP-dependent protein kinase requires a negatively charged side group at a conserved threonine. *Mol. Cell. Biol.*, **10**, 1066.

107. Adams, J. A., unpublished results.

108. Karlsson, R., Zheng, J., Xuong, N.-h., Taylor, S. S., and Sowadski, J. M. (1993) The crystal structure of the mammalian catalytic subunit of cAMP-dependent protein kinase and a di-iodinated PKI(5–24) inhibitor peptide displays an open conformation. *Acta Cryst.*, **D49**, 381.

109. Olah, G. A., Mitchell, R. D., Sosnick, T. R., Walsh, D. A., and Trewhella, J. (1993) Solution structure of the cAMP-dependent protein kinase catalytic subunit and its contraction upon binding the protein kinase inhibitor peptide. *Biochemistry*, **32**, 3649.

110. Parello, J. and Timmins, P. A., personal communication.

111. McDonald, R. C., Steitz, T. A., and Engelman, D. M. (1979) Yeast hexokinase in solution exhibits a large conformational change upon binding glucose or glucose-6-phosphate. *Biochemistry*, **18**, 338.

112. Kong, C.-T. and Cook, P. F. (1988) Isotope partitioning in the adenosine 3'-5'-monophosphate dependent protein kinase reaction indicates a steady-state random kinetic mechanism. *Biochemistry*, **27**, 4795.

113. Cook, P. F., Neville, M. E., Vrana, K. E., Hartl, F. T., and Roskoski, J. R. (1982) Adenosine cyclic 3',5'- monophosphate dependent protein kinase: kinetic mechanism for the bovine skeletal muscle catalytic subunit. *Biochemistry*, **21**, 5794.

114. Whitehouse, S., Feramisco, J. R., Casnellie, J. E., Krebs, E. G., and Walsh, D. A. (1983) Studies on the kinetic mechanism of the catalytic subunit of the cAMP-dependent protein kinase *J. Biol. Chem.*, **258**, 3693.

115. Ho, M.-f., Bramson, H. N., Hansen, D. E., Knowles, J. R., and Kaiser, E. T. (1988) Stereochemical course of the phospho group transfer catalyzed by cAMP-dependent protein kinase. *J. Amer. Chem. Soc.*, **110**, 2680.

116. Granot, J., Mildvan, A. S., Bramson, H. N., and Kaiser, E. T. (1980) Magnetic resonance measurements of intersubstrate distances at the active site of protein kinase using substitution-inert cobalt (III) and chromium (III) complexes of adenosine 5'-(β, γ-methylenetriphosphate). *Biochemistry*, **19**, 3537.

117. Yoon, M.-Y. and Cook, P. F. (1987) Chemical mechanism of the adenosine cyclic 3',5'-monophosphate dependent protein kinase from pH studies. *Biochemistry*, **26**, 4118.

118. Adams, J. A. and Taylor, S. S. (1992) The energetic limits of phosphotransfer in the catalytic subunit of cAMP-dependent protein kinase as measured by viscosity experiments. *Biochemistry*, **31**, 8516.

119. Clegg, C. H., Ran, W., Uhler, M. D., and McKnight, G. S. (1989) A mutation in the catalytic subunit of protein kinase A prevents myristylation but does not inhibit biological activity. *J. Biol. Chem.*, **264**, 20140.

120. Duronio, R. J., Jackson-Machelski, E., Heuckeroth, R. O., Olins, P., Devine, C. S., Yonemoto, W. *et al.* (1990) Protein N-myristoylation in *Escherichia coli*: reconstitution of a eukaryotic protein modification in bacteria. *Proc. Natl. Acad. Sci. USA*, **87**, 1506.

121. Yonemoto, W., McGlone, M. L., and Taylor, S. S. (1993) N-myristylation of the catalytic subunit of cAMP-dependent protein kinase conveys structural stability. *J. Biol. Chem.*, **268**, 2348.

122. Kamps, M. P., Buss, J. E., and Sefton, B. M. (1985) Mutation of NH$_2$-terminal glycine of p60^{src} prevents both myristoylation and morphological transformation. *Proc. Natl. Acad. Sci. USA*, **82**, 4625.

123. Towler, D. A., Gordon, J. I., Adams, S. P., and Glaxer, L. (1988) The biology and enzymology of eukaryotic protein acylation. *Ann. Rev. Biochem.*, **57**, 69.

124. Toner-Webb, J., van Patten, S. M., Walsh, D. A., and Taylor, S. S. (1992) Autophosphorylation of the catalytic subunit of cAMP-dependent protein kinase. *J. Biol. Chem.*, **267**, 25174.

125. Yonemoto, W., Garrod, S. M., Bell, S. M., and Taylor, S. S. (1993) Identification of phosphorylation sites in the recombinant catalytic subunit of cAMP-dependent protein kinase. *J. Biol. Chem.*, **268**, 18626.

126. Herberg, F. W., Bell, S. M., and Taylor, S. S. (1993) Expression of the catalytic subunit of cAMP-dependent protein kinase in *E. coli*: multiple isozymes reflect different phosphorylation states. *Prot. Eng.*, **6**, 771.

127. Kinzel, V., Hotz, A., Konig, N., Gagelmann, M., Pyerin, W., Reed, J. *et al.* (1987) Chromatographic separation of two heterogeneous forms of the catalytic subunit of cyclic AMP-dependent protein kinase holoenzyme type I and type II from striated muscle of different mammalian species. *Arch. Biochem. Biophys.*, **253**, 341.

128. Steinberg, R. A. (1991) A kinase-negative mutant of S49 mouse lymphoma cells is defective in posttranslational maturation of catalytic subunit of cyclic AMP-dependent protein kinase. *Mol. Cell. Biol.*, **11**, 705.

129. Steinberg, R. A., van Daalen Wetters, T., and Coffino, P. (1978) Kinase-negative mutants of S49 mouse lymphoma cells carry a trans-dominant mutation affecting expression of cAMP-dependent protein kinase. *Cell*, **15**, 1351.

130. Solomon, M. J., Lee, T., and Kirschner, M. W. (1992) Role of phosphorylation in p34^{cdc2} activation: identification of an activating kinase. *Mol. Biol. Cell*, **3**, 13.

131. Her, J.-H., Wu, J., Rall, T. B., Sturgill, T. W., and Weber, M. J. (1991) Sequence of pp42/MAP kinase, a serine/threonine kinase regulated by tyrosine phosphorylation. *Nucleic Acids Res.*, **19**, 3743.

132. Ahn, N. G., Segar, R., Bratlien, R. L., Diltz, C. D., Tonks, W. K., and Krebs, E. G. (1991) Multiple components in an epidermal growth factor-stimulated protein kinase cascade. *In vitro* activation of a myelin basic protein/microtubule-associated protein 2 kinase. *J. Biol. Chem.*, **266**, 4220.

133. Zhang, B., Tavare, J. M., Ellis, L., and Roth, R. A. (1991) The regulatory role of known tyrosine autophosphorylation sites of the insulin receptor kinase domain. An assessment by replacement with neutral and negatively charged amino acids. *J. Biol. Chem.*, **266**, 990.

134. Lee, M. G. and Nurse, P. (1987) Complementation used to clone a human homologue of the fission yeast cell cycle control gene cdc2. *Nature*, **327**, 31.

135. Knighton, D. R., Cadena, D. L., Zheng, J., Ten Eyck, L. F., Taylor, S. S., Sowadski, J. M., and Gill, G. N. (1993) Structural features that specify tyrosine kinase activity deduced from homology modeling of the EGF receptor. *Proc. Natl. Acad. Sci. USA*, **90**, 5001.

136. Orr, A. C. and Newton, A. C. (1994) Intrapeptide regulation of protein kinase C. *J. Biol. Chem.*, **269**, 8383.

137. Takeya, T. and Hanafusa, H. (1983) Structure and sequence of the cellular gene homologous to the RSV *src* gene and the mechanism for generating the transforming virus. *Cell*, **32**, 881.

138. Hofmann, F. (1980) Apparent constants for the interaction of regulatory and catalytic subunit of cAMP-dependent protein kinase I and II. *J. Biol. Chem.*, **255**, 1559.

139. Mutzel, R., Lacombe, M.-L., Simon, M.-N., De Gunzburg, J., and Veron, M. (1987) cAMP-dependent protein kinase from *Dictyostelium discoideum*: cloning and cDNA sequence of the regulatory subunit. *Proc. Natl. Acad. Sci. USA*, **84**, 6.

140. Toda, T., Cameron, S., Sass, P., Zoller, M. J., Scott, J. D., McMullen, B. *et al.* (1987) Cloning and characterization of *BCY1*, a locus encoding a regulatory subunit of the catalytic cAMP-dependent protein kinase in *Saccharomyces cerevisiae. Mol. Cell. Biol.*, **7**, 1371.

2 | Structural aspects: pseudosubstrate and substrate interactions

BRUCE E. KEMP, MAREE C. FAUX, ANTHONY R. MEANS, COLIN HOUSE, TONY TIGANIS, SHU-HONG HU, and KEN I. MITCHELHILL

1. Introduction

This chapter describes the structural aspects of protein substrate recognition and pseudosubstrate regulation of members of the protein kinase family. Substrate specificity studies have increased our understanding of how protein kinases recognize their substrates. They have also led to the development of synthetic peptide reagents for assaying and in some instances affinity purifying protein kinases.

Many protein kinases exist in latent, inactive forms and require the binding of a specific allosteric regulator or phosphorylation by another protein kinase to exhibit activity. In a number of cases the protein kinase is inactive because part of its structure blocks access to the active site and acts as an autoinhibitor. This region is called a pseudosubstrate because it binds in the active site. Activation of the protein kinase is achieved by removing the pseudosubstrate from the active site and allowing protein substrate access. This form of regulation has been termed *intrasteric* regulation to emphasize that it is occurring within the active site as opposed to acting at another site, and that the pseudosubstrate interaction mimics some aspects of the substrate's structure (1).

Both pseudosubstrate regulation and substrate recognition are intimately inter-related and our understanding of these has developed in parallel. Intrasteric regulation encompasses any change in the active site region that modifies the enzyme's kinetic properties. This may include phosphorylation at the active site (*intrasteric* phosphorylation) as is the case for the bacterial α-ketoglutarate dehydrogenase (2), phosphorylation in the ATP-binding loop for $p34^{cdc2}$ (3) and in the substrate-anchoring loop containing Thr197 in several protein kinases (4) including the cAMP-dependent protein kinase (PKA; 5). Phosphorylation events that occur away from the active site are referred to as *extrasteric*-phosphorylation events.

The rapid increase in the discovery of new protein kinases by recombinant DNA techniques has highlighted the need to understand the underlying features of substrate recognition and pseudosubstrate regulation. Increasingly, protein kinases may be studied without initially isolating native enzyme but by using expressed enzyme and in these instances insights into the mechanism of substrate specificity and regulation are of great benefit. While pseudosubstrate regulation of protein kinases appears to be a common mechanism, it is by no means universal. Many other protein kinases, including receptor-linked protein kinases, appear to be activated by dimerization events with no obvious involvement of pseudosubstrate sequences. The pp60^{c-src} family of protein kinases are regulated by interactions between the amino-terminal SH2 domains and phosphotyrosine on the carboxy-terminal end of the catalytic domain. Precisely how the interactions block enzyme activity is not known but it seems unlikely to involve a pseudosubstrate mechanism.

2. Substrate recognition

It was recognized early in the study of phosphorylase phosphorylation by phosphorylase kinase, that a specific serine residue (Ser14) was the target of the phosphorylation reaction. Furthermore, a chymotryptic peptide containing this site could be phosphorylated by phosphorylase kinase (6). Similarly, phosphorylation of proteolytic fragments of myelin basic protein by PKA was observed (7). The next major advance in understanding substrate recognition was the identification of residues in the local primary sequence around the phosphorylation site that acted as specificity determinants for protein kinases. This was shown with genetic variants of exogenous substrates such as β-casein B (8) as well as with synthetic peptides (9–11). In some cases synthetic peptides such as LRRASLG (kemptide) were phosphorylated with kinetic parameters similar to natural substrates. These findings indicated that the local phosphorylation site sequence could contain all the features required for productive substrate recognition.

However, synthetic peptides corresponding to some local phosphorylation sites were not good substrates indicating that higher orders of structure may also play a role (12, 13). The crystal structure of PKA and extensive mutation experiments have now greatly extended our earlier insights into substrate recognition. Because of the conservation of specific residues among the members of the Ser/Thr specific protein kinase family it is expected that many features of the secondary structure will be conserved between the different protein kinases. This expectation has been borne out in the newly reported cyclin dependent kinase 2 crystal structure (14) that retains many of the secondary structural features of PKA except for the β strand 9 and the B α-helix. As the number of available structures expands, greater insight will be gained into the range of ways substrates and pseudosubstrates interact with protein kinases. The substrate recognition features of specific protein kinases are discussed below under their individual headings.

3. Pseudosubstrate regulation

The concept that protein kinases can be regulated by pseudosubstrates originated from development to studies on PKA where it was proposed that the protein inhibitor, PKI, acted as a competitive substrate antagonist due to the presence of arginine residues analogous to those in protein substrates (15). It was also apparent that the type I regulatory subunit (R_I) could be displaced by Arg-rich substrates or polyarginine (16, 17), implying that the regulatory subunit functioned by masking the active site (18, 19). The proposal that PKI was acting as a substrate antagonist has since been confirmed by the solution of the crystal structure of PKA-inhibitor peptide complex. This contained the inhibitor peptide bound to the active site with electrostatic interactions similar to those expected for peptide substrates. The regulatory subunits of PKA also contain pseudosubstrate sequences (reviewed in 20) responsible for inhibiting the enzyme's catalytic subunit and while the corresponding crystal structures are not yet available there is compelling indirect evidence in favour of them binding to the active site (20). Studies of other protein kinases and their substrate specificity requirements together with their cDNA-derived primary sequences have made it possible to recognize additional examples of pseudosubstrate regulation suggesting that this is a common regulatory mechanism. These have included the myosin light chain kinases, calmodulin-dependent protein kinase II, and protein kinase C family (1).

3.1 Overview and models

Pseudosubstrate inhibitory sequences are found in several forms. The pseudosubstrate sequence can occur in a regulatory subunit of the enzyme that dissociates upon activation, as for PKA family. The inhibitor protein PKI also inhibits PKA by reversibly binding to the free catalytic subunit. However, in the majority of examples known, the pseudosubstrate is contiguous with the catalytic domain. Examples of different pseudosubstrate locations in protein kinases are given in Fig. 1. The myosin light chain kinase family have pseudosubstrate sequences carboxy-terminal to their catalytic domains whereas the protein kinase C family have their pseudosubstrate sequences amino-terminal to the catalytic domain adjacent to the lipid-binding regulatory motif (1). Some protein kinases exist as multimeric complexes ranging from dimers, such as cGMP-dependent protein kinase, to the dodecamer complexes of calmodulin-dependent protein kinase II. While it is theoretically possible for the pseudosubstrate sequence of one monomer to inhibit another in a *trans* interaction, it appears that all inhibition is *cis* with the pseudosubstrate sequences folding back into the active site of their parent catalytic domain. The connecting sequence between the pseudosubstrate sequence and the catalytic domain is often sensitive to proteolysis leading to removal of the pseudosubstrate sequence and generation of a constitutively active catalytic core (1).

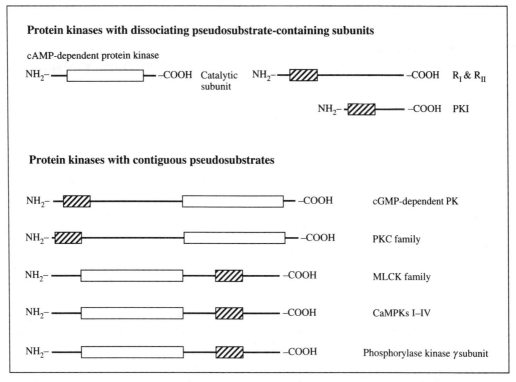

Fig. 1 Location of pseudosubstrate regulatory sequences. Protein kinase domain structures are shown with the catalytic core indicated by an open box. Pseudosubstrate regulatory sequences are shown with hatched boxes. R_I and R_{II} are regulatory subunits of cAMP-dependent protein kinase; PKI, protein kinase inhibitor; PKC, protein kinase C; MLCK, myosin light chain kinase; CaMPKs I–IV, calmodulin-dependent protein kinase isoforms I to IV.

3.2 Pseudosubstrate identification

The identification of a pseudosubstrate sequence in a newly sequenced protein kinase depends on whether it is a member of a protein kinase subfamily with an already identified pseudosubstrate motif, such as protein kinase C, or if it is one for which little is known about its properties. There are a series of considerations that aid in identification of pseudosubstrate sequences:

● Is there evidence of latent autoinhibited protein kinase activity?

● What are the substrate specificity requirements?

● Are there any amino acid sequences present in the enzyme that mimic features of the substrate specificity determinants?

● Does the corresponding synthetic peptide inhibit enzyme activity?

● Does deletion of the putative pseudosubstrate sequence render the enzyme constitutively active?

- Does an anti-pseudosubstrate sequence antibody activate the enzyme?
- What is the three dimensional structure of the autoinhibited enzyme?

PKA is the only example of a pseudosubstrate regulated protein kinase where almost all of these questions have been answered experimentally. Specific examples of these approaches are given below. The identification of pseudosubstrate sequences for PKA, protein kinase C and the myosin light chain kinase was greatly aided by a detailed knowledge of the substrate specificity of these enzymes gained by studies with synthetic peptides. The pseudosubstrate sequences have frequently been found adjacent to or overlapping allosteric regulatory sites such as the calmodulin-binding sequence in the calmodulin-regulated protein kinase family. In other instances the pseudosubstrate sequence is adjacent to sites of phosphorylation (1). Since the sequence homology between the pseudosubstrate sequence and the substrate can often be little more than a resemblance with no convincing significance in terms of homology alone, it is important to take advantage of as many clues as possible that are afforded by the enzyme's regulation. Given the structure of the extended substrate binding groove revealed in the PKA crystal structure it seems reasonable that autoinhibitory sequences of other protein kinases could bind to the active site without sharing all or possibly any of the specific contacts of the substrate. In this instance the term pseudosubstrate would be used loosely emphasizing location of binding rather than specific interactions.

Thus far all pseudosubstrate sequences have been found outside the catalytic domain so that only residues distal to either the amino-terminal or carboxy-terminal regions of the catalytic domains need to be considered. From structural considerations it seems unlikely that a pseudosubstrate sequence could be located closer than approximately 10 residues from either end of the catalytic domain but there is probably no upper limit on how distal a position may be in the linear sequence. The PKA crystal structure revealed that the extended substrate-binding groove between the two lobes was large raising the possibility that a pseudosubstrate sequence could bind in this region without making precisely the same contacts as the substrate, analogous to an antibody/protein, antibody/anti-idiotype antibody interaction. Indeed the crystal structure of twitchin (21) reveals that the autoinhibitory sequence can make extensive contacts across the entire surface of the active site including interactions with critical catalytic residues as well as those associated with substrate binding.

3.2.1 Pseudosubstrate inhibitory peptides

Synthetic peptides corresponding to the pseudosubstrate sequences for several protein kinases are very potent inhibitors with K_i values in the low to sub nM range (Table 1). However, this is not always the case, as the corresponding synthetic pseudosubstrate sequences in the cGMP-dependent protein kinase and PKA regulatory subunits have K_i values in the mM range. Pseudosubstrate inhibitor peptides have been used extensively as specific inhibitors in both *in vitro* assays and permeabilized cell systems (22). Generally, high concentrations of the pseudo-

Table 1 Synthetic pseudosubstrate inhibitor peptides

Protein kinase	Sequence	K_i/IC_{50} (μM)	Reference
cAMP-PK	TTYADFIASGRTGRRN**A**AIHD	0.003	145
cGMP-PK (α form)	PRTTRAQ**G**ISAEP	2100	Unpublished[a]
(β form)	PRTKRQ**A**ISAEP	900	Unpublished[a]
Protein kinase C	RFARKG**A**LRQKNV	0.13	141
Cam-II PK	MHRQETVDCLKKFNARRKLKG**A**ILTTMLA	2.7	111
	LKKFNARRKLKG**A**ILTTMLA	2.4	109
smMLCK (787–807)	SKDRMKKYMARRKWQKTGH**A**V	0.012	50
(774–807)	LQKDTKNMEAKKLSKDRMKKYMARRKWQKTGH**A**V	0.0003	50
skMLCK	KRRWKKNFI**A**G	3.2	70

[a] W. Chen, B. Michell, K. I. Mitchelhill, P. J. Robinson, B. E. Kemp.

substrate inhibitors are required in crude systems, possibly due to non-specific interactions with other cellular components. Caution must be exercised in these applications because of the possibility that the particular pseudosubstrate inhibitor can inhibit other protein kinases besides the target enzyme (23). Remarkably, even a polar pseudosubstrate peptide may gain access to the cytosol but very high extracellular concentrations are required to achieve this (24) and it may not be generally applicable. Recent attempts to overcome the problem of impermeability of the plasma membrane have exploited the use of myristoylated pseudosubstrate peptides for protein kinase C (25). Another alternative approach has been to use a pseudosubstrate gene construct to permit intracellular expression of the PKI (26).

3.3 cAMP-dependent protein kinase

The structure and function of this protein kinase are described in detail in Chapter 1 (27). The holoenzyme consists of two cAMP-binding regulatory subunits and two protein kinase catalytic subunits. There appear to be at least four genes encoding the regulatory subunit isoforms and three separate genes encoding the catalytic subunits (28). The catalytic subunit consists of two lobes with the amino-terminal sequence forming the smaller ATP-binding domain and the carboxy-terminal sequence forming the larger substrate-binding domain. Catalysis occurs in the extended cleft between the lobes. The pseudosubstrate sequences of the PKA regulatory subunits as well as that of the inhibitor proteins are given in Table 2. Compelling evidence supporting the identification of these regions has come from a variety of studies including proteolysis, chemical modification, site-directed mutagenesis, and protein crystallography (20, 29; see Chapter 1).

3.3.1 Substrate recognition

PKA is responsible for transducing the cAMP second messenger signal in cells and phosphorylates a variety of proteins. Inspection of the phosphorylation site

Table 2 cAMP-dependent protein kinase pseudosubstrate sequences

Protein kinase	Sequence	Reference
cAMP-PK R I subunit (88–107)		
Bovine, human, pig	VVKGRRRRG**A**ISAEVYTEED	20
Mouse	VVKARRRRG**G**VSAEVYTEED	146
Dictyostelium	NNITRKRRG**A**ISSEPLGDKP	147
cAMP-PK R II subunit (86–105)[a]		
Bovine	PGRFDRRV**S**VCAET	20
Pig	PSKFTRRV**S**VCAET	20
Rat	PAKFTRRV**S**VCAET	148
Yeast (136–155)	HFNAQRRT**S**VSGET	149
PKI (5–26) rabbit muscle	TTYADFIASGRTGRRN**A**IHDIL	150
PKI (6–27) rat testis	SVISSFASSARAGRRN**A**LPDIQ	151

[a] The core pseudosubstrate sequence RRVS$_{95}$VCAE is conserved for bovine, mouse, pig and rat; Ser95 is autophosphorylated.

sequences of the known substrates of this protein kinase show that they all contain an Arg residue on the amino-terminal side of the phosphorylatable residue. The most common arrangements are Arg–Xaa–Ser, Arg–Arg–Xaa–Ser, Arg–Xaa–Xaa–Ser, and Lys–Arg–Xaa–Xaa–Ser. Early specificity studies based on synthetic peptides indicated that the configuration, Arg–Arg–Xaa–Ser was preferred and the notion that this was the consensus phosphorylation site sequence developed, but it is important to note that less than a third (12/46) of the natural phosphorylation site sequences have the Arg–Arg–Xaa–Ser recognition motif (30). The three dimensional crystal structure of PKA (5) has provided an entirely new basis for understanding substrate recognition. From the crystal structure of the catalytic subunit inhibitor/peptide complex (31) it was possible to infer which acidic residues make important electrostatic contacts with the Arg guanidino groups in the peptide and protein substrates. Moreover, these assignments were confirmed independently by mutation studies (32–34). The substrate–active site interactions are summarized in Fig. 2.

The crystal structure has also provided insight into non-electrostatic interactions between the substrate and the enzyme. Earlier studies with synthetic peptides had indicated that a hydrophobic residue on the carboxyl side of the Ser (P + 1 position) was preferred. The crystal structure has shown that Leu198, Pro202 and Leu205 on the enzyme formed a hydrophobic pocket between the β strand 9 structure and the F α-helix. Depending on the substrate or regulatory subunit, there may be additional interactions with the catalytic subunit. A mutant form of yeast PKA with Thr241 (equivalent to mammalian residue Thr197) changed to Ala causes a 100-fold decrease in affinity for the regulatory subunit but similar catalytic activity (35). Mutation of Lys233, Asp237, Lys257, and Lys261 decrease the association of the regulatory subunit some 260-fold but only cause approximately a 2-fold

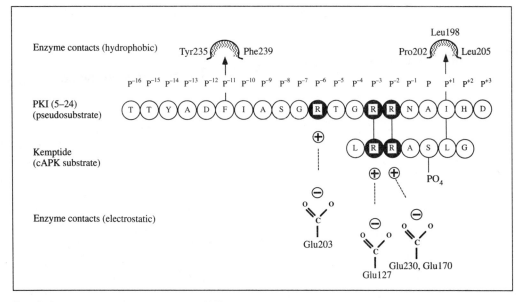

Fig. 2 Contact points between the cAMP-dependent protein kinase active site and its substrate/pseudosubstrate. The substrate and pseudosubstrate peptides are aligned according to the position of the substrate phosphorylation site (P). Other residues are assigned a position relative to Ser (PO$_4$). Basic residues making electrostatic contacts are shaded.

change in the rate of phosphorylation of kemptide (34). These findings emphasize that alteration to the protein kinase surface outside the primary electrostatic and hydrophobic interactions mentioned above may modulate protein interactions.

3.3.2 Pseudosubstrates

The regulatory subunits of PKA, as well as the protein kinase inhibitors, contain pseudosubstrate sequences that are responsible for inhibition of the catalytic subunit (Table 2). The regulatory subunit R$_I$ sequences and the protein kinase inhibitor sequences contain the motif Arg–Arg–Xaa–(Ala/Gly)–Yaa, where Yaa is typically a hydrophobic residue, Val or Ile, and Xaa is variable. This motif is the same as the substrate recognition sequence with Ala or Gly replacing the phosphorylated residue. The regulatory subunit R$_{II}$ sequences are distinct in that they contain a phosphorylatable residue, Ser. The R$_{II}$ subunit becomes phosphorylated at this site in the presence of cAMP and this leads to a reduction in the affinity for the catalytic subunit. It seems reasonable that the R$_{II}$ pseudosubstrate sequence is located in the active site in a way that precludes phosphorylation unless cAMP is bound. How this is achieved is not yet known.

Studies on the structure/function relationships of the protein kinase inhibitor of PKA using synthetic peptides together with the three dimensional structure of the enzyme have given new insight into the interactions of the pseudosubstrate inhibitor and the enzyme. In particular, they have shown how the inhibitor peptide interacts with such high affinity to the catalytic subunit (36). There are five important contacts between the inhibitor peptide PKI(5–22), T_5TYADFIASGRTGRRNAI$_{22}$ and the enzyme. The substrate recognition determinants Arg18, Arg19, and Ile22 mimic substrate binding interactions whereas Arg15 and Phe10 provide additional interactions so that the inhibitor peptide binds with a K_i in the nM range whereas substrates generally bind in the μM range (31). A similar pattern is also seen with the myosin light chain kinase where multiple contacts result in a strong interaction with this enzyme's pseudosubstrate (see below). The Ile22 acts as an important hydrophobic anchor and substitution at this position with Gly causes a 100-fold decrease in potency. Recently, small-angle X-ray scattering and Fourier transform infrared spectroscopy have been employed to provide evidence that binding of pseudosubstrate peptide PKI(5–22) is accompanied by rotation of the small and large lobes to close the cleft (37). This movement occurs around a conserved glycine between the lobes.

The potency of the PKI peptide can be further enhanced by the use of non-biological amino acids; for example, substitution of Phe10 with naphthylalanine (36). The complex interaction of the PKI(5–22) peptide with the catalytic subunit emphasizes that multiple contacts are required to ensure potent inhibition. Thus in attempts to design pseudosubstrate inhibitors based on substrate sequences for other protein kinases it will be important to engineer additional contacts to those of the substrate to ensure high affinity binding of the inhibitor.

3.4 cGMP-dependent protein kinase

In contrast to PKA, the cGMP-dependent protein kinase has its cGMP-binding regulatory domain fused with its catalytic domain (38; Fig. 1). Two genes for cGMP-dependent protein kinase have now been described for both *Drosophila* and mammalian sources which are also subject to alternative splicing (39–41). The cGMP-dependent protein kinase plays a prominent role in smooth muscle function and is under the control of nitric oxide and atriopeptin. Studies on the cGMP-dependent protein kinase regulation in the laboratories of Corbin (42) and Gill (43) were important in the initial development of the concept that this enzyme was regulated by a pseudosubstrate mechanism. Bovine cGMP-dependent protein kinase is autophosphorylated at Thr58 (38). Whereas the analogous Ser95 in the PKA R_{II} subunit is believed to occupy the active site it now seems unlikely that Thr58 does so in the cGMP-dependent protein kinase. Inspection of the range of pseudosubstrate sequences suggested that, in fact Gly62 is likely to occupy the phosphate acceptor site (1; Table 3). In the β-isoform of the cGMP-dependent protein kinase, Ala78 occupies the corresponding position to Gly62 (Table 3). We have found that substitution of the corresponding pseudosubstrate peptides with Ser (β-isoform Ala78 or the α-isoform Gly62) renders them substrates at these positions, albeit weak. If Gly62 occupies the active site then autophosphorylation on Thr58 would be a secondary event following release of the pseudosubstrate from the active site in the α-form. Activation of the enzyme by cGMP leads to

Table 3 cGMP-dependent protein kinase pseudosubstrate sequences

Protein kinase	Sequence	Reference
bcGMPK a (54–69)	GPRT**T**RAQ**G**ISAEP	38[a]
hcGMPK b (70–83)	GEPRTKRQ**A**ISAEP	40
mcGMPKII (115–128)	HSRRGAKA**G**VSAEP	41
DrcGMPKI (137–150)	MPAAIKKQ**G**VSAES	39
DrcGMPKII (474–485)	QNFRQRAL**G**ISAEP	39

[a] Thr58 is autophosphorylated and was thought to occupy the active site of the enzyme. Inspection of the aligned sequences suggests Gly62 occupies the active site as the pseudosubstrate (1). Recent proteolysis studies (Francis and Corbin, unpublished) provide direct evidence supporting the assignment of the pseudosubstrate sequence and distinguish it from the site of autophosphorylation.

autophosphorylation on Thr58 whereas up to six sites can be phosphorylated when cAMP is used to activate the enzyme (44). This suggests that an extended sequence surrounding the pseudosubstrate sequence is accessible for autophosphorylation rather than any of these residues being in the active site prior to activation.

3.4.1 Substrate recognition

The substrate specificity of the cGMP-dependent protein kinase has been studied by Glass and his colleagues using synthetic peptides corresponding to the phosphorylation site in histone, $R_{29}KRS*RKE_{35}$(45, 46). The peptide, RKRS*RAE is phosphorylated with a K_m of 29 μM and V_{max} of 20 μmol min^{-1} mg^{-1}. The basic residues were shown to be important specificity determinants (reviewed in 46) except for Lys34 which is a negative determinant. Bovine cGMP-specific cGMP-binding phosphodiesterase is also a substrate for the cGMP-dependent protein kinase (47). The synthetic peptide corresponding to the local phosphorylation site sequence in the phosphodiesterase, RKIS*ASEFDRPLR, is phosphorylated with a K_m of 68 μM and a V_{max} of 11 μmol min^{-1} mg^{-1}. Interestingly, the presence of a phenylalanine residue at the P + 4 position appeared to contribute to the selectivity of this and other peptides for the cGMP-dependent protein kinase rather than for PKA (48). We have found that a peptide corresponding to the amino-terminal 31 residues of phospholamban, MDKVQYLTRSAIRRAS*TIEMPQQARQNLQNL, is a good substrate for this enzyme with a K_m of 7 μM and a V_{max} of 20 μmol min^{-1} mg^{-1} (W. Chen *et al.*, unpublished).

3.4.2 Pseudosubstrates

Thus far attempts to develop pseudosubstrate peptide inhibitors of cGMP-dependent protein kinase based on the pseudosubstrate sequence in the enzyme (Table 3) have been disappointing and the short peptides tested were poor inhibitors (49). However, given the new insight into the features responsible for ensuring high potency for the PKA inhibitor (31) and the pseudosubstrate inhibitor of smooth muscle myosin light chain kinase (50) it may be possible to improve on pseudo-substrate inhibitor potency for this enzyme. Longer peptides incorporating multiple contacts are likely to be more effective.

3.5 Myosin light chain kinase family

The myosin light chain kinase family of protein kinases consists of structurally and functionally diverse proteins. It includes the largest known vertebrate protein, titin (3000 kDa) and the nematode protein, twitchin (700 kDa) (51) through to smooth muscle myosin light chain kinase (130–150 kDa), skeletal muscle myosin light chain kinase (87 kDa) and *Dictyostelium discoideum* myosin light chain kinase (34 kDa, 301 residues). Twitchin and titin contain a myosin light chain kinase catalytic domain near their respective carboxy-terminii but are largely made up of class I and class II motifs in super repeats (51). Smooth muscle myosin light chain kinase has three class II and one class I motif arranged around the catalytic domain, in the same juxtaposition as that found in twitchin and titin (Fig. 3).

The smooth muscle myosin light chain kinase has a well defined role in regulating the initiation of smooth muscle contraction (52, 53). Following an increase in cytosolic Ca^{2+}, the Ca^{2+}/calmodulin complex binds to the myosin light chain kinase at a specific binding site within residues 796–813 (54) and activates it to cause phosphorylation at Ser19 on the 20 kDa regulatory myosin light chains (Fig. 4). This, in turn, leads to stimulation of the actin-activated myosin MgATPase activity and muscle contraction. Myosin light chain kinase also occurs in non-muscle cells

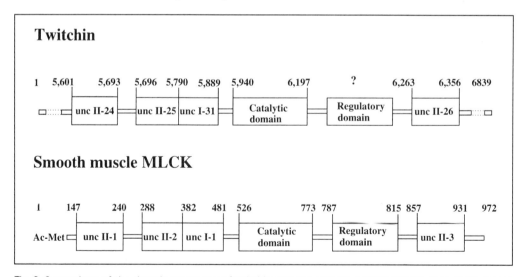

Fig. 3 Comparison of the domain structures of twitchin and the smooth muscle myosin light chain kinase (MLCK). Amino acids are numbered. The consensus motifs unc I and unc II on either side of the catalytic domain correspond to the fibronectin type III-like domain and the immunoglobulin-like domain respectively. Twitchin is the product of the *C. elegans* unc-22 gene.

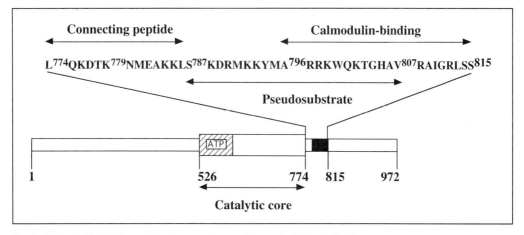

Fig. 4. Calmodulin-binding region in smooth muscle myosin light chain kinase.

where it is thought to participate in a number of functions including capping, cellular motility, organelle movement and spindle function in mitotic cells.

Skeletal muscle myosin light chain kinase is smaller, 87 kDa, and is also regulated by calmodulin but does not have the class I and class II motifs characteristic of its smooth muscle counterpart. Initiation of contraction in striated muscle is mediated via troponin on the thin filament with phosphorylation of the regulatory light chains serving to increase the rate of force development with a shift in the force–pCa relationship (55).

All of the members of the myosin light chain kinase family are autoinhibited but there is considerable variation in the mechanism of activation. The *Dictyostelium* isoform is thought to be activated by phosphorylation on its carboxy-terminal tail near the pseudosubstrate sequence (56). Neither titin nor twitchin contain calmodulin-binding motifs and their mechanisms of regulation are currently unknown. Both smooth muscle (54, 57) and skeletal muscle myosin light chain kinase (58) are regulated by calmodulin.

3.5.1 Substrate recognition

One of the striking features of the myosin light chain kinases is their restrictive substrate specificity. They do not phosphorylate other exogenous proteins and show tissue specificity with the myosin light chain source. For instance, smooth muscle myosin light chain kinase will not phosphorylate skeletal muscle myosin light chains (59), but the skeletal muscle enzyme phosphorylates both skeletal and smooth muscle myosin light chains. Synthetic peptides have been used extensively to study the specificity of the smooth muscle myosin light chain kinase (60–63). The 23-residue peptide MLC(1–23), S_1SKRAKAKTTKKRPQRATSNVFS$_{23}$ was phosphorylated with a K_m of 2.7 μM on Ser19. Although the K_m obtained with the peptide substrate was similar to the myosin light chains (approximately 7 μM), the V_{max} (3.6 μmol min^{-1} mg^{-1}) was 10-fold lower than for the myosin light chains.

The core sequence MLC(11–23) K_{11}KRPQRATSNVFA$_{23}$ was also phosphorylated with similar kinetics (61). All four basic residues, Lys11, Lys12, Arg13, and Arg16 are important specificity determinants. Moreover, the distance between the basic residues and the Ser19 phosphorylation site was critical. Shifting Arg16 to position 15 in the peptide sequence completely switched the site of phosphorylation from Ser19 to Thr18 (62). The complex spatial requirements and the number of basic residues required makes it apparent why the smooth muscle myosin light chain kinase has such restricted specificity. The reason for the 10-fold lower V_{max} for peptide phosphorylation compared to myosin light chain phosphorylation by the smooth muscle myosin light chain kinase has not been adequately resolved. The hydrophobic residues on the carboxy-terminal side of Ser19, Val21, and Phe22 contribute to the V_{max} for peptide phosphorylation (63) but other factors must be involved in phosphorylation of the parent protein. Interestingly, skeletal muscle myosin light chain kinase phosphorylated the myosin light chain peptides with V_{max} values similar to the parent protein (13). The importance of the basic residues as specificity determinants for the smooth muscle myosin light chain kinase was also supported by chemical modification studies which showed that phosphorylation of the myosin light chains was suppressed when Arg and Lys were modified (64).

The skeletal muscle myosin light chain kinase phosphorylates its corresponding myosin light chain peptide P_1KKAKRRAAEGSSNVF$_{17}$ with a K_m of 2 μM and a V_{max} of 1.0 μmol^{-1} min^{-1} mg. In the peptide, Glu10 appears to suppress phosphorylation because substitution with Arg increases the V_{max} 7-fold. The smooth muscle myosin light chain peptide K_{11}KRPQRATSNVFS$_{23}$ is a better substrate with a K_m of 11.6 μM and a V_{max} of 30.5 μmol^{-1} min^{-1} mg (13). In this case the reciprocal substitution of Arg16 with Glu reduces the V_{max} to 1.9 μmol^{-1} min^{-1} mg and increases the K_m to 592 μM. The peptide studies indicate that the skeletal muscle myosin light chain kinase also requires basic residues and that in the case of phosphorylation of the skeletal muscle myosin light chains the negative influence of Glu10 is presumably masked.

A number of studies have addressed the question of which residues on the myosin light chain kinase are responsible for recognition of the basic residue specificity determinants. Herring et al. (65) found that a monoclonal antibody that mapped to the region skMLCK(235–319), on the amino-terminal side of the catalytic core, competitively inhibited myosin light chain phosphorylation. Mutagenesis of aspartates 269 and 270 to asparagine within this region in rabbit skeletal muscle myosin light chain kinase resulted in a 10-fold increase in the apparent K_m for myosin light chain phosphorylation but only a 5-fold increase with peptide substrate. It was proposed that these residues form part of the light chain substrate-binding site (66). It was somewhat surprising that Asp residues would be involved in substrate recognition since longer side chain amino acids tend to be favoured in protein/protein electrostatic interactions. Chimeric kinases in which sequences in this region were exchanged between smooth and skeletal muscle myosin light chain kinase were inactive (67). It is not clear whether these forms of the enzyme were inherently inactive or could not be activated. If the latter were the case this would suggest that the amino-terminal region was more important for regulation than substrate recognition. Although mutagenesis of chicken skeletal muscle myosin light chain kinase at Ala494 for Glu in this region caused a 10-fold increase in the apparent K_m (67), mutagenesis of single acidic residues in the corresponding region in the smooth muscle myosin light chain kinase had no significant effect on the apparent K_m (68).

The most important insight into our understanding of the myosin light chain kinase substrate recognition came from the crystal structure of the PKA/inhibitor peptide complex (31). Molecular modelling of the smooth muscle myosin light chain kinase catalytic core on the coordinates of the PKA structure revealed potential acidic residue recognition sites (50) illustrated in Fig. 5. Both skeletal and smooth muscle myosin light chain kinases contain Glu residues corresponding to Glu170, Glu127, and Glu203 in PKA that are important for substrate recognition (31). Mutation of the myosin light chain kinase residues corresponding to Glu170 and Glu127 resulted in a 100- and 35-fold increase in the apparent K_m respectively (68) indicating that the primary substrate recognition occurs in the large lobe of the catalytic core. However, determination of the crystal structure of a myosin light chain kinase will be required to reveal the details of substrate recognition as well as the inhibitory mechanism of the pseudosubstrate sequence, and perhaps explain the earlier results (65–67) that lead to the proposal that residues amino-terminal to the catalytic core were important.

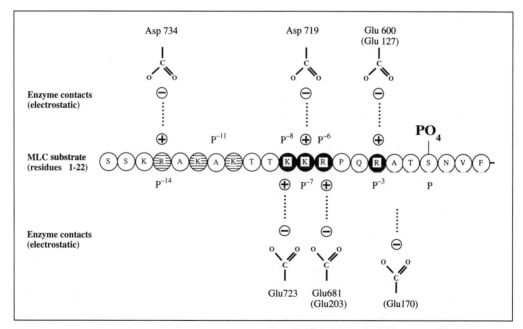

Fig. 5 Hypothetical contact points between the smooth muscle myosin light chain kinase (MLCK) active site and its substrate. Basic residues making electrostatic contacts are shaded; see also legend to Fig. 2.

3.5.2 Pseudosubstrates

Studies on the regulation of the myosin light chain kinase family have played a major role in the development of the pseudosubstrate regulatory model for the calmodulin-dependent protein kinases (reviewed in ref. 1). Initially it was recognized (69) that the basic residues present in the calmodulin-binding domain (58) resembled closely the number and juxtaposition of the basic residues identified as specificity determinants in the myosin light chain kinase (62).

substrate smMLC(1–22) $SSKR\,AKAK\,TT\,KKR\,P\,QRA\,TSNV^{22}-$

sm MLCK(787–808) $-SKDRMKKYMARRKWQKTGHAV^{807}-$

Synthetic peptides corresponding to this region were found to be inhibitors of the smooth muscle myosin light chain kinase (69). Similar results were also found for the skeletal muscle enzyme where the peptide KRRWKKNFIAVSAANRFG, related to the skeletal muscle myosin light chain kinase (skMLCK(577–594), inhibited a calmodulin-independent chymotryptic fragment of the enzyme βC35 (70). The kinetic studies with the intact smooth muscle enzyme were complicated due to the peptides acting as both substrate and calmodulin antagonists so that the initial IC_{50} values underestimated the potency of the peptides as substrate inhibitors and it was not until more thorough kinetic studies (71) and the use of calmodulin-independent fragments of the myosin light chain kinase were used (72) that the K_i values in the nanomolar range were determined. Detailed peptide studies as well as partial proteolysis experiments have shown that the pseudosubstrate inhibitor and calmodulin-binding functions overlap but are not identical (1). These studies have shown that the most potent calmodulin antagonist activity is contained within the sequence $A_{796}RRKWQKTGHAVRAIGRLSS_{815}$ as originally proposed by Lucas *et al.* (54). Carboxy-terminal extension of peptides beyond Val807 did not enhance their substrate antagonist activities (69, 72, 73). However, amino-terminal extension of the pseudosubstrate peptide greatly increased the potency as a substrate antagonist. Whereas the smMLCK(787–807) peptide has a K_i of 12 nM the longer peptide smMLCK(774–807) has a K_i of 0.3 nM (Table 2). The amino-terminal extension is not itself an inhibitor but greatly boosts the potency of the pseudosubstrate sequence (50).

The pseudosubstrate peptide, smMLCK(787–807) also acts as a potent ATP antagonist with an IC_{50} of 149 nM (50, 73). Similar ATP antagonism is also seen with the corresponding pseudosubstrate peptide from the calmodulin-dependent protein kinase II (74). We cannot exclude the possibility that the extended pseudosubstrate peptide naturally binds to the small lobe and modulates ATP binding. Nevertheless, this is surprising since active site labelling of the skeletal muscle myosin light chain kinase by 5'-*p*-fluorosulphonylbenzoyl adenosine (FSBA) is relatively insensitive to the presence of calmodulin, indicating that access to the ATP-binding site is not blocked by the pseudosubstrate inhibitor sequence in the native enzyme (75). This may indicate that the pseudosubstrate peptide does not mask the ATP-binding site in the small lobe or that the free pseudosubstrate peptide is able to bind to the active site in ways not available to its tethered counterpart. Alternatively, the on and off rate may be high for the pseudosubstrate sequence so that even if it masks the ATP-binding site FSBA may gain access and the pseudosubstrate sequence re-bind, but not necessarily in the same conformation. This vast number of options underscores the need for definitive three dimensional information.

Tyr794 in the myosin light chain kinase pseudosubstrate peptide also influences peptide substrate antagonist potency since substitution with Ala decreases potency 30-fold (50). This residue corresponds to Phe10 at the important $P - 11$ position in the inhibitor of PKA. The Ala794 substitution also strongly reduces ATP antagonist activity of the pseudosubstrate peptide. Paradoxically, Trp800 appears to be important for both optimal pseudosubstrate inhibition as well as playing a crucial role in the binding to calmodulin and the activation event (76).

A variety of approaches have been used to test the pseudosubstrate hypothesis and these will be discussed below to illustrate their strengths and shortcomings.

3.5.3 Proteolysis

Both skeletal and smooth muscle myosin light chain kinase (77) can be activated by partial proteolysis. Tryptic digestion of smooth muscle myosin light chain kinase proceeds with the initial generation of an intermediate 64 kDa fragment that is inactive: subsequent digestion gives rise to a constitutively active 61 kDa fragment (78). The 64 kDa fragment resulted from cleavage at Arg808 in the calmodulin-binding sequence adjacent to the pseudosubstrate sequence (79). Similar inactive fragments can also be generated by either thermolytic cleavage at Val807 or Lys-C endoproteinase cleavage at Lys802 (80) as shown in Fig. 6.

These results indicated that only a portion of the pseudosubstrate sequence is required for inhibition (up to Lys802) and that TGHAV$_{807}$ was not essential. The carboxy-terminal residue of the constitutively active 61 kDa fragment is Lys779, amino-terminal to the beginning of the pseudosubstrate sequence at Ser807 (81). Trypsin also cleaves the myosin light chain kinase on the amino-terminal side of the catalytic domain at Thr283 but the generation of the constitutively active 61 kDa fragment can be attributed entirely to the removal of residues on the carboxy-terminal side of the catalytic core (81). The peptide D$_{777}$TK$_{779}$NMEAKKLS$_{787}$KDRMKK$_{793}$ containing the sequence between Lys779 and Ser787 was a 10 000-fold weaker inhibitor of the myosin light chain kinase than the pseudosubstrate peptide 787–807, indicating that activation of the 64 kDa fragment of the myosin light chain kinase was due to the removal of the pseudosubstrate sequence smMLCK(787–807) (80). Ikebe et al. (82) reported a different result suggesting that cleavage of the enzyme at Lys793 generated an inactive fragment: however, this result is inconsistent with both peptide inhibition (81) and mutation studies (83). A 35 kDa chymotryptic fragment of the skeletal enzyme, skMLCK(256–584) that does not bind calmodulin was isolated (84). The cleavage site is within the calmodulin binding sequence of this enzyme. Interestingly, it retains most of the

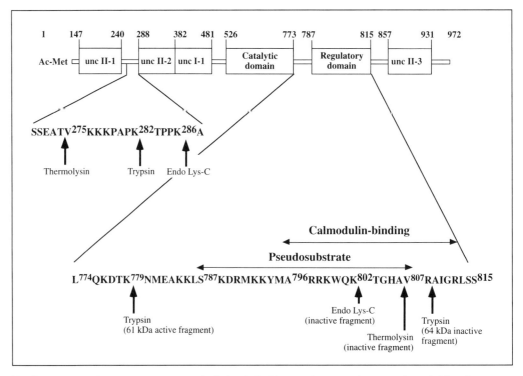

Fig. 6 Topology of smooth muscle myosin light chain kinase. Protease cleavage sites yielding fragments discussed in the text are indicated by arrows.

pseudosubstrate sequence but becomes increasingly active during purification (84). This inherent instability made it difficult to recognize that the carboxy-terminal sequence of the 35 kDa fragment was responsible for autoinhibition of the intact skeletal muscle enzyme (84).

The instability of the skeletal muscle myosin light chain kinase is also apparent from observations that addition of calmodulin to the enzyme in the absence of substrates resulted in temperature- and time-dependent inactivation (85). The instability resulted from the removal of the pseudosubstrate/calmodulin-binding sequence (85) and could be prevented by the addition of either substrate or inhibitor peptides. Subsequently, it was found that the constitutively active 61 kDa fragment of smooth muscle myosin light chain kinase was also very susceptible to thermal inactivation although, in this case, the native enzyme is not inactivated by calmodulin binding (86). Detailed structure/function studies revealed that the pseudosubstrate peptide 787–807, SKDRMKKYMARRKWQKTGHAV, protected the enzyme fragment from inactivation. The short peptide, RRKWQK, was equipotent (73 nM $PC_{0.5}$) at protecting the enzyme and all four basic residues as well as the tryptophan were important. Substitution of Trp800 in the full length peptide did not alter its potency for protecting the enzyme from the thermal inactivation suggesting that multiple interactions were involved (86). It is now apparent that the overlapping pseudosubstrate/calmodulin-binding sequence has multiple functions including a stabilizing structural role as well as its regulatory role. This may be a common feature of pseudosubstrate sequences because mutation studies with protein kinase C have shown that deletion of the pseudosubstrate sequence rendered the enzyme unstable (87).

3.5.4 Mutation analysis

The role of the pseudosubstrate sequence in regulating the myosin light chain kinase has been investigated by several laboratories using mutagenesis. Four approaches have been taken:

- Truncation mutagenesis to extend the information gained from proteolysis experiments.
- Point mutations to investigate the contribution of individual residues.
- Sequence inversions and exchanges to form chimeric molecules.
- Insertion of a phosphorylatable residue in the pseudosubstrate sequence.

Truncation mutagenesis studies of smooth muscle myosin light chain kinase by Ito *et al.* (83) showed that the pseudosubstrate sequence could be truncated up to Trp800 before there was substantial release from autoinhibition. Deletion of the residues $Y_{794}MARRKW_{800}$ resulted in activation of the enzyme. Truncation of the non-muscle myosin light chain kinase up to Glu1067 (corresponding to Glu782 in smMLCK), deleting AKKL as well as the pseudosubstrate sequence also resulted in constitutively active enzyme (88) as expected from the proteolysis studies. Based on the knowledge of the substrate specificity of the myosin light chain kinase one would have expected that mutation of the basic residues in the pseudosubstrate sequence would have rendered the enzyme constitutively active. This does not occur with either Ala substitutions or charge reversal of these basic residues (88–90). Thus, while truncation mutagenesis and proteolysis experiments yielded data consistent with the pseudosubstrate model point mutation, experiments did not. This paradox indicates that the pseudosubstrate sequence is sufficient to cause autoinhibition but key residues within it are not essential in the context of the full length enzyme.

The model of the myosin light chain kinase pseudosubstrate structure (50) provides important clues as to why individual residues can appear to make such a small contribution to autoinhibition. The connecting peptide pseudosubstrate sequence smMLCK (774–807) extends 34 residues across the surface of the kinase into the extended substrate binding groove making a number of electrostatic and hydrophobic contacts.

We do not know the binding energy between the pseudosubstrate sequence and the enzyme catalytic core and this complicates the interpretation of mutation experiments. Activity measurements comparing inactive and active forms of the enzyme are limited and if there is several-fold redundancy in the degree of inhibition this means that substitution of important residues may have strong effects on the binding energy but not cause the degree of inhibition to go below 100% and generate constitutively active enzyme. Truncation mutations are the most powerful because they more readily identify the minimum structural requirement for autoinhibition. Where an active site is being masked there is also the possibility of a steric inhibition component due to the overall location of the sequence backbone that may be relatively insensitive to point mutations.

The most crucial point of the pseudosubstrate hypothesis is that the pseudosubstrate is located in the enzyme's active site substrate binding groove. In order to test this directly, Bagchi *et al.* (91) replaced the pseudosubstrate/calmodulin-binding sequence with S_{787}KDRMKKYMAAAAWQRATSNVRAIGRLSS incorporating the phosphorylation site sequence RATSNV within the pseudosubstrate/calmodulin-binding sequence. This mutant underwent autophosphorylation in the absence of calmodulin supporting the view that this sequence was located in the active site region. However, whether the substrate/pseudosubstrate mutant precisely mimics the state of autoregulation in the native enzyme is not entirely clear. Moreover, calmodulin inhibited the autophosphorylation reaction as would be expected during activation of the enzyme by removal of the pseudosubstrate sequence from the active site. In the case of twitchin, a member of the myosin light chain kinase family with 50% identity in the catalytic domain, we have recently obtained the crystal structure and this shows unequivocally that the autoinhibitory carboxy-terminal extension occupies the active site and functions by an *intrasteric* mechanism (21).

3.5.5 Mechanism of activation by calmodulin

Calmodulin serves as an intracellular Ca^{2+} receptor and transduces Ca^{2+} transients that frequently are generated in cells in response to hormones or growth factors. The calmodulin molecule binds 4 moles of Ca^{2+} with high affinity and each Ca^{2+} is coordinated by a helix–loop–helix structure that Kretsinger and Nockolds have termed an EF-hand or calmodulin-fold (92). The Ca^{2+}-binding sites are arranged in pairs and each pair is located in a globular domain that comprises most of each half of this bilaterally symmetrical protein (93). The three-dimensional structure of Ca^{2+}/calmodulin revealed a dumbbell-shaped molecule in which each terminal globular domain containing two Ca^{2+} binding sites was separated by a solvent exposed eight-turn α-helix (94). The Ca^{2+}-binding sites in the carboxy-terminal lobe of vertebrate calmodulin have a slightly higher affinity for Ca^{2+} than do the pairs of sites in the amino-terminal lobe (95). Calcium binding to either member of a pair of sites greatly influences binding to the other whereas one pair of sites is not markedly dependent on the other pair (96).

In terms of the mechanism activation of target enzymes by calmodulin, the most is known about smMLCK. The calmodulin-binding region was identified by cyanogen bromide fragmentation of chicken smooth muscle myosin light chain kinase (54) between amino acids Ser787 and Ser813 (81) overlapping the autoregulatory domain. By the use of synthetic peptide analogues it was shown that the autoregulatory domain could be roughly divided into three with the amino-terminal two-thirds (787–807) containing the pseudosubstrate region and the carboxy-terminal two-thirds (796–813) being responsible for calmodulin binding. Bagchi *et al.* (76) expressed a 40 kDa fragment of smMLCK in bacteria that contained the entire catalytic and autoregulatory domain (Asp450–Ser815). Deletion of three residues from the carboxy-terminus of the calmodulin-binding region, Gly811, Arg812, and Leu813, was sufficient to completely prevent calmodulin-binding and consequently calmodulin-dependent activation of the enzyme. Subsequently, it was found that mutation of any one of these three amino acids would also eliminate calmodulin-binding (76). Araki and Ikebe (97) reported that a monoclonal antibody specific for amino acids between Arg808 and Ser815 activates the myosin light chain kinase.

Extensive mutagenesis studies within the calmodulin-binding sequence have examined the relationship between calmodulin-binding and activation (76, 88, 91). Generally it has been found that residues in the autoregulatory sequence that encompasses the pseudosubstrate sequence (residues 796–807) are less important for calmodulin-dependent activation than are residues 808–813 with the exception of Trp800. Mutation of Trp800 to Ala completely eliminates calmodulin-binding (76). Many other calmodulin-dependent enzymes have a hydrophobic residue at an identical position in the calmodulin-binding domains such as the plasma membrane Ca^{2+}/calmodulin-dependent ATPase (98). Interestingly, chimeric enzymes where the calmodulin-binding sequences of skeletal muscle myosin light chain kinase or the multifunctional calmodulin-dependent protein kinase (CaM kinase) have been substituted into the smooth muscle myosin light chain kinase reveal calmodulin-dependent activity (67, 88). Despite considerable variation between the three calmodulin domains, all share a hydrophobic residue at the position normally occupied by Trp800. Shoemaker *et al.* (88) reported that inversion of the calmodulin-binding sequence in non-muscle myosin light chain kinase does not disrupt calmodulin dependency. In the inverted sequence an Ile occupies the position of Trp800 in the normal sequence. Thus, in all reports to date, calmodulin-dependent activation of myosin light chain kinase isoforms requires a hydrophobic residue in the fourth position (equivalent to Trp800).

A structural basis for calmodulin-binding to myosin light chain kinase has now been provided by the solution of the crystal structure of vertebrate Ca^{2+}/calmodulin complexed to a synthetic peptide corresponding to smMLCK(796–813) (99). The calmodulin in the complex forms a tunnel diagonal to its long axis that completely engulfs the 18 amino acid peptide which is helical along its entire length. To accommodate the peptide, calmodulin has undergone a bend of 100°, due to flexibility of residues 73–77 within the central helix, and a 120° twist between the two globular lobes of calmodulin. The close proximity of the two lobes of calmodulin creates a long arc of hydrophobic residues that is contiguous with the hydrophobic patch in each lobe. This arc interacts with the hydrophobic face of the helical peptide. The orientation has the amino-terminal lobe of calmodulin interacting with the carboxy-terminus of the peptide and vice versa as was previously suggested from NMR experiments (100). The peptide is anchored into position by Leu813 in the amino-terminal lobe with Trp800 in the carboxy-terminal lobe. The only residue in the peptide that extensively interacts with the bend and the helices that occur immediately proximal and distal to the bend is Arg812. A considerable number of these contacts are hydrogen bonds and charge-coupling interactions. Thus Arg812 serves as the pivotal residue to orchestrate the bend in calmodulin. Overall there are about 185 contacts between calmodulin and the peptide with 80% of these <4 Å, associations being van der Waals interactions.

A high resolution structure of calmodulin bound to a calmodulin-binding peptide based on that present in skeletal muscle myosin light chain kinase has been solved by multi-dimensional NMR (101). Comparison of this structure and the crystal structure of the Ca^{2+}/calmodulin/smMLCK peptide complex reveals considerable similarities. Particularly relevant to the question of how calmodulin might activate its target enzymes is that both structures are arranged amino- to carboxy- and carboxy- to amino- and the essential hydrophobic residues are conserved. Thus, the critical importance of hydrophobic residues flanking the peptide has been confirmed in two high resolution structures.

The pseudosubstrate of myosin light chain kinase seems to be arranged in the active site so that the carboxy-terminal residues that are critical for calmodulin binding (808–813) would be available for calmodulin binding, whereas the amino-terminal residues (787–807) are much more intimately associated with the catalytic core of the enzyme (50). Indeed, the molecular modelling predicts that amino acids 801–813 would probably not exist as an α-helix in the autoinhibited state of the enzyme. We can predict from the data available that the amino-terminal portion of calmodulin might initially bind to the carboxy-terminal end of the autoregulatory region of smooth muscle myosin light chain kinase with Arg812 and Leu813 playing prominent roles. This initial interaction would result in the bend in calmodulin and a conformational change in the carboxy-terminal portion of the calmodulin-binding domain that would result in helix formation. Generation of the helix would remove the pseudosubstrate from the active site such that the carboxy-terminal lobe of calmodulin could twist and engulf Trp800. We would expect that the surface of the amino-terminal lobe of calmodulin would not be involved in direct binding to the calmodulin-binding residues of the enzyme but may participate in removal of the pseudosubstrate. This is consistent with the observation that substitution of only three amino acids that exist in a cluster on the external surface of Ca^{2+}/calmodulin domain I, converts the regulatory protein from agonist to potent antagonist (K_i light chain = 30 nM) (102). Validation of these putative interactions requires solution of a three-dimensional structure of Ca^{2+}/calmodulin complexed to smooth muscle myosin light chain kinase. Interestingly the details of the mechanism of calmodulin activation of the myosin light chain kinase may be distinct from those for phosphorylase kinase (103).

3.6 Calmodulin-dependent protein kinase II family

Early studies on neuronal protein phosphorylation by Greengard and colleagues led to the identification of a series of calmodulin-dependent protein kinases (reviewed in 104): calmodulin-dependent protein kinase I–IV. However, it is now recognized that many of these are found in non-neuronal tissues and cell types.

These enzymes are thought to play a central role in transduction of Ca^{2+} signals in cells. Both calmodulin-dependent protein kinases I and II were identified as synapsin kinases. The most thoroughly studied member of the family is the calmodulin-dependent protein kinase II also known as the multifunctional calmodulin-dependent protein kinase. Calmodulin-dependent protein kinase III is an elongation factor-2 kinase (105) and this enzyme does not readily phosphorylate other substrates (105). Calmodulin-dependent protein kinase IV also phosphorylates synapsin and has been purified from rat cerebellum. There are multiple isoforms (α, β, β', γ, δ) of the calmodulin-dependent protein kinase II as well as calmodulin-dependent protein kinase IV. Since the most extensive studies of the substrate recognition and pseudosubstrate regulation have been done on the calmodulin-dependent protein kinase II it will be discussed in greatest detail. This enzyme is especially abundant in brain, constituting approximately 0.3% of the total brain protein but it has a wide tissue distribution. Typically the mammalian calmodulin-dependent protein kinase II isoforms have subunits of 50–62 kDa that occur in multimers (10–12 subunits) to yield a native mass of 300–700 kDa. The enzyme is made up of an amino-terminal catalytic domain, a central regulatory sequence and a carboxy-terminal domain responsible for association of the subunits into multimers. Upon activation by Ca^{2+}/calmodulin the enzyme becomes auto-phosphorylated, which leads to a reduction in calmodulin dependency.

3.6.1 Substrate recognition

A large number of protein substrates have been identified for the calmodulin-dependent protein kinase II (104) but of these synapsin I, pyruvate kinase, phenyl-alanine hydroxylase, tyrosine hydroxylase, phospholamban and MAP-2 are among the most likely substrates *in vivo*. This enzyme typically recognizes the phosphoryl-ation site sequence motif Arg–Xaa–Xaa–Ser/Thr. Although approximately three quarters of the known phosphorylation site sequences conform to this motif (104), sometimes Lys is the nearest amino acid and in several cases there is no nearby basic residue in the primary sequence. The specificity requirements of calmodulin-dependent protein kinase II has been studied using synthetic peptides correspond-ing to the phosphorylation site at Ser7 in glycogen synthase, $PLSRTLS_7VSS$. Phosphorylation of this peptide is dependent on the presence of Arg4 (106) since the Leu analog is not phosphorylated and the Lys analog has a 10-fold higher K_m. Two synthetic peptide analogues of this sequence have been used widely to study the calmodulin-dependent protein kinase, PLRRTLSVAA (K_m 3.5 μM, V_{max} 11.3 μmol^{-1} min^{-1} mg^{-1}) (106) and PLARTSVAGLPGKK, 'syntide' (K_m 12 μM, V_{max}, 2.75 μmol min^{-1} mg^{-1}) (107). Calmodulin-dependent protein kinase II prefers the Arg specificity determinant in the P-3 position and placing it in different positions leads to reduced or no phosphorylation of model peptides (106). Model-ling studies (108) suggest that the P-3 Arg makes an electrostatic contact with Glu96 in the catalytic core (equivalent to Glu127 in PKA).

3.6.2 Pseudosubstrates

Following the observation that the smooth muscle myosin light chain kinase contained a pseudosubstrate autoinhibitory sequence we proposed that calmodulin-dependent protein kinase II also contained a pseudosubstrate sequence overlapping its calmodulin-binding sequence (Table 1; 69, 109). The pseudosubstrate sequence originally proposed, corresponding to residues 290–309 in the sequence $M_{281}HRQET_{286}VDCL_{290}KKFNARRKLKGAILTTMLA_{309}$, is adjacent to the major autophosphorylation site at Thr286 (109). The calmodulin-binding sequence extends from Ala295 to Ala309 (104, 110). There have been a large number of studies directed at understanding the function of autophosphorylation and locating the autoinhibitory sequence. Since this area has been exhaustively reviewed (e.g. 104), we will concentrate on a few aspects that highlight the problems and limitations of the approaches used to investigate the mechanism of autoinhibition.

Several groups have observed that increasing the length of peptide amino-terminal to Leu290 increases the potency of the peptide inhibitors (74, 111). This extension includes the autophosphorylation site at Thr286. The extended peptide 281–309 was a competitive inhibitor with ATP in contrast to $L_{290}KKFNARRKLKGAILTTMLA_{309}$ that was competitive with respect to peptide substrate. This lead to the suggestion of a dual interaction between the ATP and protein substrate binding elements of the catalytic core (112). Extension of the myosin light chain kinase pseudosubstrate peptide to include the connecting residues also enhanced ATP antagonism (see Section 4.3.2). The difficulty with the dual interaction hypothesis is that the extension that enhances ATP antagonism is on the amino-terminal side of the pseudosubstrate sequence distal from the ATP-binding site. The possibility that the autoinhibitory sequence is oriented in the active site in the reverse direction is structurally implausible.

A number of site-directed mutagenesis studies have been undertaken but truncation mutagenesis studies have been the most illuminating (108). We found that truncation of the calmodulin-dependent protein kinase II to Lys294 resulted in an inactive form of the protein kinase. This was unexpected because the truncated form of the enzyme only retained $L_{290}KKFN$, five of the 15 pseudosubstrate residues in the proposed pseudosubstrate sequence. Surprisingly, substitution of these residues with AAAL only led to partial (7%) Ca^{2+}-independent activity. Deletion of these residues resulted in 38% Ca^{2+}-independent activity. Thus, depending on the type of mutation, one could conclude that the sequence $L_{290}KKFN$ was sufficient for autoinhibition of the calmodulin-dependent protein kinase II but not essential in the context of the remaining sequence being present. This paradox reflects the problem of measuring enzyme activity versus the binding energy for the autoinhibitory peptide for the catalytic core. Deletion of residues 283–289, RQETVDC, resulted in 36% Ca^{2+}-independent activity, equivalent to deletion of $L_{290}KKFN$. We interpret this deletion as having shifted the frame of the remaining pseudosubstrate sequence, rather than it being directly involved in contacts within the active site. The outcome of the mutation studies serves to strongly emphasize the need for comprehensive comparisons including truncation, deletion and substitution mutations to provide reliable insight into the mechanism, but above all, the need for direct structural information.

3.6.3 Calmodulin-dependent protein kinase I

This enzyme was identified based on its ability to phosphorylate synapsin I. It has now been cloned from rat brain and has a mass of 37.5 kDa. This enzyme phosphorylates the synapsin site 1 with the sequence RRRLSDS and synthetic peptide studies indicate that all three Arg residues are important for recognition (113). The putative calmodulin-binding sequence in the calmodulin-dependent protein kinase I is shown in Table 4. This sequence contains KKNF similar to KKFN that is important in autoregulation of calmodulin-dependent protein kinase II and it seems reasonable that it would have a similar function.

3.6.4 Calcium-dependent protein kinase

A soybean calcium-dependent protein kinase has been cloned (114) with a mass of 57 kDa that contains a catalytic domain fused to a calmodulin-like sequence with four calcium-binding motifs (EF hands). This configuration makes the enzyme directly dependent on Ca^{2+} concentration and independent of calmodulin. A putative autoregulatory sequence is also present between the catalytic core and the calmodulin-like domain (K_{383}LKKMALRVIA). The substrate specificity and regulation of this enzyme have yet to be studied.

3.7 Phosphorylase kinase

Phosphorylase kinase is a calcium-dependent protein kinase that regulates glycogen breakdown via phosphorylation of phosphorylase b. The enzyme is a complex

---

Table 4 Calmodulin-dependent protein kinase pseudosubstrate sequences

Protein kinase	Sequence	Reference
Cam-I PK	HQSVSEQIKKNFAKSKWKQ**A**FNRTA	113
Cam-II PK	LKKFNARRKLKG**A**ILTT	109
	RQETVDCLKKFNARRKLKG**A**ILTT	74
	KKFN	108
Cam-IV PK	MDTAQKKLQEFNARRKLKA**A**VKAVVASS	152
smMLCK	SKDRMKKYMARRKWQKTGH**A**V	69
skMLCK	KRRWKKNFI**A**G	70
Phosphorylase kinase γ[a]	GKFKVICLTVLASVRIYYQYRRVKP	
	LRRLIDAYAFRIYGHWVKKGQQQNR	115

[a] Phosphorylase kinase has two calmodulin-binding sequences at the COOH-terminal end of the γ subunit but it is not known which region is critical for autoinhibition.

heterododecamer $(\alpha\beta\gamma\delta)_4$ where γ is the catalytic subunit and δ is a tightly bound calmodulin subunit. The enzyme is calcium-dependent and can be further activated by a second molecule of calmodulin or troponin C. The γ subunit is auto-inhibited and synthetic peptide studies have suggested that its carboxy-terminal sequence contains two calmodulin-binding sequences (115). It has a restricted substrate range and although it will phosphorylate troponin I, neuronal proteins (116) and glycogen synthase *in vitro*, phosphorylase *b* may be its major substrate *in vivo*.

3.7.1 Substrate recognition

The substrate specificity of phosphorylase kinase was among the earliest to be investigated with synthetic peptides (6, 117). The peptide $K_1RKQIS_{14}VRGLAG_{20}$ corresponding to the phosphorylation site in phosphorylase is a substrate with a K_m of 270 μM and a V_{max} of 3.7 μmol min^{-1} mg^{-1} (118) compared to a K_m of 18 μM and V_{max} of 4.1 μmol min^{-1} mg^{-1} for phosphorylase (119). With this peptide as substrate, Arg16, Lys9, and Lys11 were shown to influence the kinetics of phosphorylation (120) but not Arg10 (117, 121). The surrounding hydrophobic residues also appear to contribute to the rate of peptide phosphorylation (117, 118) but in the case of neurogranin the phosphorylation site is IQASF (116) so that this is not an absolute requirement. The glycogen synthase peptide PLSRTLSVSS with only the single Arg is readily phosphorylated at Ser7 by phosphorylase kinase (118) so that multiple basic residues also may not be necessary in every instance. The troponin I and troponin T peptides corresponding to their respective local phosphorylation site sequences are poor substrates (118). Recently, melittin has been shown to act as a poor substrate for phosphorylase kinase as well as acting as an inhibitor (122). The K_m for melittin as substrate was approximately 10 μM, similar to that with phosphorylase.

3.7.2 Pseudosubstrates

Since the γ subunit of phosphorylase kinase is autoregulated it seems likely that it will contain an inhibitory sequence on its carboxy-terminal tail similar to the other members of the calmodulin-dependent protein kinase subfamily. Constitutively active fragments of phosphorylase kinase γ have been generated by proteolysis (123) as well as recombinant protein expression (124). Cleavage with chymotrypsin at Phe298 resulted in a calcium independent 34 kDa fragment with 6-fold higher activity than the γ subunit calmodulin complex (123). Previously, we speculated that the sequence $V_{332}IRDPYALRPLRRLIDAYAFRI_{353}$ may act as a pseudosubstrate for phosphorylase kinase based on analogy with other pseudosubstrate regulated protein kinases (1). However, recent evidence has been presented indicating that a sequence $K_{420}RNPGSQKRFPSNCGRD_{436}$ derived from the β-subunit is responsible for autoinhibition (125). While this proposal does not appear to make structural sense at this time, it is clear that direct structural information is required for this complicated holoenzyme.

3.8 Protein kinase C family

Protein kinase C was initially identified as a soluble, proteolytically activatable brain histone kinase (126). The discovery that this enzyme was regulated by phospholipids and diacylglycerol lead to the realization that it is centrally involved in a major signal transduction system. Protein kinase C, composed of a single polypeptide chain of 77–83 kDa, has been found in all animal cells and tissues examined and is the primary receptor for the phorbol ester class of tumour promoters. The cloning of protein kinase C revealed it to consist of a subfamily of related gene products as well as isoforms arising from differential splicing that fall into two major groups (α β_I, β_{II}, γ and δ, ϵ, ζ, η, θ; see Chapter 3; 127–130). The former group are Ca^{2+}-dependent whereas the latter group are Ca^{2+}-independent. The δ, ϵ, ζ, η, and θ isoforms have amino-terminal extensions varying from 100 to 130 residues compared to the α β_I, β_{II}, γ isoforms. These latter isoforms do not phosphorylate histone readily and their activities and specificities have only been revealed with the availability of recombinant enzyme. The isoforms of protein kinase C show distinct tissue distributions as well as varying subcellular localizations.

3.8.1 Substrate recognition

The substrate specificity of protein kinase C has been the subject of a number of studies. Prior to the cloning and sequencing of protein kinase C and the discovery of the complex isoenzyme pattern it was recognized that protein kinase C phosphorylated Ser and Thr residues in a wide variety of phosphorylation site sequences (30). Commonly used exogenous protein substrates included: histones H1 and HIIIS, myelin basic protein and protamine. Initial synthetic peptide studies revealed that the motif Arg–Xaa–Xaa–Ser/Thr–Xaa–Arg was phosphorylated with excellent kinetics (131, 132; Table 5). It is clear, however, that in the context of protein substrates many sites do not conform to this motif. Recent studies with purified isoenzymes have revealed a number of important specificity differences and indicated a higher level of complexity than was apparent with the initial studies using purified enzyme containing multiple isoforms. Initially it was hoped that peptide substrates modelled on the protein kinase C isoenzyme pseudo-substrates would be specific for the individual isoenzymes (133, 134). Using a

Table 5 Protein kinase C peptide substrate sequences

Substrate	Sequence	Reference
MARCKS	FKK**S***FKL	153, 154
Glycogen synthase	PLSRTL**S***VAKK	131, 132
EGF-receptor	VRKR**T***LRRL	132, 135

* Phosphorylated residue; reviewed in refs 12, 127.

series of synthetic analogues of the sequence GGRLARALS$_9$VAAG one study found that the protein kinase C-α and -β isoforms have a requirement for a basic residue at the P-3 position but the kinetics of peptide phosphorylation are improved by the inclusion of an additional basic residue at the P$+3$ position (135). The γ-isoform had a requirement for basic residues on the carboxy-terminal side of Ser9 in this peptide sequence. Interestingly, deletion of the pseudosubstrate sequence from the η-isoform generated a constitutively active form of the enzyme that was able to phosphorylate histone with 30-fold lower K_m (136). Using rat brain as a source, Nakabayashi *et al.* (137) also observed that protein kinase C had altered phosphorylation site specificity for myosin light chains upon proteolysis to the constitutively active M form. Chimeric protein kinase C isoenzymes in which the pseudosubstrate sequences form the α- and η-isoforms were exchanged revealed that the residues on the carboxy-side of the pseudosubstrate Ala161 were critical for ensuring a high K_m for histone phosphorylation (138).

3.8.2 Pseudosubstrates

Studies on the substrate specificity of protein kinase C with synthetic peptides indicated that basic residues were important specificity determinants and that peptide substrates with the sequence Arg–Xaa–Xaa–Ser/Thr–Xaa–Arg were most readily phosphorylated (131, 132). With this sequence motif in mind it was possible to recognize a putative pseudosubstrate sequence within protein kinase C (139). The sequence R$_{19}$FARKGALRQKNV$_{31}$ located at the start of the C1 region resembled a phosphorylation site motif (Table 6). This pseudosubstrate sequence is located ~20 residues upstream of the cysteine-rich Zn^{2+}-finger motifs suggesting that the two motifs are structurally interdependent (140). Synthetic peptides corresponding to the pseudosubstrate sequence were found to act as potent competitive inhibitors of protein kinase C (139, 141). Subsequently it was shown that an anti-peptide antibody to this region caused complete activation of the enzyme in the absence of its allosteric regulators calcium and phospholipids (142). It is presumed that under resting conditions the pseudosubstrate sequence binds to

Table 6 Protein kinase C pseudosubstrate sequences[a]

Isoform	Sequence	Zn^{2+} finger	Ca^{2+}-dependence
PKC α	R^{19}FARKG**A**LRQKNVHEVK^{35}	I and II	C2
PKC βI and II	R^{19}FARKG**A**LRQKNVHEVK^{35}	I and II	C2
PKC γ	L^{18}FCARKG**A**LRQKVVHEVK^{34}	I and II	C2
PKC δ	R^{141}MNRRG**A**IKQAKIHYIK^{157}	I and II	
PKC ϵ	P^{153}RKRQG**A**VRRRVHQVNG^{169}	I and II	
PKC η	T^{154}RKRQR**A**MRRRVHQING^{170}	I and II	
PKC θ	L^{142}HQRRG**A**IKQAKVHHVK^{158}	I and II	
PKC ζ	S^{113}IYRRG**A**RRWRKLYRAN^{129}	I	

[a] See reviews (127, 140).

the active site and renders the enzyme inactive. Activation of the enzyme by the binding of the cofactors phospholipid and diacylglycerol causes the pseudo-substrate sequence to dissociate from the active site.

Both proteolysis and mutation experiments have provided supporting evidence for the hypothesis that the pseudosubstrate sequence is important for maintaining protein kinase C in an inactive form in the absence of its cofactors. Proteolysis of protein kinase C by trypsin removes the regulatory domain by cleaving in the hinge region contained within the V3 sequence generating the constitutively active M form of the enzyme (126). Orr *et al.* (143) found that the binding of phospholipid to the recombinant β_{II} form of protein kinase C exposed Arg19 within the pseudo-substrate sequence $S_{16}TVR_{19}FARKGA_{25}LRQKNVHEVKN_{36}$ to cleavage by endo-proteinase Arg-C. This elegant experiment demonstrated that the pseudosubstrate sequence was masked in the inactive form of the enzyme but became exposed following cofactor binding. Since Arg27 was not cleaved it suggests that there may be only partial removal of the pseudosubstrate sequence. The work of Dekker *et al.* (136) also suggests that cofactor binding weakens the pseudosubstrate/active site interaction and is followed by substrate/pseudosubstrate competition. The en-hanced susceptibility of the pseudosubstrate sequence to proteolysis following protein kinase C activation is similar to the situation with the myosin light chain kinase where the binding of calmodulin modulates the protease cleavage pattern (see Section 3.3). Support for the concept that the pseudosubstrate sequence does in fact bind to the active site comes from the observation that the nearby Ser16 and Thr17 are autophosphorylation sites (127). Mutagenesis experiments have also provided supporting evidence that the pseudosubstrate sequence is critical for regulation. Initial attempts to isolate recombinant protein kinase C with the pseudosubstrate sequence deleted were complicated by problems of stability (87), nevertheless cotransfection of this form of the enzyme with a phorbol ester re-porter gene showed that the enzyme was transiently constitutively active (87). Mutation of Ala25 to Glu within the pseudosubstrate sequence only partially activated protein kinase C but did lead to enhanced susceptibility to proteolysis (87). Recently, Dekker *et al.* (136) prepared a series of mutants of protein kinase Cη expressed in COS cells. Again substitution of the pseudosubstrate Ala (Ala161 to Glu) partially activated the recombinant protein kinase Cη. Deletion of the protein kinase η pseudosubstrate sequence 155–171 resulted in constitutively active en-zyme. This latter form of the enzyme also exhibited enhanced specificity for histone (136).

Detailed structure–function studies have been undertaken with the protein kinase C pseudosubstrate peptide. Arg22 was found to have the strongest in-fluence on the pseudosubstrate inhibitor peptide potency with the Ala22 analogue having a 600-fold higher IC_{50} value of 81 μM (141). Other basic residues also have an important influence as do uncharged residues such as Leu26 and Gly24. Protein kinase C shares the same hydrophobic cluster of residues corresponding to the P+1 binding-site in the PKA structure and Leu26 in the pseudosubstrate peptide would be expected to bind to it. The results of studies with synthetic peptides indicate that the potency of the pseudosubstrate peptide inhibitor results from the contribution of a number of residues and not just from the basic residues.

The contact residues on the protein kinase C catalytic core that bind the pseudo-substrate sequence have not been positively identified. Prior to the solving of the PKA structure we attempted to locate the putative pseudosubstrate binding region by inspection of the protein kinase catalytic core sequence (144). It was anticipated that a corresponding peptide may act as a pseudosubstrate antagonist and activate the enzyme. A 29-residue peptide, L_{530}LYEMLAGQAPFEGEDEDELFQSIMEHNV$_{558}$ was found to act as a potent activator of the enzyme with a K_a of approximately 10 μM (144). A brain preparation of protein kinase C was activated 59% compared to the activity observed with phosphatidylserine and was partially Ca^{2+}-dependent. It was proposed that this was part of the protein substrate binding region and that it activated protein kinase C by binding to the pseudosubstrate sequence and blocking its association with the active site. In the light of the subsequent PKA crystal structure, this interpretation seemed less plausible because the acidic residues present in the catalytic core responsible for recognition of the substrate basic residues were not derived from a short segment of the linear sequence. Although shorter segments of the acidic peptide L_{530}LYEMLAGQAPFEGEDEDELFQSIMEHNV$_{558}$ were poor activators of protein kinase C, suggesting that the full length sequence was important and implying specificity, polyaspartic acid peptides were also capable of activating protein kinase C which made the interpretation equivocal. However, recent modelling studies by Newton and colleagues (unpublished) have recognized that the acidic activator sequence would be expected to occur in the region corresponding to the G-helix of PKA and would therefore be in close proximity to the putative location of the pseudosubstrate sequence in the protein kinase C active site. If direct structural studies confirm this then it will be interesting to compare the similarity between pseudosubstrate and substrate binding for protein kinase C.

4. General conclusions

The first major steps have now been made in identifying the structural basis for substrate and pseudosubstrate regulation of protein kinases. The rapidly expanding number of protein kinase crystal structures is dramatically enriching our understanding of the regulation of protein kinases. It seems likely that there will be a diversity of interactions involving the autoinhibitory sequences and the active site. Such *intrasteric* mechanisms (1) can be expected to be just as varied as the *allosteric* mechanisms that have been described since the discovery that AMP regulated phosphorylase by an *allosteric* mechanism, which was also the first enzyme shown to be regulated by protein phosphorylation.

References

1. Kemp, B. E. and Pearson, R. B. (1991) Intrasteric regulation of protein kinases and phosphatases. *Biochim. Biophys. Acta*, **1094**, 67.

2. Hurley, J. H., Dean, A. M., Sohl, J. L., Koshland, D. E. J., and Stroud, R. M. (1990) Regulation of an enzyme by phosphorylation at the active site. *Science*, **249**, 1012.

3. Norbury, C. and Nurse, P. (1992) Animal cell cycles and their control. *Ann. Rev. Biochem.*, **61**, 441.

4. Cazaubon, S. M. and Parker, P. J. (1993) Identification of the phosphorylated region responsible for the permissive activation of protein kinase C. *J. Biol. Chem.*, **268**, 17559.

5. Knighton, D. R., Zheng, J., Eyck, L. F. T., Ashford, V. A., Xuong, N.-H., Taylor, S. S., and Sowadski, J. M. (1991) Crystal structure of the catalytic subunit of cyclic adenosine monophosphate-dependent protein kinase. *Science*, **253**, 407.

6. Nolan, C., Novoa, W. B., Krebs, E. G., and Fischer, E. H. (1964) Further studies on the site phosphorylated in the phosphorylase *b* to *a* reaction. *Biochemistry*, **3**, 542.

7. Daile, P. and Carnegie, P. R. (1974) Peptides from myelin basic protein as substrates for adenosine 3',5'-cyclic AMP-monophosphate dependent protein kinase. *Biochem. Biophys. Res. Commun.*, **61**, 852.

8. Kemp, B. E., Bylund, D. B., Huang, T. S., and Krebs, E. G. (1975) The substrate specificity of the cyclic AMP-dependent protein kinase. *Proc. Natl. Acad. Sci. USA*, **72**, 3448.

9. Kemp, B. E., Brenjamini, E., and Krebs, E. G. (1976) Synthetic peptide substrates and inhibitors of the cyclic AMP-dependent protein kinase. *Proc. Natl. Acad. Sci. USA*, **73**, 1038.

10. Daile, P., Carnegie, P. R., and Joung, J. D. (1975) Synthetic substrate for the cyclic AMP-dependent protein kinase. *Nature*, **257**, 416.

11. Zetterqvist, O., Ragnarsson, U., Humble, E. E. J., Berglund, L., and Engstrom, L. (1976) The minimum substrate of cyclic AMP-stimulated protein kinase, as studied by synthetic peptides representing the phosphorylatable site of pyruvate kinase (type L) of rat liver. *Biochem. Biophys. Res. Commun.*, **70**, 696.

12. Kemp, B. E. and Pearson, R. B. (1991) Design and use of peptide substrates for protein kinases. In *Methods in enzymology* (ed. T. Hunter and B. M. Sefton), p. 121. Academic Press, San Diego.

13. Michnoff, C. H., Kemp, B. E., and Stull, J. T. (1986) Phosphorylation of synthetic peptides by skeletal muscle myosin light chain kinases. *J. Biol. Chem.*, **261**, 8320.

14. Bondt, H. L. D., Rosenblatt, J., Jones, H. D., Morgan, D. O., and Kim, S. H. (1993) Crystal structure of cyclin-dependent kinase 2. *Nature*, **363**, 595.

15. Demaille, J. G., Peters, K. A., and Fischer, E. H. (1977) Isolation and properties of the rabbit skeletal muscle protein inhibitor of adenosine 3',5'-monophosphate dependent protein kinases. *Biochemistry*, **16**, 3080.

16. Corbin, J. D., Keely, S. L., and Park, C. R. (1975) The tissue distribution and dissociation of the cyclic adenosine 3',5'-monophosphate-dependent protein kinases in adipose, cardiac and other tissues. *J. Biol. Chem.*, **250**, 218.

17. Rosen, O. M., Erlichman, J., and Rubin, C. S. (1973) Molecular characterization of cyclic AMP-dependent protein kinases derived from bovine heart and human erythrocytes. In *International symposium of metabolic interconversion of enzymes* (ed. E. H. Fischer, *et al.*), p. 143. Springer–Verlag, Berlin.

18. Witt, J. J. and Roskoski, R. J. (1975) Bovine brain adenosine 3',5'-monophosphate dependent protein kinase. Mechanism of regulatory subunit inhibition of the catalytic subunit. *Biochemistry*, **14**, 4503.

19. Corbin, J. D., Sugden, P. H., West, L., Flockhart, D. A., Lincoln, T. M., and McCarthy, D. (1978) Studies on the properties and mode of action of the purified regulatory

subunit of bovine heart adenosine 3':5' monophosphate-dependent protein kinase. *J. Biol. Chem.*, **253**, 3997.

20. Taylor, S. S., Buechler, J. A., and Yonemoto, W. (1990) cAMP-dependent protein kinase: framework for a diverse family of regulatory enzymes. *Annu. Rev. Biochem.*, **59**, 971.

21. Hu, S.-H., Parker, M. W., Lei, J. Y., Wilce, M. C. J., Benian, G. M., and Kemp, B. E. (1994) Insights into autoregulation from the crystal structure of twitchin kinase. *Nature*, **369**, 581.

22. Kemp, B. E., Pearson, R. B., and House, C. M. (1991) Pseudosubstrate based peptide inhibitors. In *Methods in enzymology* (ed. T. Hunter and B. M. Sefton), p. 287. Academic Press, San Diego.

23. Smith, M. K., Colbran, R. J., and Soderling, T. R. (1990) Specificity of autoinhibitory domain peptides for four protein kinases. *J. Biol. Chem.*, **265**, 1837.

24. Buchler, W., Walter, U., Jastorff, B., and Lohmann, S. M. (1988) Catalytic subunit of cAMP-dependent protein kinase is essential for cAMP-mediated mammalian gene expression. *FEBS Lett.*, **228**, 27.

25. Eichholtz, T., Bont, D. B. d., Widt, J. d., Liskamp, R. N., and Pleogh, H. L. (1993) A myristoylated pseudosubstrate peptide, novel protein kinase C inhibitor. *J. Biol. Chem.*, **268**, 1982.

26. Grove, J. R., Price, D. J., Goodman, H. M., and Avruch, J. (1987) Recombinant fragment of protein kinase inhibitor blocks cyclic AMP-dependent gene transcription. *Science*, **238**, 530.

27. Walsh, D. A., Glass, D. B., and Mitchell, R. D. (1992) Substrate diversity of the cAMP-dependent protein kinase: regulation based upon multiple binding interactions. *Curr. Opin. Cell. Biol.*, **4**, 241.

28. Scott, J. D. (1991) Cyclic nucleotide-dependent protein kinases. *Pharmacol. Ther.*, **50**, 123.

29. Taylor, S. S., Knighton, D. R., Zheng, J., Sowadski, J. M., Gibbs, C. S., and Zoller, M. J. (1993) A template for the protein kinase family. *TIBS*, **18**, 84.

30. Pearson, R. B. and Kemp, B. E. (1991) Protein kinase phosphorylation site sequences and consensus phosphorylation site specificity motifs. In *Methods in enzymology* (ed. T. Hunter and B. M. Sefton), p. 62. Academic Press, San Diego.

31. Knighton, D. R., Zheng, J., Eyck, L. F. T., Ashford, V. A., Xuong, N.-H., Taylor, S. S., and Sowadski, J. M. (1991) Structure of a peptide inhibitor bound to the catalytic subunit of cyclic adenosine monophosphate-dependent protein kinase. *Science*, **253**, 414.

32. Gibbs, C. S. and Zoller, M. J. (1991) Rational scanning mutagenesis of a protein kinase identifies functional regions involved in catalysis and substrate interactions. *J. Biol. Chem.*, **266**, 8923.

33. Gibbs, C. S. and Zoller, M. J. (1991) Identification of electrostatic interactions that determine site specifity of the cAMP-dependent protein kinase. *Biochemistry*, **30**, 5329.

34. Gibbs, C. S., Knighton, D. R., Sowadsky, J. M., Taylor, S. S., and Zoller, M. J. (1992) Systematic mutational analysis of cAMP-dependent protein kinase identifies un-regulated catalytic subunits and defines regions important for the recognition of the regulatory subunit. *J. Biol. Chem.*, **267**, 4806.

35. Levin, L. R., Kuret, J., Johnson, K. E., Powers, S., Cameron, S., Michaeli, T. *et al.* (1988) A mutation in the catalytic subunit of cAMP dependent protein kinase that disrupts regulation. *Science*, **240**, 68.

36. Walsh, D. A. and Glass, D. B. (1991) Utilization of the inhibitor protein of adenosine cyclic monophosphate-dependent protein kinase, and peptides derived from it, as tools to study adenosine cyclic monophosphate-mediated cellular processes. *Methods Enzymology*, **201**, 304.

37. Olah, G. A., Mitchell, R. D., Sosnick, T. R., Walsh, D. A., and Trewhella, J. (1993) Solution structure of the cAMP-dependent protein kinase catalytic subunit and its contraction upon binding the protein kinase inhibitor peptide. *Biochemistry*, **32**, 3649.

38. Takio, K., Wade, R. D., Smith, S. B., Krebs, E. G., Walsh, K. A., and Titani, K. (1984) Guanosine cyclic 3',5'-phosphate dependent protein kinase, a chimeric protein homologous with two separate protein kinase families. *Biochemistry*, **23**, 4207.

39. Kalderon, D. and Rubin, G. M. (1989) cGMP-dependent protein kinase genes in *Drosophila*. *J. Biol. Chem.*, **264**, 10738.

40. Wernet, W., Flockerzi, V., and Hofmann, F. (1989) The cDNA of the two isoforms of bovine cGMP-dependent protein kinase. *FEBS Lett.*, **251**, 191.

41. Uhler, M. D. (1993) Cloning and expression of a novel cGMP-dependent protein kinase from mouse brain. *J. Biol. Chem.*, **268**, 13586.

42. Lincoln, T. M., Flockhart, D. A., and Corbin, J. D. (1978) Studies on the structure and mechanism of activation of the cGMP-dependent protein kinase. *J. Biol. Chem.*, **253**, 6002.

43. Gill, G. N. (1977) A hypothesis concerning the structure of cAMP- and cGMP-dependent protein kinase. *J. Cyclic Nucleotide Res.*, **3**, 153.

44. Aitken, A., Hemmings, B. A., and Hofmann, F. (1984) Identification of residues on cyclic GMP-dependent protein kinase that are autophosphorylated in the presence of cyclic AMP and cyclic GMP. *Biochim. Biophys. Acta*, **790**, 219.

45. Glass, D. B. and Krebs, E. G. (1979) Comparison of the substrate specificity of adenosine 3':5'-monophosphate- and guanosine 3'-5'-monophosphate-dependent protein kinase. *J. Biol. Chem.*, **254**, 9728.

46. Glass, D. B. (1990) Substrate specificity of the cyclic GMP-dependent protein kinase. In *Peptides and protein phosphorylation* (ed. B. E. Kemp), p. 209. CRC Press, Boca Raton, Florida.

47. Thomas, M. K., Francis, S. H., and Corbin, J. D. (1990) Characterization of a purified bovine lung cGMP-binding cGMP phosphodiesterase. *J. Biol. Chem.*, **265**, 14964.

48. Colbran, J. L., Francis, S. H., Leach, A. B., Thomas, M. K., Jiang, H., McAllister, L. M., and Corbin, J. D. (1992) A phenylalanine in peptide substrates provides for selectivity between cGMP- and cAMP-dependent protein kinases. *J. Biol. Chem.*, **267**, 9589.

49. Glass, D. B. and Smith, S. B. (1983) Phosphorylation by cyclic GMP-dependent protein kinase of a synthetic peptide corresponding to the autophosphorylation site in the enzyme. *J. Biol. Chem.*, **258**, 14797.

50. Knighton, D. R., Pearson, R. B., Sowadski, J. M., Means, A. R., Eyck, L. F. T., Taylor, S. S., and Kemp, B. E. (1992) Structural basis of the intrasteric regulation of myosin light chain kinases. *Science*, **258**, 130.

51. Benian, G. M., Kiff, J. E., Neckelmann, N., Moerman, D. G., and Waterston, R. H. (1989) Sequence of an unusually large protein implicated in regulation of myosin activity in *C. elegans*. *Nature*, **342**, 45.

52. Adelstein, R. S. and Conti, M. A. (1975) Phosphorylation of platelet myosin increases actin-activated myosin ATPase activity. *Nature*, **256**, 597.

53. Hoar, P. E., Kerrick, W. G., and Cassidy, R. L. (1979) Chicken gizzard: relation between calcium phosphorylation and contraction. *Science*, **204**, 503.

54. Lucas, T. J., Burgess, W. H., Prendergast, F. G., Lau, W., and Watterson, D. M. (1986) Calmodulin binding domains: characterisation of a phosphorylation and calmodulin binding site from myosin light chain kinase. *Biochemistry*, **25**, 1450.

55. Sweeney, H. L., Bowman, B. F., and Stull, J. T. (1993) Myosin light chain phosphorylation in vertebrate striated muscle: regulation and function. *Am. J. Physiol.*, **253**, C1085.

56. Tan, J. L. and Spudich, J. A. (1991) Characterization and bacterial expression of the *Dictyostelium* myosin light chain kinase. *J. Biol. Chem.*, **266**, 16044.

57. Dabrowska, R., Sherry, J. M. F., Aromatorio, D. K., and Hartshorne, D. J. (1978) Modulator protein as a component of the myosin light chain kinase from chicken gizzard. *Biochemistry*, **17**, 253.

58. Blumenthal, D. K., Takio, K., Edelman, A. M., Charbonneau, H., Titani, K., Walsh, K. A., and Krebs, E. G. (1985) Identification of the calmodulin-binding domain of skeletal muscle myosin light chain kinase. *Proc. Natl. Acad. Sci. USA*, **82**, 3187.

59. Kemp, B. E. and Stull, J. T. (1990) Myosin light chain kinases. In *Peptides and protein phosphorylation* (ed. B. E. Kemp), p. 115. CRC Press, Bocan Raton, Florida.

60. Kemp, B. E., Pearson, R. B., and House, C. (1982) Phosphorylation of a synthetic heptadecapeptide by smooth muscle myosin light chain kinase. *J. Biol. Chem.*, **257**, 13349.

61. Kemp, B. E., Pearson, R. B., and House, C. (1983) Role of basic residues in the phosphorylation of synthetic peptides by myosin light chain kinase. *Proc. Natl. Acad. Sci. USA*, **80**, 7471.

62. Kemp, B. E. and Pearson, R. B. (1985) Spatial requirements for location of basic residues in peptide substrates for smooth muscle myosin light chain kinase. *J. Biol. Chem.*, **260**, 3355.

63. Pearson, R. B., Misconi, L. Y., and Kemp, B. E. (1986) Smooth muscle myosin kinase requires residues on the COOH-terminal side of the phosphorylation site: peptide inhibitors. *J. Biol. Chem.*, **261**, 25.

64. Pearson, R. B. and Kemp, B. E. (1986) Chemical modification of lysine and arginine residues in the myosin light chain inhibits phosphorylation. *Biochim. Biophys. Acta*, **870**, 312.

65. Herring, B. P., Stull, J. T., and Gallagher, P. J. (1990) Domain characterization of rabbit skeletal muscle myosin light chain kinase. *J. Biol. Chem.*, **265**, 1724.

66. Herring, B. P., Fitzsimons, D. P., Stull, J. T., and Gallagher, P. J. (1990) Acidic residues comprise part of the myosin light chain-binding site on skeletal muscle myosin light chain kinase. *J. Biol. Chem.*, **265**, 16588.

67. Leachman, S. A., Gallagher, P. J., Herring, B. P., McPhaul, M. J., and Stull, J. T. (1992) Biochemical properties of chimeric skeletal and smooth muscle myosin light chain kinases. *J. Biol. Chem.*, **267**, 4930.

68. Herring, B. P., Gallagher, P. J., and Stull, J. T. (1992) Substrate specificity of myosin light chain kinase. *J. Biol. Chem.*, **267**, 25945.

69. Kemp, B. E., Pearson, R. B., Guerriero, V., Bagchi, I. C., and Means, A. R. (1987) The calmodulin binding domain of chicken smooth muscle myosin light chain kinase contains a pseudosubstrate sequence. *J. Biol. Chem.*, **262**, 2542.

70. Kennelly, P. J., Edelman, A. M., Blumenthal, D. K., and Krebs, E. G. (1987) Rabbit skeletal muscle myosin light chain kinase. *J. Biol. Chem.*, **262**, 11958.

71. Ikebe, M. (1990) Mode of inhibition of smooth muscle myosin light chain kinase by synthetic peptide analogs of the regulatory site. *Biochem. Biophys. Res. Commun.*, **168**, 714.

72. Foster, C. J., Johnston, S. A., Sunday, B., and Gaeta, F. C. A. (1990) Potent peptide inhibitors of smooth muscle myosin light chain kinase: mapping of the pseudo-substrate and calmodulin binding domains. *Archiv. Biochem. Biophys.*, **280**, 397.

73. Pearson, R. B., Hunt, J. T., Mitchelhill, K. I., and Kemp, B. E. (1991) Myosin light chain kinase autoregulatory pseudosubstrate prototope. *Peptide Research*, **4**, 147.

74. Colbran, R. J., Smith, M. K., Schworer, C. M., Fong, Y. L., and Soderling, T. R. (1989) Regulatory domain of calcium/calmodulin-dependent protein kinase II. *J. Biol. Chem.*, **264**, 4800.

75. Kennelly, P. J., Colburn, J. C., Lorenzen, J., Edelman, A. M., Stull, J. T., and Krebs, E. G. (1991) Activation mechanism of rabbit skeletal muscle myosin light chain kinase. *FEBS Lett.*, **286**, 217.

76. Bagchi, I. C., Huang, Q., and Means, A. R. (1992) Identification of amino acids essential for calmodulin binding and activation of smooth muscle myosin light chain kinase. *J. Biol. Chem.*, **267**, 3024.

77. Tanaka, T., Naka, M., and Hidaka, H. (1980) Activation of myosin light chain kinase by trypsin. *Biochem. Biophys. Res. Commun.*, **92**, 313.

78. Ikebe, M., Stepinska, M., Kemp, B. E., Means, A. R., and Hartshorne, D. J. (1987) Proteolysis of smooth muscle myosin light chain kinase. *J. Biol. Chem.*, **260**, 13828.

79. Pearson, R. B., Wettenhall, R. E. H., Means, A. R., Hartshorne, D. J., and Kemp, B. E. (1988) Autoregulation by enzyme pseudosubstrate prototopes: myosin light chain kinase. *Science*, **241**, 970.

80. Pearson, R. B., Ito, M., Morrice, N. A., Smith, A. J., Condron, R., Wettenhall, R. E. H. *et al.* (1991) Proteolytic cleavage sites in smooth muscle myosin light chain kinase and their relation to structural and regulatory domains. *Eur. J. Biochem.*, **200**, 723.

81. Olson, N. J., Pearson, R. B., Needleman, D., Hurwitz, M. Y., Kemp, B. E., and Means, A. R. (1990) Regulation and structural motifs of chicken gizzard myosin light chain kinase. *Proc. Natl. Acad. Sci. USA*, **87**, 2284.

82. Ikebe, M., Maruta, S., and Reardon, S. (1989) Location of the inhibitory region of smooth muscle myosin light chain kinase. *J. Biol. Chem.*, **264**, 6967.

83. Ito, M., Guerriero, V. J., Chen, X., and Hartshorne, D. J. (1991) Definition of the inhibitory domain of smooth muscle myosin light chain kinase by site-directed muta-genesis. *Biochemistry*, **30**, 3498.

84. Edelman, A. M., Takio, K., Blumenthal, D. K., Scott Hanson, R., Walsh, K. A., Titani, K., and Krebs, E. G. (1985) Characterization of the calmodulin-binding and catalytic domains in skeletal muscle myosin light chain kinase. *J. Biol. Chem.*, **260**, 11275.

85. Kennelly, P. J., Starovasnik, M. A., Edelman, A. M., and Krebs, E. G. (1990) Modula-tion of the stability of rabbit skeletal muscle myosin light chain kinase through the calmodulin-binding domain. *J. Biol. Chem.*, **265**, 1742.

86. Faux, M. C., Mitchelhill, K. I., Katsis, F., Wettenhall, R. E. H., and Kemp, B. E. (1993) Chicken smooth muscle myosin light chain kinase is acetylated on its NH$_2$-terminal methionine. *Mol. Cell. Biochem.*, **127/128**, 81.

87. Pears, C. J., Kour, G., House, C., Kemp, B. E., and Parker, P. J. (1991) Mutagenesis of the pseudosubstrate site of protein kinase C leads to activation. *Eur. J. Biochem.*, **194**, 189.

88. Shoemaker, M. O., Lau, W., Shattuck, R. L., Kwiatkowski, A. P., Matrisian, P. E.,

Guerra-Santos, L. *et al.* (1990) Use of DNA sequence and mutant analysis and antisense oligonucleotides to examine the molecular basis of nonmuscle myosin light chain kinase autoinhibition, calmodulin recognition, and activity. *J. Cell Biol.*, **111**, 1107.

89. Herring, B. P. (1991) Basic residues are important for Ca^{2+}/calmodulin binding and activation but not autoinhibition of rabbit skeletal muscle myosin light chain kinase. *J. Biol. Chem.*, **266**, 11838.

90. Fitzsimons, D. P., Herring, B. P., Stull, J. T., and Gallagher, P. J. (1992) Identification of basic residues involved in activation and calmodulin binding of rabbit smooth muscle myosin light chain kinase. *J. Biol. Chem.*, **267**, 23903.

91. Bagchi, I. C., Kemp, B. E., and Means, A. R. (1992) Intrasteric regulation of myosin light chain kinase: the pseudosubstrate prototope binds to the active site. *Molec. Endocrinol.*, **6**, 621.

92. Kretsinger, R. H. and Nockolds, C. E. (1973) Carp muscle calcium-binding protein. II. Structure determination and general description. *J. Biol. Chem.*, **248**, 3313.

93. Moncrief, N. D., Kretsinger, R. H., and Goodman, M. (1990) Evolution of EF-hand calcium-modulated proteins. I. Relationships based on amino acid sequences. *J. Mol. Evol.*, **30**, 522.

94. Babu, Y. S., Sack, J. S., Greenbough, T. J., Bugg, C. E., Means, A. R., and Cook, W. J. (1985) Three-dimensional structure of calmodulin. *Nature*, **315**, 37.

95. Klee, C. B. and Vanaman, T. C. (1982) Calmodulin. *Adv. Prot. Chem.*, **35**, 213.

96. Starovasnik, M. A., Su, D.-R., Beckingham, M. A., and Klevit, R. E. (1992) A series of point mutations reveal interactions between the calcium-binding sites of calmodulin. *Protein Sci.*, **1**, 245.

97. Araki, Y. and Ikebe, M. (1991) Activation of smooth muscle myosin light chain kinase activity by a monoclonal antibody which recognizes the calmodulin-binding region. *Biochem. J.*, **275**, 679.

98. Vorherr, T., James, P., Krebs, J., Enyedi, A., McCormick, D. J., Penniston, J. T., and Carafoli, E. (1990) Interaction of calmodulin with the calmodulin binding domain of the plasma membrane Ca^{2+} pump. *Biochemistry*, **29**, 355.

99. Meador, W. E., Means, A. R., and Quiocho, F. A. (1992) Target enzyme recognition by calmodulin: 2.4 Å structure of a calmodulin-peptide complex. *Science*, **257**, 1251.

100. Roth, S. M., Schneider, D. M., Strobes, L. A., VanBerkum, M. F. A., Means, A. R., and Wand, A. J. (1991) Structure of smooth muscle myosin light chain kinase calmodulin-binding domain peptide bound to calmodulin. *Biochemistry*, **30**, 10078.

101. Ikura, M., Clore, G. M., Gronenborn, A. M., Zhu, G., Klee, C. B., and Bax, A. (1992) Solution structure of a calmodulin-target peptide complex by multidimensional NMR. *Science*, **256**, 632.

102. VanBerkem, M. F. A. and Means, A. R. (1991) Three amino acid substitution in domain I of calmodulin prevent the activation of chicken smooth muscle myosin light chain kinase. *J. Biol. Chem.*, **266**, 21488.

103. Farrar, Y. J., Lukas, T. J., Craig, T. A., Watterson, D. M., and Carlson, G. M. (1993) Features of calmodulin that are important in the activation of the catalytic subunit of phosphorylase kinase. *J. Biol. Chem.*, **268**, 4120.

104. Hanson, P. I. and Shulman, H. (1992) Neuronal Ca^{2+}/calmodulin-dependent protein kinases. *Ann. Rev. Biochem.*, **61**, 559.

105. Mitsui, K., Brady, M., Palfrey, H. C., and Nairn, A. C. (1993) Purification and characterization of calmodulin-dependent protein kinase III from rabbit reticulocytes and rat pancreas. *J. Biol. Chem.*, **268**, 13422.

106. Pearson, R. B., Woodgett, J. R., Cohen, P., and Kemp, B. E. (1985) Substrate specificity of a multifunctional calmodulin-dependent protein kinase. *J. Biol. Chem.*, **260**, 14471.

107. Hashimoto, Y. and Soderling, T. R. (1987) Calcium calmodulin-dependent protein kinase II and calcium phospholipid-dependent protein kinase activities in rat tissues assayed with a synthetic peptide. *Arch. Biochem. Biophys.*, **252**, 418.

108. Cruzalegui, F. H., Kapiloff, M. S., Morfin, J.-P., Kemp, B. E., Rosenfeld, M. G., and Means, A. R. (1992) Regulation of intrasteric inhibition of the multifunctional calcium/calmodulin-dependent protein kinase. *Proc. Natl. Acad. Sci. USA*, **89**, 12127.

109. Payne, M. E., Fong, Y.-L., Ono, T., Colbran, R. J., Kemp, B. E., Soderling, T. R., and Means, A. R. (1988) Calcium/calmodulin-dependent protein kinase II: characterization of distinct calmodulin binding and inhibitory domains. *J. Biol. Chem.*, **263**, 7190.

110. Hanley, R. M., Means, A. R., Ono, T., Kemp, B. E., Burgin, K. E., Waxman, N., and Kelley, P. T. (1987) Functional analysis of a complementary DNA for the 50-kilodalton subunit of calmodulin kinase II. *Science*, **237**, 293.

111. Colbran, R. J., Fong, Y. L., Schworer, C. M., and Solderling, T. R. (1988) Regulatory interactions of the calmodulin-binding, inhibitory, and autophosphorylation domains of Ca^{2+}/calmodulin-dependent protein kinase II. *J. Biol. Chem.*, **263**, 18145.

112. Smith, M. K., Colbran, R. J., Bickey, D. A., and Soderling, T. R. (1992) Functional determinants in the autoinhibitory domain of calcium/calmodulin-dependent protein kinase II. *J. Biol. Chem.*, **267**, 1761.

113. Picciotto, M. R., Czernik, A. J., and Nairn, A. C. (1993) Calcium/calmodulin-dependent protein kinase I. *J. Biol. Chem.*, **268**, 26512.

114. Harper, J. G., Sussman, M. R., Schaller, G. E., Putnam-Evans, C., Charbonneau, H., and Harmon, A. C. (1991) A calcium-dependent protein kinase with a regulatory domain similar to calmodulin. *Science*, **252**, 951.

115. Dasgupta, M., Honeycutt, T., and Blumenthal, D. K. (1989) The γ-subunit of skeletal muscle phosphorylase kinase contains two noncontiguous domains that act in concert to bind calmodulin. *J. Biol. Chem.*, **264**, 17156.

116. Paudel, H. K., Zwiers, H., and Wang, J. H. (1993) Phosphorylase kinase phosphorylates the calmodulin-binding regulatory regions of neuronal tissue-specific proteins B-50 (Gap-43) and neurogranin. *J. Biol. Chem.*, **268**, 6207.

117. Tessmer, G. W., Skuster, J. R., Tabatabai, L. B., and Graves, D. J. (1977) Studies on the specificity of phosphorylase kinase using peptide substrates. *J. Biol. Chem.*, **252**, 5666.

118. Kemp, B. E. and John, M. (1981) Synthetic peptide substrates for protein kinases. In *Protein phosphorylation* (ed. O. M. Rosen and E. G. Krebs), Vol. 8, p. 331. Cold Spring Harbor Laboratory, Cold Spring Harbor.

119. Yuan, C.-J., Huang, C.-Y., and Graves, D. J. (1993) Phosphorylase kinase, a metal ion-dependent dual specificity kinase. *J. Biol. Chem.*, **268**, 17683.

120. Tabatabai, L. B. and Graves, D. J. (1978) Kinetic mechanism and specificity of the phosphorylase kinase reaction. *J. Biol. Chem.*, **253**, 2196.

121. Viriya, J. and Graves, D. J. (1979) Phosphorylation of synthetic peptide analogs of the phosphorylatable site of phosphorylase *b* with phosphorylase kinase. *Biochem. Biophys. Res. Commun.*, **87**, 17.

122. Paudel, H. K., Xu, Y.-H., Jarrett, H. W., and Carlson, G. M. (1993) The model calmodulin-binding peptide melittin inhibits phosphorylase kinase by interacting with its catalytic center. *Biochemistry*, **32**, 11865.

123. Harris, W. R., Malencik, D. A., Johnson, C. M., Carr, S. A., D. Roberts, G., Byles, C. A. *et al.* (1990) Purification and characterization of catalytic fragments of phosphory-

lase kinase γ subunit missing a calmodulin-binding domain. *J. Biol. Chem.*, **265**, 11740.

124. Huang, C. Y. F., Yuan, C.-J., Livanova, N. B., and Graves, D. J. (1993) Expression, purification, characterization, and deletion mutations of phosphorylase kinase γ subunit: identification of an autoinhibitory domain in the γ subunit. *Mol. Cell Biochem.* **127/128**, 7.

125. Sanchez, V. E. and Carlson, G. M. (1993) Isolation of an autoinhibitory region from the regulatory beta-subunit of phosphorylase kinase. *J. Biol. Chem.*, **268**, 17889.

126. Inoue, M., Kishimoto, A., Takai, Y., and Nishizuka, Y. (1977) Studies on a cyclic nucleotide-independent protein kinase and its proenzyme in mammalian tissues. II Proenzyme and its activation by calcium-dependent protease from rat brain. *J. Biol. Chem.*, **252**, 7610.

127. Hug, H. and Sarre, T. F. (1993) Protein kinase C isozymes: divergence in signal transduction. *Biochem. J.*, **291**, 329.

128. Azzi, A., Boscoboinik, D., and Hensey, C. (1992) The protein kinase C family. *Eur. J. Biochem.*, **92**, 547.

129. Clemens, M. J., Trayner, I., and Menaya, J. (1992) The role of protein kinase C isoenzymes in the regulation of cell proliferation and differentiation. *J. Cell Sci.*, **103**, 881.

130. Stabel, S. and Parker, P. (1991) Protein kinase C. *Pharmacol. Ther.*, **51**, 71.

131. Woodgett, J. R., Gould, K. L., and Hunter, T. (1986) Substrate specificity of protein kinase C. *Eur. J. Biochem.*, **161**, 177.

132. House, C., Wettenhall, R. E. H., and Kemp, B. E. (1987) Influence of basic residues on the substrate specificity of protein kinase C. *J. Biol. Chem.*, **262**, 772.

133. Marais, R. M. and Parker, P. J. (1989) Purification and characterisation of bovine brain protein kinase C isotypes α, β and γ. *Eur. J. Biochem.*, **182**, 129.

134. Schaap, D. and Parker, P. J. (1990) Expression, purification and characterisation of protein kinase C-ε. *J. Biol. Chem.*, **265**, 7301.

135. Marais, R. M., Nguyen, O., Woodgett, J. R., and Parker, P. J. (1990) Studies on the primary sequence requirements for PKC-α, -β₁ and -γ peptide substrates. *FEBS Lett.*, **277**, 151.

136. Dekker, L. V., McIntyre, P., and Parker, P. J. (1993) Mutagenesis of the regulatory domain of rat protein kinase C-η. *J. Biol. Chem.*, **268**, 19498.

137. Nakabayashi, N., Sellers, J. R., and Huang, K.-P. (1991) Catalytic fragment of protein kinase C exhibits altered substrate specificity toward smooth muscle myosin light chain. *FEBS Lett.*, **294**, 144.

138. Dekker, L. V., McIntyre, P., and Parker, P. J. (1993) Altered substrate selectivity of PKC-η pseudosubstrate site mutants. *FEBS Lett.*, **329**, 129.

139. House, C. and Kemp, B. E. (1987) Protein kinase C contains a pseudosubstrate prototope in its regulatory domain. *Science*, **238**, 1726.

140. Gschwendt, M., Kittstein, W., and Marks, F. (1991) Protein kinase C activation by phorbol esters: do cysteine-rich regions and pseudosubstrate motifs play a role? *TIBS*, **16**, 167.

141. House, C. and Kemp, B. E. (1990) Protein kinase C pseudosubstrate prototope: structure–function relationships. *Cellular Signalling*, **2**, 187.

142. Makowske, M. and Rosen, O. M. (1989) Complete activation of protein kinase C by an antipeptide antibody directed against the pseudosubstrate prototope. *J. Biol. Chem.*, **264**, 16155.

143. Orr, J. W., Keranen, L. M., and Newton, A. C. (1992) Reversible exposure of the pseudosubstrate domain of protein kinase C by phosphatidylserine and diacylglycerol. *J. Biol. Chem.*, **267**, 15263.

144. House, C., Robinson, P. J., and Kemp, B. E. (1989) A synthetic peptide analog of the putative substrate-binding motif activates protein kinase C. *FEBS Lett.*, **249**, 243.

145. Walsh, D. A., Angelos, K. L., Patten, S. M. V., Glass, D. B., and Garetto, L. P. (1990) The inhibitor protein of the cAMP-dependent protein kinase. In *Peptides and protein phosphorylation* (ed. B. E. Kemp), p. 43. CRC Press, Boca Raton, Florida.

146. Clegg, C. H., Cadd, G. G., and McKnight, G. S. (1988) Genetic characterization of a brain specific form of the type I regulatory subunit of cAMP-dependent protein kinase. *Proc. Natl. Acad. Sci. USA*, **85**, 3703.

147. Mutzel, R., Lacombe, M. L., Simon, M. N., Gunzburg, J. D., and Vernon, M. (1987) cAMP-dependent protein kinase from *Dictyostelium discoideum*: cloning and cDNA sequence of the regulatory subunit. *Proc. Natl. Acad. Sci. USA*, **84**, 6.

148. Scott, J. D., Glaccum, M. B., Zoller, M. J., Uhler, M. D., Helfman, D. M., McKnight, G. S., and Krebs, E. G. (1987) The molecular cloning of a type II regulatory subunit of the cAMP-dependent protein kinase from rat skeletal muscle and mouse brain. *Proc. Natl. Acad. Sci. USA*, **84**, 5192.

149. Toda, T., Cameron, S., Sass, P., Zoller, J. D., Scott, B., McMullen, M. *et al.* (1987) Cloning and characterization of BCY1, a locus encoding a regulatory subunit of the cyclic AMP-dependent protein kinase in *Saccharomyces cerevisiae*. *Mol. Cell. Biol.*, **7**, 1371.

150. Scott, J. D., Fischer, E. H., Demaille, J. G., and Krebs, E. G. (1985) Identification of an inhibitory region of the heat-stable protein inhibitor of the cAMP-dependent protein kinase. *Proc. Natl. Acad. Sci. USA*, **82**, 4379.

151. Patten, S. M. V., Ng, D. C., Th'ng, J. P. H., Angelos, K. L., Smith, A. J., and Walsh, D. A. (1991) Molecular cloning of a rat testis form of the inhibitor protein of cAMP-dependent protein kinase. *Proc. Natl. Acad. Sci. USA*, **88**, 5383.

152. Ohmstede, C. A., Jensen, K. F., and Sahyoun, N. (1989) Ca^{2+}/calmodulin-dependent protein kinase enriched in cerebellar granule cells. Identification of a novel neuronal calmodulin-dependent protein kinase. *J. Biol. Chem.*, **264**, 5866.

153. Graff, J. M., Stumpo, D. J., and Blackshear, P. J. (1989) Characterization of the phosphorylation sites in the chicken and bovine myristoylated alanine-rich C kinase substrate protein, a prominent cellular substrate for protein kinase C. *J. Biol. Chem.*, **264**, 11912.

154. Chakravarthy, B. R., Bussey, A., Whitfield, J. F., Sikorsky, M., Williams, R. E., and Durkin, J. P. (1991) The direct measurement of protein kinase C (PKC) activity in isolated membranes using a selective peptide substrate. *Anal. Biochem.*, **196**, 144.

3 | Protein kinase C and its relatives

PETER J. PARKER

1. Introduction

Before plumbing the murky depths of the protein kinase C (PKC) family and its relatives, it would be useful to take a brief backward glance and set the scene. The identification of PKC by Nishizuka and colleagues was initially based upon the ability of cellular Ca^{2+}-activated proteases to cleave the inactive holoenzyme *in vitro* and so derive a catalytically active fragment, often referred to as PKM (Fig. 1) (1, 2). As a physiological means of control, proteolysis remains a controversial issue; however, the subsequent demonstration that the holoenzyme could be reversibly

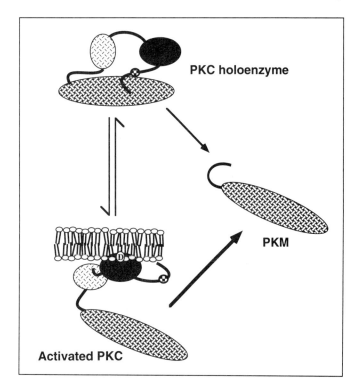

Fig. 1 Proteolytic and reversible activation of PKC. The inactive PKC holoenzyme can be proteolysed *in vitro* to generate a catalytically active fragment (PKM). The enzyme can also be activated reversibly in the presence of diacylglycerol (D), phospholipid, and Ca^{2+} (for α, β, γ proteins, see text). The activated form of the enzyme is more susceptible to proteases than the holoenzyme as indicated.

activated by a combination of Ca^{2+} (μM), phospholipid, and the neutral lipid diacylglycerol (DAG) (3, 4), provided a cornerstone to a signalling pathway involved in a plethora of biological events. The critical factor in this reversible activation process was considered to be DAG since this permitted activation within the physiological range of Ca^{2+} concentrations. That this means of regulation pertained *in vivo* was suggested by the use of membrane penetrating short chain diacylglycerols that appear to be able to activate PKC in intact cells (5). These observations, in conjunction with those made earlier on agonist-induced inositol lipid turnover (reviewed in 6), suggested the general pathway outlined in Fig. 2. This signal transduction pathway is dependent upon the ability of particular cell surface receptors to stimulate the activity of phospholipase C, that in turn act upon the lipid phosphatidylinositol 4,5 bisphosphate (PtdIns4,5P$_2$) yielding DAG (and 1,4,5 inositol trisphosphate) and consequent PKC activation.

While it is now well established that agonist-induced increases in cellular DAG can be derived from sources other than PtdIns4,5,P$_2$, such as phosphatidylcholine, strategies employed in coupling diverse receptors to this particular inositol lipid pathway are better understood and will serve to illustrate the coupling mechanisms that probably exist for a variety of other as yet poorly defined DAG generating pathways. The triggering of this pathway is common to many distinct agonists (reviewed in 7). As such, the activation of PKC has been implicated in the control of many diverse cellular events and investigation into the specific functions of PKC and the means of receptor activation have been intense. The structural similarity between the PKC and Raf proteins (see below) has argued for a similar cascade of events in the control of Raf, although to date such a cascade has not been established (see Chapter 4).

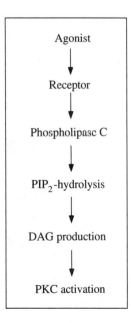

Fig. 2 Physiological activation of PKC. A schematic representation of a cascade of intracellular events initiated by agonist binding to its receptor at the cell surface.

Agonist

↓

Receptor

↓

Phospholipase C

↓

PIP$_2$-hydrolysis

↓

DAG production

↓

PKC activation

In the context of the above, it is the intention in this chapter to delineate mechanisms and strategies involved in signal generation and to discuss in more detail the properties and potential roles of kinase families PKC and Raf.

2. Generating the signal; phosphatidylinositol-specific phospholipase C

2.1 Phospholipase C structures

Phosphatidylinositol-specific phospholipase C (PtdIns-PLC) activities have been purified from a variety of tissue sources (reviewed in 7–9). These enzymes show an absolute specificity for inositol lipids, a preference for the polyphosphoinositides, PtdIns-4P and PtdIns-4,5P_2, an inability to hydrolyse PtdIns-3P lipids and a dependence upon Ca^{2+} within the physiological range (0.1–10 μM) (7–9). The general catalytic similarity probably reflects the existence of two conserved domains that appear to make up the catalytic cores of these proteins. Outside these conserved domains there is significant divergence that subdivides the proteins into three groups, β, γ, and δ (see Fig. 3). The structural divergence of the three subclasses of PtdIns-PLCs reflects the distinct modes of regulation. It is now clear that the PtdIns-PLCβ class are activated by G-proteins of the Gq family (10,11) and that the β$_2$ and β$_3$ enzymes can also be activated by G-protein βγ subunits (12–14); the PtdIns-PLCγ class are activated by complex formation with tyrosine kinase linked receptors and tyrosine phosphorylation (see 15). While the δ-class remain as yet unattached, it is anticipated that this PtdIns-PLC class will respond physiologically through a distinct receptor-operated regulatory pathway.

2.2 Specificity of action

The multiplicity of enzymes within each subclass suggests that either: (i) there is some specificity in respect of who is talking to whom; (ii) feedback controls may differ for these enzymes leading to differences in the longevity of responses;

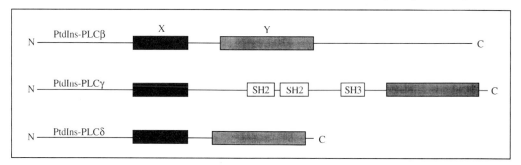

Fig. 3 Domain structure of PtdIns-PLC family. The conserved regions of PtdIns-PLC β, γ, and δ are indicated as the X and Y boxes. The src homology domains in PtdIns-PLCγ (SH2, SH3) are also indicated. The members within each subfamily (β$_{1-3}$; γ$_{1,2}$; δ$_{1-3}$) retain similar organization.

(iii) localization of the activated enzymes is distinct[1]; or (iv) any combination of the above. These distinctions are subtleties of the pathway that marginally complicate our view of the input (receptor–PLC coupling) but do not affect the rather singular output, i.e. the products of hydrolysis, namely inositol 1,4,5 trisphosphate and DAG. The basic need for the different PtdIns-PLC subclasses reflects a requirement for the coupling of different receptor types to this pathway.[2]

In respect of other sources of diacylglycerol and related lipid effectors, it is probable that a similar spectrum of controls operate in these circumstances also. For example, phospholipase D activation (producing phosphatidic acid and subsequently DAG) can be regulated by G-proteins, PKC activation and Ca^{2+} (see 16 and references therein). A distinct lipid effector, PtdIns-3,4,5P$_3$, is also produced through multiple mechanisms involving tyrosine kinase activation and G-proteins; the nature of the target for this lipid is not known (reviewed in 17). With respect to DAG, a critical, yet unresolved, issue is where and when it is formed and how this varies for different sources, e.g. PtdIns versus Ptd choline. The context of DAG production is also important in view of the synergistic effects of other lipid metabolites (recently reviewed in 18).

In the context of this chapter, it is also noteworthy that the kinase, Raf, is likely to be regulated by an effector (see below and Chapter 4). While the nature of this effector is currently unknown, it is reasonable to suppose that multiple strategies are employed in coupling receptors to its generation as described above for DAG. So in respect of PKC, and perhaps other cellular targets, distinct strategies of agonist-induced effector production permit convergence of signals.

3. Structural aspects of the protein kinase C and Raf families

3.1 Topology of the functional domains in PKC

A schematic representation of the protein kinase C (PKC) and Raf gene families is shown in Fig. 4. The degree of identity between the Raf and PKC families within the kinase domain is not as high as it is, for example, for the PKC/cAMP-dependent protein kinase (19). Nevertheless, the PKC and Raf families share regulatory domain features that suggest a significant degree of operational similarity. While these operational aspects are well documented for PKC, significantly less is understood for Raf. Here PKC will serve as a model for both families; Raf will be considered towards the end of this section. Attributes of the domains shown in

---

[1] It is important to note that agonist-induced DAG is initially likely to be restricted to the membrane in which it is produced.

[2] That such a need exists relates to the broader issue of signal integration. It is evident that individual receptors can trigger a number of events in a given cell type. The manner in which cells respond reflects the programmed way that they interpret combinations of signals to produce an appropriate response. This integration of multiple events is the basis of synergistic and antagonistic effects of different agonists.

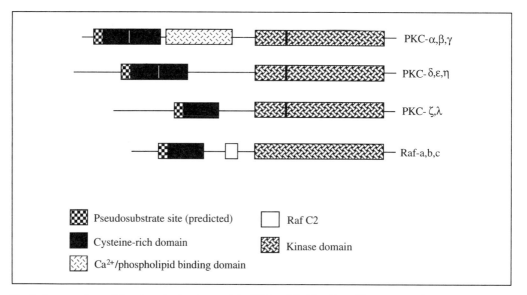

Fig. 4. A comparison of the domain structures of the PKC and Raf families. All members of these two families retain C-terminal kinase domains and all retain at least one cysteine-rich domain (the first two subclasses of PKC have two cysteine-rich domains). The Raf C2 region is a short segment conserved in c-Raf, a-Raf, and b-Raf within a proline, serine, threonine-rich putative linker region.

Fig. 4 are based upon mutagenesis, expression of individual domains, comparison between individual PKC isotypes, and protein chemistry as outlined below.

The amino-termini of the PKC family show little overall sequence similarity, and those of δ–λ are all extended relative to the α–γ group. It has been suggested that this amino-terminal extension confers a distinct specificity upon these enzymes relative to α–γ based upon comparative specificity, proteolysis and the characterization of a chimeric PKC-ε/γ construct (see 20 and references therein). However, whether this domain in fact confers this restricted substrate specificity requires further analysis.

The pseudosubstrate site, while not highly conserved across the family, is recognizable by the presence of basic amino acids either side of an alanine. This region was first noted by Kemp and colleagues (see 21) and its role as an autoinhibitory region has been confirmed through use of antisera (22) and mutagenesis (23). The presence and operation of pseudosubstrate sites in protein kinases is described in Chapter 2.

Immediately C-terminal to the pseudosubstrate site is a cysteine-rich domain that in all isotypes other than PKC-ζ and λ is tandemly repeated. Deletion studies and mutagenesis have provided evidence that this domain is responsible for DAG binding and that either one of the repeat units confers this binding property (see 24 and references therein). In fact, the overwhelming majority of studies on effector binding have been carried out with the relatively hydrophilic phorbol ester, phorbol dibutyrate (PDBu). This agent is one of a series of biologically active compounds that includes phorbol esters and some structurally related and apparently unrelated tumour promoters and irritants that are capable of substituting for DAG *in vitro* in the activation of PKC, as originally demonstrated by Castagna and colleagues (25). Since PDBu competes with DAG binding to PKC it is assumed that these sites overlap. The assignment of DAG (and PDBu) binding to this cysteine-rich repeat unit is further corroborated by the finding that a similar domain in the neural protein n-chimaerin retains this property (26).

In the α, β, and γ proteins there is a second conserved domain in the regulatory half of the protein (often referred to as the C_2 domain) that is circumstantially linked to the Ca^{2+}-dependence of phospholipid interactions. The remaining PKC isotypes that do not retain a C_2 domain display no Ca^{2+}-dependence for membrane interaction nor for kinase activity (discussed in 24). This C_2 domain shows similarity to regions of p65 (27) and $cPLA_2$ (28) both of which interact with phospholipids in a Ca^{2+}-dependent fashion. A domain containing the C_2-related region of $cPLA_2$, when expressed in *E. coli*, retains this property, providing direct evidence for this functional assignment (28).

The kinase domain of all these proteins retains the typical Ser/Thr kinase hallmarks (reviewed in 19) encompassing the bulk of the carboxy-terminal half of the proteins. The last fifty or so amino acids form a more variable region within the PKC family. This is the region of the protein that distinguishes the β_1 and β_2 gene products; these two proteins are encoded by a single gene and vary in this C-terminal region through alternative splicing that includes the last coding exon (29, 30). No clear properties have been assigned to this region, although deletion of C-terminal sequences have been correlated with both decreased (31) and increased (32) nuclear localization of another PKC isotype. Moderate (~2-fold) differences in activity have also been noted for PKC-β_1 compared to PKC-β_2 when GSK-3β is employed as a substrate (33), suggesting some role in substrate binding.

3.2 Domains in Raf

As illustrated above, the Raf family show a general structure similar to that of PKC. In particular, towards the amino-terminus of Raf is a cysteine-rich region which shows some similarity to the equivalent domain of PKC (34). Raf constructs with deletions encompassing this cysteine-rich domain display a dominant-negative influence (suggesting non-productive effector sequestration: see below). However, it is known that in Raf this domain does not bind phorbol esters. Rather, the domain has been implicated in Ras association (see below and Chapter 4, Section 4.5.4).

There is a potential pseudosubstrate site amino-terminal to the cysteine-rich domain in Raf; however, whether this plays a role similar to that assigned in the PKC proteins is not established. Nevertheless, the fact that amino-terminal deletions are activating (see 34) suggests that there are inhibitory sequences within the regulatory domain of the protein.

4. Mechanisms of activation

4.1 Turning on PKC

Due to the nature of the effector for PKC, the process of activation is minimally a two-step one. The first must involve the association of PKC with phospholipid in a non-productive complex. In the case of PKC-α, -β, -γ this is controlled by the free Ca^{2+} concentration presumably acting via the C_2 domain (see Fig. 5). At submicro-molar Ca^{2+} concentrations these PKC isotypes can form stable ternary complexes with phospholipid vesicles (see, for example, 35). In the second step the membrane (or vesicle) associated enzyme can bind DAG, inducing conformational change and consequent activation. This quaternary complex (PKC/Ca^{2+}/phospholipid/DAG) is slow to dissociate (even in the presence of Ca^{2+}-chelators); it is this phenomenon that is the basis for the translocation assay that monitors membrane-associated PKC levels. For those PKC isotypes that show no dependence upon Ca^{2+} the initial membrane association may not be controlled, but as for the Ca^{2+}-dependent forms once activated by DAG, this membrane form is also slow to dissociate (i.e. agonist-induced translocation can be demonstrated for PKC isotypes other than -α, -β, -γ; see, for example, 36).

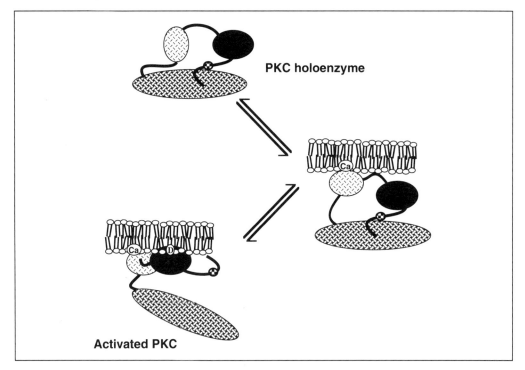

Fig. 5 Activation of Ca^{2+}-dependent PKC. The domain structure is shaded as indicated in Fig. 4 and schematically illustrates pseudosubstate site—kinase interaction in the inactive holoenzyme, Ca^{2+}-dependent binding to membranes via the C2 domain, and subsequent activation on binding diacylglycerol (D) through the cysteine-rich domain.

There is, at least *in vitro*, a third step in this process. It appears that in a stochastic fashion and as a consequence of membrane association, PKC isotypes can penetrate the lipid bilayer (37). This leads to the formation of an activated PKC that remains active even on removal of effector. However, detergent extraction re-establishes phospholipid/DAG dependence. That such phenomena operate *in vivo* is suggested by the finding that a significant proportion of the PKC expressed in, for example, brain tissue is intrinsic to the membrane fraction and can be released by neutral detergents or high pressure to yield typically dependent PKC (see 38). Interestingly, high pressure release from membranes leads to isolation of a PKC mixture containing a higher than normal content of phospholipid (38).

4.2 Activation of Raf

The mechanism via which Raf becomes activated in cells has attracted considerable attention. In view of the structural similarities to the PKC family, initial searches were aimed at isolating effector molecules. However, in many instances, physiological activation of Raf was detectable in immunocomplexes, suggesting a post-translational change. In these experiments native extracts were precipitated with Raf-selective antibodies and the washed immunoprecipitates assayed for kinase activity. The associated kinase activity has been monitored as apparent autophosphorylation (e.g. 39) or substrate phosphorylation (e.g. 40). The finding that these immunocomplexes contain increased kinase activity, following pretreatment of cells with particular agonists, indicated that these immunocomplexes contain stably activated kinase(s). It remains possible that activity reflects a Raf-associated protein kinase, although studies with inactive Raf mutants indicate that such an activity would have to be dependent upon Raf function (39).

By these assays Raf is 'activated' by a wide variety of agonists. Several of these, such as growth factors, act via the Ras protein. Using immobilized Ras, several groups have demonstrated specific binding to, and activation of, Raf via its amino-terminal domain (see Chapter 4, Section 4.5.4 for a detailed discussion). However, Raf may also be subject to Ras-independent regulation which may explain the stability of the immunoprecipitation assay. Two groups have demonstrated activation of Raf by PKC in insect cells or fibroblasts (39, 41). In the latter case, PKCα was shown to directly phosphorylate and activate Raf.

4.3 Monitoring PKC activation *in vivo*

Physiologically, the reversible activation of PKC is of primary relevance to the acute action of cellular agonists. As mentioned above, monitoring activation can be carried out by following membrane association. In intact cells, chronic activation also induces a decreased half-life (42) and in certain circumstances activation is evidenced by a decrease in the steady state level of expression (36). These procedures are open to immunological analysis and thus permit distinctions between different isotypes. This is not so for PKC-selective substrates which, while providing evidence for activation, in the absence of clear-cut specificity distinctions *in vivo*, cannot provide information on which PKC isotypes have been activated.

One other monitor of activation of PKC is phosphorylation state. Thus PKC-α has been shown to increase phosphorylation on stimulation (43), although no direct evidence that this is due to autophosphorylation has been presented. Without detailed phosphorylation site mapping and identification of autophosphorylation sites,[3] this type of analysis remains of limited value.

5. PKC and Raf substrates

An enormous variety of substrates for PKC have been described; these are assigned either based upon direct studies *in vitro* (usually with mixtures of PKC isotypes), or through identification of phosphoproteins induced *in vivo* by phorbol esters (see 45). Both approaches have limitations, thus PKC may well show promiscuity *in vitro* while phosphoproteins induced by phorbol ester treatment of cells may or may not be direct substrates for PKC.

5.1 Membrane-associated targets

There are a number of membrane-associated substrates for PKC including intrinsic transmembrane proteins such as the EGF receptor (46, 47). The phorbol ester induced phosphorylation of a number of such cell surface receptors is associated with increased rates of internalization (e.g. transferrin, 48). However, based upon site directed mutagenesis this internalization does not appear to be due to the receptor-directed phosphorylations (49) and the expectation is that PKC activation has some more fundamental regulatory role in the internalization pathway.

The membrane-localized protein c-src is also phosphorylated by PKC (50). The only documented effect of mutation of the PKC-phosphorylated residues is on an interaction between c-src and β-adrenergic receptor signalling (51). In response to β-adrenergic agonist, c-src overexpressing cells accumulate higher levels of cAMP compared with control cells and this effect requires the PKC sites. It is of interest that in contrast to the wild type protein, non-myristoylated (non-membrane associated) c-src is not phosphorylated following activation of PKC in intact cells (52). This implies that the non-myristoylated, non-membrane localized c-src is not 'seen' by membrane activated PKC. A similar analysis has been carried out for another PKC substrate p80/MARCKS; this protein is also myristoylated and associated with the particulate fraction. However, in contrast to c-src, the non-myristoylated MARCKS is still a PKC substrate *in vivo*, even though it shows a cytoplasmic localization (53). These contrasting results suggest that while the membrane localization of PKC is necessary for initial activation, this does not limit substrates exclusively to the same compartment, thus some may diffuse to PKC

---

[3] Autophosphorylation sites induced *in vitro* for PKC-β have been defined (44). It has not yet been established how these relate to basal/stimulated phosphorylation of PKC-β *in vivo*.

and/or PKC may itself relocate. This latter possibility might be effected through generation of a constitutively active catalytic fragment of PKC (see below).

5.2 Soluble substrates

There are a number of non-membrane associated proteins that will also serve as substrates for PKC *in vitro* and become phosphorylated following phorbol ester treatment *in vivo*. In few cases is it established that these are direct substrates for PKC, perhaps the most well defined being p80/MARCKS (which itself may be part of a family of related proteins; 54, 55). The sites within this protein serve as excellent substrates (low K_m) for PKC (56) and the evidence indicates direct control *in vivo*. In contrast, while the ribosomal S6 protein is phosphorylated by PKC *in vitro*, the evidence clearly implicates a kinase cascade *in vivo* (57, see Chapter 4). This latter scenario may reflect a common feature of PKC action in controlling kinase cascades that would minimally include S6 kinase, the MAP kinase pathway (reviewed in 58, 59) and glycogen synthase kinase-3 (33). It may yet transpire that a substantial contribution of PKC isotypes is in the regulation of other diffusible kinases, so permitting the membrane activation event to be propagated through the cell.

5.3 Generation of soluble, active PKC by proteolysis

Contrasting with the above, there appears to be one other mechanism for active PKC to induce substrate phosphorylation *in vivo*, and that involves proteolysis. There is direct evidence that activated PKC is susceptible to proteolysis *in vitro* and *in vivo* with the consequent production of a constitutively active catalytic domain fragment (60, 61); this would no longer be subject to membrane localization. Evidence has been provided to indicate that such a catalytic domain fragment is responsible for myosin light chain (MLC) phosphorylation in neutrophils (62). While the evidence is compelling, it is difficult to rule out a PKC controlled MLC kinase and rigorous proof of the role of proteolysis will no doubt come from mutant forms of PKC that are not subject to this mode of regulation.

5.4 Raf substrates

Unlike the plethora of identified targets for PKC (including Raf) intensive study of Raf has revealed only one likely substrate, MAP kinase kinase (MEK; 63–65). This specific issue is detailed in the context of MAP kinase elsewhere in Chapter 4 (Section 4.5.3). However, it is of interest to note that, as alluded to above for PKC, Raf likely controls the activity of other kinases. This would be consistent with the pleiotropic action of Raf as a protooncogene. As for PKC, the primary site of Raf activation is likely the membrane (via association with Ras) with the signal propagated through the cell via phosphorylation of other diffusible protein kinases.

6. Alterations in and to PKC and Raf gene expression

6.1 PKC-related genes

Members of both the PKC and Raf gene families have been identified in various species including *Drosophila* and *C. elegans*; a PKC relative has also been identified in yeast (reviewed in 66; see below and Chapter 8). As genetically tractable organisms, insights into the roles of the various gene products have been obtained and epigenetic evidence drawn up to localize the proteins at particular levels within signalling hierarchies.

In *Drosophila*, three PKC genes have been identified to date; one of these (53E) is exclusively expressed in photoreceptor cells (67, 68). A mutation (inaC) has been mapped to the PKC53E structural gene and studies on phototransduction in these mutants provides evidence for a role of PKC53E in deactivation and rapid desensitization of the light-activated signalling cascade (69). Reduced PKC activity has been documented also in a *Drosophila* learning mutant; however, this mutation does not map to a known structural gene (70). Mutations at other *Drosophila* PKC loci have not been described.

Mutations in the PKC-related *C. elegans tpa1* gene correct behavioural and developmental disorders induced by the phorbol ester TPA, suggesting that the direct action of TPA on this PKC gene product is causative in inducing these abnormalities (71). In the yeast *S. cerevisiae*, deletion of PKC1 induces a cell lysis defect as part of a cascade minimally involving PKC1 and a second protein kinase, BCK1 (72). In *Dictyostelium* a myosin heavy chain kinase has been identified as a member of the PKC family (73); the function of this PKC is evidently very specialized.

The loss of function phenotypes and the patterns of expression of PKC genes in these various organisms support the notion that individual PKC isotypes provide unique functions and furthermore, that tissue-specific PKC isotypes exist presumably adapted for specific roles. For example it is likely, though unproven, that as in *Drosophila* an eye-specific PKC exists in mammalian retina and likewise that it serves to regulate the operation of visual transduction.

6.2 Raf homologues

A Raf-related gene *D-Raf* has been identified in *Drosophila* (74) and been shown to be allelic to *1(1) pole hole* (75). The mutants in this gene show defects in proliferating tissues of the larvae and die at larval/pupal stages (74, 76). Thus *D-Raf* plays an essential role in development and this is in part associated with defective cell proliferation. Such a role is consistent with the role of activated Raf constructs as oncogenes (see 34). Recently, genetic studies in *Drosophila* have clearly placed Raf downstream of protein tyrosine kinases and p21[ras] in cellular signalling (77; see section 8.2.1). Manipulation of Raf expression in mammalian cells through the use of antisense RNA also provides evidence for a role in mitogenic signalling from

EGF and PDGF (78). A variation on this theme is the use of dominant-negative Raf constructs that interfere with the activation of c-Raf. Thus, for example, a point mutant without catalytic activity or a kinase domain-deleted construct suppress signalling pathways elicited by serum, TPA and Ras (79). The implications for the action of these mutant constructs is that c-Raf is regulated at least in part by a binding component that is limiting and can thus be titrated out. Whether this putative component is a protein or perhaps a diffusible second messenger is not clear, although the parallels that can be drawn with PKC would suggest the latter.

6.3 Overexpression of PKC genes

The introduction of PKC genes into different contexts has been shown to have interesting effects on cellular behaviour. For example, PKC-β overexpressed in a rodent fibroblast induces hypersensitivity to TPA, growth in soft agar (TPA-dependent) and a moderate tumorigenic phenotype (80). This is similar to the effect of overexpression of PKC-γ in fibroblasts (81). Induced changes in the expression of PKC also affect interaction with other signalling pathways as evidenced by susceptibility to H-*ras* transformation (82). Conversely transformation or expression of transforming genes is found to affect patterns of PKC gene expression (83, 84). Whether these represent necessary alterations with respect to the perceived phenotype is unclear; however, oncogene-induced alterations in certain cellular responses are restored on ectopic expression of PKC (83).

7. Perspectives

The PKC and Raf families evidently play critical regulatory roles in a number of biological contexts. The future will no doubt lie in firstly defining clearly the input into these signalling molecules, particularly the putative effector for Raf and where and when activation is triggered, and secondly in identifying substrates (kinase cascades?) that are responsible for propagating the cellular response. With this information in hand it may then be possible to rigorously dissect the responsibilities of individual gene family members and this will no doubt be complemented by the selective knock-out of individual genes. In the meantime, the combination of unique tools available for the study of PKC, the isolated genes, specific inhibitors and activators, will ensure continued progress in defining the biological roles of this expansive protein kinase family.

References

1. Takai, Y., Kishimoto, A., Inoue, M., and Nishizuka, Y. (1977) Studies on a cyclic nucleotide independent protein kinase and its proenzyme in mammalian tissues I. Purification and characterization of an active enzyme from bovine cerebellum. *J. Biol. Chem.*, **252**, 7603.

2. Inoue, M., Kishimoto, A., Takai, Y., and Nishizuka, Y. (1977) Studies on a cyclic nucleotide-independent protein kinase and its proenzyme in mammalian tissues II. *J. Biol. Chem.*, **252**, 7610.

3. Takai, Y., Kishimoto, A., Kikkawa, U., Mori, T., and Nishizuka, Y. (1979) Unsaturated diacylglycerol as a possible messenger for the activation of calcium-activated, phospholipid-dependent protein kinase system. *Biochem. Biophys. Res. Commun.*, **91**, 1218.

4. Kishimoto, A., Takai, Y., Mori, T., Kikkawa, U., and Nishizuka, Y. (1980) Activation of calcium- and phospholipid-dependent protein kinase by diacylglycerol, its possible relation to phosphatidylinositol turnover. *J. Biol. Chem.*, **255**, 2273.

5. Kaibuchi, K., Takai, Y., Sawamura, M., Hoshijima, M., Fujikura, T., and Nishizuka, Y. (1983) Synergistic functions of protein phosphorylation and calcium mobilization in platelet activation. *J. Biol. Chem.*, **258**, 6701.

6. Hokin, L. E. (1985) Receptors and phosphoinositide-generated second messengers. *Ann. Rev. Biochem.*, **54**, 205.

7. Meldrum, E., Parker, P. J., and Carozzi, A. (1991) The PtdIns-PLC superfamily and signal transduction. *Biochim. Biophys. Acta*, **1092**, 49.

8. Rhee, S. G., Suh, P.-G., Ryu, S-H, and Lee, S. Y. (1989) Studies of inositol phospholipid-specific phospholipase C. *Science*, **244**, 546.

9. Rhee, S. G. and Choi, K. D. (1992) Regulation of inositol phospholipid-specific phospholipase C isozymes. *J. Biol. Chem.*, **267**, 12393.

10. Taylor, S. J., Chae, H. Z., Rhee, S. G., and Exton, J. H. (1991) Activation of the β_1 isozyme of phospholipase C by α subunits of the G_q class of G proteins. *Nature*, **350**, 516.

11. Taylor, S. J. and Exton, J. H. (1991) Two α subunits of the G_q class of G proteins stimulate phosphoinositide phospholipase C-β_1 activity. *FEBS Lett.*, **286**, 214.

12. Camps, M., Carozzi, A., Schnabel, P., Scheer, A., Parker, P. J., and Gierschik, P. (1992) Isozyme-selective stimulation of phospholipase C-β_2 by G-protein $\beta\gamma$-subunits. *Nature*, **360**, 684.

13. Carozzi, A., Camps. M., Gierschik, P., and Parker, P. (1993) Activation of phosphatidylinositol lipid specific phospholipase C-β_3 by G-protein $\beta\gamma$ subunits. *FEBS Letts.*, **315**, 340.

14. Katz, A., Wu, D., and Simon, M. I. (1992) Subunits $\beta\gamma$ of heterotrimeric G protein activate $\beta2$ isoform of phospholipase C. *Nature*, **360**, 686.

15. Wahl, M. and Carpenter, G. (1992) Selective phospholipase C activation. *Bioassays*, **13**, 107.

16. Exton, J. (1990) Signalling through phosphatidylcholine breakdown. *J. Biol. Chem.*, **265**, 1.

17. Parker, P. J. and Waterfield, M. D. (1992) Phosphatidylinositol 3-kinase: A novel effector. *Cell Growth Diff.*, **3**, 747.

18. Nishizuka, Y. (1992) Intracellular signalling by hydrolysis of phospholipids and activation of protein kinase C. *Science*, **258**, 607.

19. Hanks, S. K., Quinn, A. M., and Hunter, T. (1988) The protein kinase family: conserved features and deduced phylogeny of the catalytic domains. *Science*, **241**, 42.

20. Pears, C., Schaap, D., and Parker, P. J. (1991) The regulatory domain of protein kinase C-ϵ restricts the catalytic-domain-specificity. *Biochem. J.*, **276**, 257.

21. Houslay, M. D. (1990) 'Crosstalk' and the regulation of hepatocyte adenylate cyclase activity: desensitization, G_i and cyclic AMP phosphodiesterase regulation. In *Biology of cellular transducing signals* (ed. J. Y. Vanderhoek), p. 141. Plenum Press, New York.

22. Makowske, M. and Rosen, O. M. (1989) Complete activation of protein kinase C by an antipeptide antibody directed against the pseudosubstrate prototope. *J. Biol. Chem.*, **264**, 16155.

23. Pears, C. J., Kour, G., House, C., Kemp, B. E., and Parker, P. J. (1990) Mutagenesis of the pseudosubstrate site of protein kinase C leads to activation. *Eur. J. Biochem.*, **194**, 89.

24. Parker, P. J., Kour, G., Marais, R. M., Mitchell, F., Pears, C. J., Schaap, D. *et al.* (1989) Protein kinase C—a family affair. *Mol. Cell. Endocrinol.*, **65**, 1.

25. Castagna, M., Takai, Y., Kaibuchi, K., Sano, K. Kikkawa, U., and Nishizuka, Y. (1982) Direct activation of calcium-activated, phospholipid-dependent protein kinase by tumor-promoting phorbol esters. *J. Biol. Chem.*, **257**, 7847.

26. Ahmed, S., Kozma, R., Monfries, C., Hall, C., Lim, H. H., Smith, P., and Lim, L. (1990) Human brain n-chimaerin cDNA encodes a novel phorbol ester receptor. *Biochem. J.*, **272**, 767.

27. Perin, M. S., Fried, V. A., Mignery, G. A., Jahn, R., and Sudhof, T. C. (1990) Phospholipid binding by a synaptic vesicle protein homologous to the regulatory region of protein kinase C. *Nature*, **345**, 250.

28. Clark, J. D., Lin, L. L., Kriz, R. W., Ramesha, C. S., Sultzman, L. A., Lin, A. Y. *et al.* (1991) A novel arachidonic acid-selective cytosolic PLA_2 contains a Ca^{2+}-dependent translocation domain with homology to PKC and GAP. *Cell*, **65**, 1043.

29. Coussens, L., Rhee, L., Parker, P. J., and Ullrich, A. (1987) Alternative splicing increases the diversity of the human protein kinase C family. *DNA*, **6**, 389.

30. Ono, Y., Kikkawa, U., Ogita, K., Fujii, T., Kurokawa, T., Asaoka, Y. *et al.* (1987) Expression and properties of two types of protein kinase C: alternative splicing from a single gene. *Science*, **236**, 1116.

31. James, G. and Olson, E. (1992) Deletion of the regulatory domain of protein kinase C α exposes regions in the hinge and catalytic domains that mediate nuclear targeting. *J. Cell Biol.*, **116**, 863.

32. Eldar, H., Ben-Chaim, J., and Livneh, E. (1992) Deletions in the regulatory or kinase domains of protein kinase C-α cause association with the cell nucleus. *Exp. Cell Res.*, **202**, 259.

33. Goode, N., Hughes, K., Woodgett, J. R., and Parker, P. J. (1992) Differential regulation of glycogen synthase kinase-3β by protein kinase C isotypes. *J. Biol. Chem.*, **267**, 16878.

34. Rapp, U. R., Cleveland, J. L., and Bonner, T. I. (1988) c-Raf. In *Handbook of oncogenes* (ed. P. Reddy, T. Curran, and A. Skalka), p. 213. Elsevier, Amsterdam.

35. Wolf, M., Cuatrecases, P., and Sahyoun, N. (1985) Interaction of protein kinase C with membranes is regulated by Ca^{2+}, phorbol esters and ATP. *J. Biol. Chem.*, **260**, 15718.

36. Kiley, S., Parker, P. J., Fabbro, D., and Jaken, S. (1991) Differential regulation of protein kinase C isozymes by thyrotropin-releasing hormone in GH_4C_1 cells. *J. Biol. Chem.*, **266**, 23761.

37. Bazzi, M. D. and Nelsestuen, G. L. (1988) Association of protein kinase C with phospholipid monolayers: two-stage irreversible binding. *Biochemistry*, **27**, 6776.

38. Lester, D. S. (1989) High pressure extraction of membrane-associated protein kinase C. *J. Neurochem.*, **52**, 1950.

39. Sözeri, O., Vollmer, K., Liyanage, M., Frith, D., Kour, G., Mark III, G. E., and Stabel, S. (1992) Activation of the c-Raf protein kinase by protein kinase C phosphorylation. *Oncogene*, **7**, 2259.

40. Oshima, M., Sithanandam, G., Rapp, U. R., and Guroff, G. (1991) The phosphorylation

and activation of B-Raf in PC12 cells stimulated by nerve growth factor. *J. Biol. Chem.*, **266**, 23753.

41. Kolch, W., Heidecker, G., Kochs, G., Hummel, R., Vahidl, H., Mischak, H. *et al.* (1993) Protein kinase Cα activates RAF-1 by direct phosphorylation. *Nature*, **364**, 249.

42. Young, S., Parker, P. J., Ullrich, A., and Stabel, S. (1987) Down-regulation of protein kinase C is due to an increased rate of degradation. *Biochem. J.*, **244**, 775.

43. Mitchell, F. E., Marais, R. M., and Parker, P. J. (1989) The phosphorylation of protein kinase C as a potential measure of activation. *Biochem. J.*, **261**, 131.

44. Flint, A. J., Paladini, R. D., and Koshland Jr, D. E. (1990) Autophosphorylation of protein kinase C at three separated regions of its primary sequence. *Science*, **249**, 408.

45. Woodgett, J. R., Hunter, T., and Gould, K. L. (1987) Protein kinase C and its role in cell growth. In *Cell membranes: methods and reviews* (ed. E. L. Elson, W. A. Frazier, and L. Glaser), vol. 3, p. 215. Plenum, New York.

46. Davis, R. J. and Czech, M. P. (1985) Tumor-promoting phorbol diesters cause the phosphorylation of epidermal growth factor receptors in normal human fibroblasts at threonine-654. *Proc. Natl. Acad. Sci. USA*, **82**, 1974.

47. Downward, J., Waterfield, M. D., and Parker, P. J. (1985) Autophosphorylation and protein kinase C phosphorylation of the epidermal growth factor receptor. *J. Biol. Chem.*, **260**, 14538.

48. May, W. S., Sahyoun, N., Jacobs, S., Wolf, M., and Cuatrecasas, P. (1985) Mechanism of phorbol diester-induced regulation of surface transferrin receptor involves the action of activated protein kinase C and an intact cytoskeleton. *J. Biol. Chem.*, **260**, 9419.

49. Davis, R. J. and Meisner, H. (1987) Regulation of transferrin receptor cycling by protein kinase C is independent of receptor phosphorylation at serine 24 in Swiss 3T3 fibroblasts. *J. Biol. Chem.*, **262**, 16041.

50. Gould, K. L., Woodgett, J. R., Cooper, J. A., Buss, J. E., Shalloway, D., and Hunter, T. (1985) Protein kinase C phosphorylates pp60^{src} at a novel site. *Cell*, **42**, 849.

51. Moyers, J. S., Bouton, A. H., and Parsons, S. J. (1993) The sites of phosphorylation by protein kinase C and an intact SH2 domain are required for the enhanced response to β-adrenergic agonists in cells overexpressing c-*src*. *Mol. Cell. Biol.*, **13**, 2391.

52. Buss, J. E., Kamps, M. P., Gould, K., and Sefton, B. M. (1986) The absence of myristic acid decreases membrane binding of p60^{src} but does not affect tyrosine kinase activity. *J. Virol.*, **58**, 468.

53. Graff, J. M., Gordon, J. I., and Blackshear, P. J. (1989) Myristoylated and nonmyristoylated forms of a protein are phosphorylated by protein kinase C. *Science*, **246**, 503.

54. Li, J. and Aderem, A. (1992) MacMARCKS, a novel member of the MARCKS family of protein kinase C substrates. *Cell*, **70**, 791.

55. Umekage, T. and Kato, K. (1991) A mouse brain cDNA encodes a novel protein with the protein kinase C phosphorylation site domain common to MARCKS. *FEBS Letts.*, **286**, 147.

56. Graff, J. M., Rajan, R. R., Randall, R. R., Nairn, A. C., and Blackshear, P. J. (1991) Protein kinase C substrate and inhibitor characteristics of peptides derived from the myristoylated alanine-rich C kinase substrate (MARCKS) protein phosphorylation site domain. *J. Biol. Chem.*, **266**, 14390.

57. Novak-Hofer, I. and Thomas, G. (1984) An activated S6 kinase in extracts from serum- and epidermal growth factor-stimulated Swiss 3T3 cells. *J. Biol. Chem.*, **259**, 5995.

58. Cobb, M. H., Boulton, T. G., and Robbins, D. J. (1991) Extracellular signal-regulated kinases: ERKs in progress. *Cell Reg.*, **2**, 965.

59. Posada, J. and Cooper, J. A. (1992) Molecular signal integration. Interplay between serine, threonine, and tyrosine phosphorylation. *Mol. Biol. Cell*, **3**, 583.

60. Kishimoto, A., Mikawa, K., Hashimoto, K., Yasuda, I., Tanaka, S-I., Tommaga, M. *et al.* (1989) Limited proteolysis of protein kinase C subspecies by calcium-dependent neutral protease (calpain). *J. Biol. Chem.*, **264**, 4088.

61. Young, S., Rothbard, J., and Parker, P. (1988) A monoclonal antibody recognising the site of limited proteolysis of protein kinase C. *Eur. J. Biochem.*, **173**, 247.

62. Pontremoli, S., Melloni, E., Michetti, M., Sparatore, B., Salamino, F., Sacco, O., and Horecker, B. L. (1987) Phosphorylation and proteolytic modification of specific cyto-skeletal proteins in human neutrophils stimulated by phorbol 12-myristate 13-acetate. *Proc. Natl. Acad. Sci. USA*, **84**, 3604.

63. Howe, L. R., Leevers, S. J., Gómez, N., Nakielny, S., Cohen, P., and Marshall, C. J. (1992) Activation of the MAP kinase pathway by the protein kinase Raf. *Cell*, **71**, 335.

64. Kyriakis, J. M., App, H., Zhang, X.-F., Banerjee, P., Brautigan, D., Rapp, U. R., and Avruch, J. (1992) Raf-1 activates MAP kinase-kinase. *Nature*, **358**, 417.

65. Dent, P., Haser, W., Haystead, T. A. J., Vincent, L. A., Roberts, T. M., and Sturgill, T. W. (1992) Activation of mitogen-activated protein kinase kinase by v-Raf in NIH 3T3 cells and *in vitro*. *Science*, **257**, 1404.

66. Stabel, S. and Parker, P. J. (1991) Protein kinase C. *Pharm. Therap.*, **51**, 71.

67. Rosenthal, A., Rhee, L., Yadegari, R., Paro, R., Ullrich, A., and Goeddel, D. V. (1987) Structure and nucleotide sequence of a *Drosophila melanogaster* protein kinase C gene. *EMBO J.*, **6**, 433.

68. Schaeffer, E., Smith, D., Mardon, G., Quinn, W., and Zuker, C. (1989) Isolation and characterization of two new *Drosophila* protein kinase C genes, including one specifically expressed in photoreceptor cells. *Cell*, **57**, 403.

69. Smith, D. P., Ranganathan, R., Hardy, R. W., Marx, J., Tsuchida, T., and Zuker, C. S. (1991) Photoreceptor deactivation and retinal degeneration mediated by a photoreceptor-specific protein kinase C. *Science*, **254**, 1478.

70. Choi, K.-W., Smith, R. F., Buratowski, R. M., and Quinn, W. G. (1991) Deficient protein kinase C activity in *turnip*, a *Drosophila* learning mutant. *J. Biol. Chem.*, **266**, 15999.

71. Tabuse, Y., Nishiwaki, K., and Miwa, J. (1989) Mutations in a protein kinase C homolog confer phorbol ester resistance on *Caenorhabditis elegans*. *Science*, **243**, 1713.

72. Lee, K. S. and Levin, D. E. (1992) Dominant mutations in a gene encoding a putative protein kinase (*BCK1*) bypass the requirement of a *Saccharomyces cerevisiae* protein kinase C homolog. *Mol. Cell. Biol.*, **12**, 172.

73. Ravid, S. and Spudich, J. A. (1992) Membrane-bound *Dictyostelium* myosin heavy chain kinase: a developmentally regulated substrate-specific member of the protein kinase C family. *Proc. Natl. Acad. Sci. USA*, **89**, 5877.

74. Nishida, Y., Hata, M., Ayaki, T., Ryo, H., Yamagata, M., Shimizu, K., and Nishizuka, Y. (1988) Proliferation of both somatic and germ cells is affected in the *Drosophila* mutants of *Raf* proto-oncogene. *EMBO J.*, **7**, 775.

75. Ambrosio, L., Mahowald, A. P., and Perrimon, N. (1989) Requirement of the *Drosophila* Raf homologue for *torso* function. *Nature*, **342**, 288.

76. Perrimon, N., Engstrom, L., and Mahowald, A. P. (1985) A pupal lethal mutation with paternally influenced material effect an embryonic development in *Drosophila melanogaster*. *Dev. Biol.*, **110**, 480.

77. Dickson, B., Sprenger, F., Morrison, D., and Hafen, E. (1992) Raf functions down-stream of Ras1 in the *sevenless* signal transduction pathway. *Nature*, **360**, 600.

78. Kizaka-Kondoh, S., Sato, K., Tamura, K., Nojima, H., and Okayama, H. (1992) Raf-1 protein kinase is an integral component of the oncogenic signal cascade shared by epidermal growth factor and platelet-derived growth factor. *Mol. Cell. Biol.*, **12**, 5078.

79. Bruder, J. T., Heidecker, G., and Rapp, U. R. (1992) Serum-, TPA-, and Ras-induced expression from Ap-1/Ets-driven promoters requires Raf-1 kinase. *Genes Dev.*, **6**, 545.

80. Housey, G. M., Johnson, M. D., Hsiao, W. L. M., O'Brian, C. A., Murphy, J. P., Kirschmeier, P., and Weinstein, I. B. (1988) Overproduction of protein kinase C causes disordered growth control in rat fibroblasts. *Cell*, **52**, 343.

81. Persons, D. A., Wilkison, W. O., Bell, R. M., and Finn, O. J. (1988) Altered growth regulation and enhanced tumorigenicity of NIH 3T3 fibroblasts transfected with protein kinase C-I DNA. *Cell*, **52**, 447.

82. Hsiao, W.-L. W., Housey, G. M., Johnson, M. D., and Weinstein, I. B. (1989) Cells that overproduce protein kinase C are more susceptible to transformation by an activated H-*ras* oncogene. *Mol. Cell. Biol.*, **9**, 2641.

83. Bernards, R. (1991) N-*myc* disrupts protein kinase C-mediated signal transduction in neuroblastoma. *EMBO J.*, **10**, 1119.

84. Borner, C., Guadagno, S. N., Hsiao, W. W.-L., Fabbro, D., Barr, M., and Weinstein, I. B. (1992) Expression of four protein kinase C isoforms in rat fibroblasts. *J. Biol. Chem.*, **267**, 12900.

4 | S6 kinases and MAP kinases: sequential intermediates in insulin/ mitogen-activated protein kinase cascades

JOHN M. KYRIAKIS and JOSEPH AVRUCH

1. Introduction

The characterization of signal transduction pathways that mediate cellular responses to insulin (1), growth factors (2) and cytokines (3) is an area of intense study. These agents act through cell surface receptors that are themselves transmembrane tyrosine kinases, or are functionally but non-covalently coupled to intracellular tyrosine kinases. Activation of these receptors leads to a rapid increase in the tyrosine phosphorylation of the receptors themselves, as well as a small subset of intracellular proteins.

Within minutes after the initial tyrosine phosphorylation, there follows a much more widespread and quantitatively extensive increase in Ser/Thr phosphorylation (4, 5). This response is now known to reflect the activation of a relatively large number of Ser/Thr protein kinases (6), arrayed in several parallel cascades, which are presumed to transmit, amplify and diversify the signals initiated via tyrosine phosphorylation. This chapter describes the characterization of two families of enzymes from among this array of mitogen-activated (Ser/Thr) kinases, the S6 kinases, p70 and p85 Rsk, and the MAP kinases, in particular the entities now known as Erk-1 and -2, and the distinct but homologous p54 SAP kinase subfamily. We summarize what is known concerning: (i) biochemical properties and structure of these enzymes; (ii) identity of their physiological substrates, basis for their substrate specificity and nature of their physiological role in elaborating the cellular response to insulin and mitogens; and (iii) molecular mechanism by which their activity is regulated and identity of the molecules that couple these protein (Ser/Thr) kinases to the signals initiated at the receptor.

1.1 Initial steps in insulin/mitogen signal transduction

Although receptor tyrosine autophosphorylation was initially viewed as an in-
cidental concomitant of the activation of the receptor-associated protein tyrosine
kinase activity, it is now clear that in many instances, it is the first indispensable,
intracellular event in signal transduction. Receptor autophosphorylation participates in signalling through two mechanisms: in some systems, receptor tyrosine
phosphorylation is required to activate the tyrosine kinase activity toward exogenous, non-receptor substrates; this is a property of the insulin/IGF-1 receptor
kinase family (7) (but not the EGF/PDGF families), as well as the Src tyrosine
kinase family (8) (see Chapter 7). Interestingly, the latter appear also to require a
prior tyrosine dephosphorylation at another site to disinhibit their capacity to carry
out the activating tyrosine autophosphorylation (9). A second and perhaps more
generally important mechanism utilizes the receptor tyrosine phosphate residues,
when situated in a specific sequence context, as recognition elements for the
binding of polypeptides with complementary *src* homology domains, type 2 (SH2
domains) (10).[1] The binding of signal transduction molecules through their SH2
domains to receptor phosphotyrosine residues facilitates their activation and
signalling function by several means: in some instances, the binding *per se* is able to
activate the catalytic function of SH2-containing protein complexes (as appears to
be the case with phosphatidylinositol 3-kinase; see 11). Such binding may also
facilitate the preferential phosphorylation and activation of the signal transduction
molecule by the receptor tyrosine kinase (as occurs with PLCγ; see 12). In addition,
this docking function serves to bring signalling molecules to the region of the inner
leaflet of the plasma membrane where they have immediate access to their
membrane-localized substrates, e.g. phosphatidylinositol lipids (for PLCγ and
phosphatidylinositol 3-kinase) or Ras (for RasGAP).

Thus, tyrosine phosphorylation serves, in general, a somewhat different role in
signal transduction as compared to Ser/Thr phosphorylation. Tyrosine phosphate
is more analogous in function to the soluble second messenger molecules that are
generated as an initial intracellular product of the activation of signal transduction
pathways by extracellular stimuli, and provide a transient, localized chemical
signal that recruits signal transduction molecules into an activated state. In this
fashion, multiple, parallel pathways of intracellular signal transduction are initiated concurrently; these pathways are mediated in subsequent steps largely
through the actions of Ser/Thr kinases. Insulin/mitogen-stimulated Ser/Thr phosphorylation, although less extensively characterized than tyrosine phosphorylation, is presumed to serve a regulatory role much like that served by Ser/Thr
phosphorylation in the 'classical' second messenger-activated pathways: a reversible
covalent post-translational modification that alters the catalytic or other functional
states of the protein substrate by superseding the functional state specified by
ligand modulators. This achieves immediate and direct regulation of the key

---

[1] However, an apparent SH2 domain in the c-Abl tyrosine kinase recognizes sequences that contain P-Ser
but not P-Tyr (11).

enzymes, transporters, transcription factors and structural proteins that mediate the multiple changes in cell structure and physiology necessary to accomplish such events as cell division, differentiation and possibly apoptosis.

This distinction between the functional roles of Tyr and Ser/Thr phosphorylation may be useful but is not absolute. For example, interferons activate a set of nuclear transcription factors by inducing tyrosine phosphorylation of these polypeptides, which directly modulates their function (13; see also the Jak kinases, Chapter 7). However, it is clear that post-translationally introduced P-Ser/P-Thr residues are present at ~100-fold greater abundance than P-Tyr, even in cells transformed by unregulated tyrosine kinase oncogenes, so that a fundamental difference in the functional roles for the two classes of phosphorylated hydroxyamino acids seems likely.

2. Discovery of insulin/growth factor-activated Ser/Thr kinases

The concept that protein (Ser/Thr) phosphorylation serves as the dominant bio-chemical mechanism for conveying extracellular signals to diverse targets within the cell interior gained acceptance with the characterization of the cellular role of the cAMP-dependent protein kinase (14; see Chapter 1). The elucidation of this enzyme's broad range of physiological substrates, and its function as the (almost) universal mediator of cAMP's numerous actions validated the idea that any cell activity is potentially susceptible to regulation by Ser/Thr phosphorylation. However, even as information accumulated establishing the central role of cAMP-dependent protein kinase in the response to numerous extracellular stimuli, paral-lel work demonstrated that some agonists, operating through receptors not linked directly to the regulation of adenylyl cyclase, nevertheless also elicited rapid, reversible changes in the activity of intracellular enzymes that persisted after cell disruption, consistent with changes in protein function caused by Ser/Thr phos-phorylations. One large class of such cAMP-independent agonists, exemplified by α_1 adrenergic agents and certain peptide hormones like vasopressin and angio-tensin II, was ultimately shown to initiate signal transduction pathways involving Ca^{2+} (15) and/or diacylglycerol (DAG)-regulated Ser/Thr kinases (16) as the major effector mechanisms, through the generation of the dual second messengers inositol 1,4,5 trisphosphate and DAG (16, 17; see Chapter 3).

Insulin, in addition to its rapid actions on nutrient and ion transport that are largely independent of cAMP, also rapidly induces stable alterations in the activity of several intracellular enzymes, usually in a direction opposite to that caused by activation of the cAMP-dependent protein kinase (5). This is best illustrated by glycogen synthase, an enzyme that is fully active in the dephospho-state but when multiply phosphorylated at specific sites is essentially inactive in the absence of the allosteric activator, glucose-6-phosphate. High levels of glucose-6-phosphate will, however, fully activate phosphorylated glycogen synthase. Insulin causes an increase in glucose-6-phosphate independent glycogen synthase activity within minutes by promoting the dephosphorylation of multiple Ser/Thr residues on this enzyme (18, 19).

The opposing effects of insulin and cAMP on metabolic enzymes such as glycogen synthase led to the view that activation of the insulin receptor, in addition to acting in the plane of the membrane to alter transport functions, was likely also to generate intracellular signals that led to dephosphorylation of regulatory proteins at Ser/Thr residues, perhaps through inhibition of Ser/Thr kinases, such as cAMP-dependent protein kinase, or by activation of Ser/Thr phosphatase(s) (20). Unexpectedly, direct examination of the effects of insulin on ^{32}P incorporation into cellular proteins in intact rat adipose cells (21–23) and subsequently rat hepatocytes (24) demonstrated that although insulin antagonized some of the increased phosphorylation elicited by cAMP, insulin's major effect was to increase ^{32}P incorporation into a subset of cellular proteins through a cAMP-independent mechanism. The polypeptides identified initially as substrates for this modification were the lipogenic enzymes, ATP-citrate lyase (25, 26) and acetyl-CoA carboxylase (27, 28), findings undoubtedly biased by the high abundance of these polypeptides in lipogenic tissues. Comparable studies of overall protein phosphorylation, carried out with cultured cells (29) and *Xenopus* oocytes (30), confirmed increased ^{32}P incorporation as the predominant response to insulin; in these cells, however, the most consistently detected and usually most abundant insulin-stimulated ^{32}P-polypeptide was a 31 kDa band identified as the 40S ribosomal subunit protein S6. As purified polypeptide growth factors and cytokines became available, studies examining the effects of agents such as IGF-1, EGF and PDGF on cellular protein phosphorylation established that increases in (Ser/Thr) phosphorylation of S6 is one of the most rapid and consistent intracellular biochemical responses to mitogens (31). The working hypothesis as to the biochemical mechanism underlying this phenomenon proposed that insulin/mitogens activated one or more protein (Ser/Thr) kinases that served as signalling intermediates catalysing the phosphorylation of S6 and other cellular proteins, altering their function to facilitate the enzymatic and structural changes required for DNA synthesis and cell division, as well as the metabolic adaptations necessary to support these activities. In the case of insulin, Ser/Thr dephosphorylation at regulatory sites on crucial metabolic enzymes was already known to be a central aspect of the hormone's action *in vivo*; it was therefore necessary to define the relationship of insulin stimulated Ser/Thr phosphorylation to the well-established regulatory dephosphorylation (5).

Direct evidence that an increased 40S S6 kinase activity was present in cell-free extracts of insulin-treated cells was first provided by Rosen and colleagues, as measured by the phosphorylation of endogenous S6 (32), or S6 on added 40S ribosomal subunits (33). Little progress in purification of these 40S S6 kinases was achieved, however, because of the unreliable recovery of mitogen-stimulated S6 kinase activity, until Novak-Hofer and Thomas demonstrated that inclusion of EGTA and β-glycerophosphate in the extraction buffer permitted reproducible recovery of an EGF/serum-stimulated 40S S6 kinase activity from 3T3 cells (34).

This buffer had been adapted from one employed to extract maturation promoting factor (MPF) activity from stimulated *Xenopus* oocytes (35), as it had been observed that impure isolates of MPF catalysed 40S phosphorylation (31). Although MPF and S6 kinase were shown to be separate entities, the extraction conditions were soon adopted by many groups who demonstrated activation of S6 kinase activity in cultured cells and *Xenopus* oocytes by the wide array of stimuli that had been shown to stimulate S6 phosphorylation *in situ*. Furthermore, comparable extraction conditions later enabled the detection of additional, novel insulin/mitogen protein (Ser/Thr) kinases, that were active on other protein substrates, such as MAP-2 or casein.

3. Insulin/mitogen-stimulated S6 kinases
3.1 General considerations

Once suitable conditions for the extraction of activated 40S S6 kinases were defined, early efforts at purification employed fractionated extracts from mitogenically stimulated cultured cells. A dominant peak of S6 kinase activity in these chromatograms (eluting broadly on anion-exchange chromatography near 0.25 M NaCl) and several minor peaks were usually evident (36, 37). It is now clear that this dominant peak is attributable to the molecular entity known as the p70 S6 kinase, primarily the (502 amino acid) p70 αII polypeptide (38). A second, molecularly distinct and independently regulated subfamily of mitogen-stimulated 40S S6 kinase activity, attributable to the Rsk polypeptides, can also be reliably detected upon anion-exchange chromatography of these extracts as a minor peak of kinase activity eluting between 0.05–0.1 M NaCl (36–39). The relative 40S kinase activity ordinarily detected in these peaks does not reflect accurately the relative abundance of the p70 versus Rsk polypeptides, because the usual conditions of 40S kinase assay employ 40S subunits at concentrations far below the K_ms for both the p70 S6 kinase (K_m 40S, ~0.5 μM; ref. 40) and the Rsk enzymes (K_m 40S, 5–10 μM; ref. 41). Given the difference between the substrate affinities of the two enzymes, the p70 S6 kinase would catalyse 40S phosphorylation at an approximately 10-fold higher rate than the Rsk enzymes, were both present at equal catalytic capacity. Thus, as with Rsk, some of the other minor peaks of 40S S6 kinase activity detected in chromatograms of crude extracts of mammalian cells may also represent as yet uncharacterized mitogen-activated protein kinases, although the likelihood of such enzymes contributing to 40S phosphorylation *in situ* is low.

Many protein Ser/Thr kinases phosphorylate ribosomal protein S6 *in vitro*. Nearly all of these enzymes, e.g. cAMP-dependent protein kinase, protein kinase C and even the Rsk enzyme, *Xenopus* S6 kinase II, also phosphorylate synthetic peptide substrates corresponding to the carboxy-terminal region of S6, the segment that contains all the *in vivo* phosphorylation sites, but at a much higher rate than they phosphorylate intact S6 on 40S subunits. The Rsk enzyme *Xenopus* S6 kinase II, for example, exhibits a 10-fold higher affinity (lower K_m) for such synthetic peptides as compared to the intact 40S subunit (42). The p70 S6 kinase is an exception in this regard, exhibiting at best equal and usually much lower affinity for any of the synthetic S6 peptide substrates than for the intact 40S subunit itself (40). This suggests that except for the p70 S6 kinase, the structure of S6 or its surroundings in the 40S subunit restricts access to protein kinases that are potentially capable of interacting strongly with the S6 polypeptide *per se*. Consequently, the use of a peptide substrate to survey cell extracts rather than 40S ribosomal subunits, even if the sequence of the peptide is derived from S6, allows a substantial relaxation of the specificity for 40S kinases, and results in a broadening in the array of protein kinases detected. A series of studies by Krebs and coworkers systematically applied such an approach to the detection of mitogen-activated kinases, employing several model substrates including a short peptide derived from the region surrounding the initial phosphorylation sites on S6 (RRLSSLRA) (37, 43–45). Although this peptide is avidly phosphorylated by the Rsk enzymes, it is a rather poor substrate for p70 S6 kinase (K_m ~0.2 mM; ref. 40), which prefers a block of three arginine residues amino-terminal to the phosphorylation site. The resulting patterns of peptide kinase activity underestimate the contribution of the p70 S6 kinase and accentuate the contribution of kinases with overlapping specificity. Nevertheless, inspections of these chromatograms, as obtained from a variety of ecchinoderm, amphibian and mammalian extracts, all indicate that the Rsk and p70 S6 kinases are the dominant mitogen-stimulated protein kinases that phosphorylate S6-derived (i.e. basic context) sequences. A third enzyme, chromatographically distinct from both p70 and Rsk, preferentially phosphorylates Kemptide (LRRASLG) compared to RRLSSLRA, and may correspond to MAPKAP kinase 2 (46; see Section 4.4.1 (ii), or to an apparently novel, insulin-activated 85 kDa kinase, the molecular structure of which is not yet available (47).

3.2   The Rsk S6 kinases

The first enzyme to be purified based on its ability to catalyse phosphorylation of 40S S6 was obtained from *Xenopus* eggs. Treatments of *Xenopus* oocytes arrested in prophase of their first meiotic division with progesterone, insulin, or by micro-injection of MPF, v-Abl, v-Src, or insulin receptor proteins, all led to a marked increase in S6 phosphorylation *in situ*, which increased further as oocytes completed their meiotic divisions and matured into unfertilized eggs (48). Extracts of mature unfertilized *Xenopus* eggs contain 40S S6 kinase at a relatively high specific activity. Chromatography of such extracts on DEAE Sephacel separated two peaks of S6 kinase activity, the second of which was further purified to near homogeneity and termed S6 kinase (S6K) II (49). This purified kinase is a 92 kDa polypeptide, verified by partial renaturation of activity from an SDS-PAGE (50). The S6 kinase activity in the DEAE peak eluting earlier (termed S6K I) proved far more labile and has surrendered only recently to purification as a 90 kDa protein (42). A polyclonal antibody raised to the purified, denatured *Xenopus* S6K II, although not reactive with *Xenopus* S6K I, did immunoprecipitate a portion of the increased S6 kinase activity from extracts of insulin and progesterone-treated of oocytes, as well as from extracts of chick embryo fibroblasts activated by serum or by expression of a temperature-sensitive v-*src* (48). These findings verified that a mitogen-stimulated S6 kinase of broad expression had been purified, and impelled efforts at molecular cloning.

The rapid success in the purification and characterization of the *Xenopus* Rsk enzymes, as compared to similar efforts in avian and mammalian systems, is striking. This may be partly attributable to a high level of S6 kinase in the egg, perhaps related to the intrinsic biology of this system. Unfertilized eggs contain a superabundance of ribosomal subunits presumably because much of the protein synthesis that occurs during the first several cell divisions after fertilization occurs without substantial gene transcription, and is generated by expression of maternally derived mRNA already present in the egg prior to fertilization (48). The more surprising aspect of *Xenopus* system in contrast to mammalian systems is the recovery of the Rsk family of S6 kinases, rather than the p70 S6 kinase; the latter is clearly the dominant 40S S6 kinase present during the G_0–G_1–S phase of the mammalian cell cycle (see Section 3.4.3). Moreover, it appears that p70 S6 kinase, although present, is in a low activity state in the unfertilized egg and contributes relatively little of the total 40S kinase (51). Thus, in the oocyte and egg a different 40S phosphorylating system appears to be operating than observed early after mitogen stimulation of growth factor-deprived mammalian cells. Whether this reflects the arrest of the oocyte cell cycle at the G_2/M boundary, or some other unique feature, is presently unknown.

Jones *et al.* (52), employing probes based on tryptic peptide sequence from *Xenopus* S6K II, isolated several *Xenopus* cDNAs that encoded a novel protein kinase, designated S6 kinase α. Polyclonal antiserum raised to the recombinant S6 kinase α polypeptide immunoprecipitated both S6K I and S6K II from extracts of *Xenopus* oocytes (53). The deduced S6 kinase α polypeptide was 84 kDa, and contained two complete kinase catalytic domains; the amino-terminal catalytic domain was most closely related (~35–40% identity) to the protein kinase C family, whereas the second, more carboxy-terminal domain, was most closely related to those of the calcium/calmodulin family of protein kinases, especially phosphorylase kinase. Utilizing low stringency hybridization, homologues to *Xenopus* S6 kinase α (renamed ribosomal S6 kinase, *rsk*α) were cloned from avian, mouse and rat sources (54); these *rsk* cDNAs encoded polypeptides of 84–86% identity to *Xenopus rsk*α. Moreover, a second mouse cDNA (*rsk*[mo]-2) was isolated whose polypeptide sequence was 79% identical to *Xenopus* Rskα, but only 84% identical to murine Rskα; Rsk[mo]-2 is clearly the product of a second *rsk*-like gene (54). The mRNAs corresponding to the two mouse *rsk* sequences exhibit a generally similar pattern of tissue-specific expression with the important exception that *rsk*[mo]-1 expression is low in heart and brain, whereas *rsk*[mo]-2 expression is relatively high in those two terminally differentiated, postmitotic tissues (54).

Present evidence indicates that the polypeptides encoded by murine *rsk*-1 and *rsk*-2 cDNAs are homologous to the *Xenopus* S6K I and S6K II polypeptides, respectively. A *Xenopus rsk* cDNA corresponding to S6K II, however, has yet to be isolated. Despite the lack of direct evidence concerning the relation of *Xenopus* Rskα to *Xenopus* S6K I and II, compelling evidence as to the essential identity of *Xenopus* S6K II and Rsk-2 has been provided by Cohen and associates, who isolated a 91 kDa protein kinase from rabbit skeletal muscle, which they named insulin-stimulated protein kinase-1 (ISPK-1) (55). This rabbit skeletal muscle enzyme is cross-reactive with highly specific polyclonal anti-*Xenopus* S6K II antiserum that reacts with purified *Xenopus* S6K II but not *Xenopus* S6K I or *Xenopus* Rskα. Moreover, peptide sequence obtained from rabbit ISPK-1 indicates that this kinase is identical to Rsk^{mo}-2 (56).

In summary, two Rsk isoforms have been identified at the protein and cDNA levels. Rskα (=Rsk-1) is either equivalent to *Xenopus* S6K I, or is a third Rsk, more closely related to S6K I than S6K II. Rsk-2 is the homologue of *Xenopus* S6K II; a full-length cDNA corresponding to the Rsk-2 isoform has not yet been reported. The nomenclature of the Rsks is given in Table 1.

3.3 The p70 S6 kinases

Contemporaneous with the purification of the S6 kinases from *Xenopus* and the subsequent molecular cloning of the *rsk* genes, several groups pursued the isolation of insulin/mitogen-stimulated S6 kinase from avian and mammalian sources. These efforts were impeded by the enzyme's low abundance, marked susceptibility to proteolysis and inhibition by transition metals (presumably acting through sensitive SH groups on the enzyme), and by the need to activate the enzyme *in situ* prior to extraction. Although extensive purification was achieved by several groups using cultured cells, yield and purity were insufficient for peptide sequence analysis. A more abundant source of enzyme was provided by the regenerating rat liver. S6 phosphorylation had first been identified as the basis for the 10-fold increase of ribosome phosphorylation which occurs in rat liver undergoing regeneration after a 70% hepatectomy (57). Nemenoff *et al.* (36) showed that S6 kinase activity in liver extracts also increased 6- to 8-fold within 2 h of partial hepatectomy and remained elevated for 36 h, thereafter falling rapidly toward baseline. An even more convenient source for purification was provided by livers from rats injected with the protein synthesis inhibitor cycloheximide. The ability of cycloheximide and puromycin to increase hepatic S6 phosphorylation, first observed by Gressner and Wool (58), was subsequently shown to occur in all tissues *in vivo* except for brain (59); cycloheximide also increased S6 phosphorylation when added directly to cultured cells (60). Not surprisingly, cycloheximide induced a 4- to 8-fold increase in hepatic S6 kinase activity, evident within 30 min of injection (61). The S6 kinase activated *in situ* after partial hepatectomy or cycloheximide treatment exhibited chromatographic and enzymatic properties indistinguishable from S6 kinase activity stimulated by insulin treatment of cultured (H4) hepatoma cells, i.e. coelution from anion and cation exchange columns and upon gel filtration as well as a requirement for EGTA during extraction. The cycloheximide-stimulated S6 kinase was purified

Table 1 Summary of known S6 and MAP kinase nomenclature[a]

Names	Molecular weight kDa	Substrates	Regulation
Rsk[(mo)2], S6 kinase II ISPK-1	85	40S subunits, PP1-G subunit, c-Fos, p67[SRF], lamin C	p42/44 MAPKs
Rsk[(mo)1], Rskα	90	Likely the same as Rsk2	p42/44 MAPKs
S6 kinase-I (possibly the same as Rsk1)	90	40S subunits	?
p70 S6 kinase	70 (αII) 93 (αI) (SDS-PAGE); 56 (αII) 59 (αI) (cDNA)	40S subunits	p42/44 MAPKs(?) Cdks(?), Rapamycin-sensitive species
p42 MAPK, Erk2	42	Rsks 1 and 2, MAPKAP, p70, S6 kinase, c-Jun, c-Myc, C-Fos, p62[TCF], ATF-2, EGF receptor, MAP2, tau, c-Raf-1	MAPKK/MEK
p44 MAPK, Erk1	44	same as p42 MAPK	MAPKK/MEK
Erk3	63	?	?
p54 SAPK-α[b]	54	p70 S6 kinase, c-Jun ATF-2, EGF receptor, MAP2	?
p54 SAPKs-β-γ	54, 46[c]	?	?

[a] Names used correspond to kinases identified by purification or molecular cloning, therefore ambiguities are present. These are discussed in the text. Molecular weights correspond to those seen upon SDS-PAGE, except where indicated, or where the kinase has been characterized by cloning only. Kinases with multiple isoforms encoded by a single gene are listed under one name. Substrates listed are presumed physiological substrates.
[b] Refers to the isoform purified from rat liver and characterized enzymatically; other p54 SAPKs have been characterized by molecular cloning only.
[c] Splice variants.

~50 000-fold from rat liver and characterized by two groups (61, 62). A crucial step in the purification was the use of a synthetic peptide for affinity chromatography corresponding to the sequences surrounding the phosphorylation sites on S6. This single step gave 30- to 100-fold purification with high recovery (61). Purified S6 kinase exhibited a major polypeptide at 70 kDa which copurified with a ladder of slightly smaller polypeptides at 68 and 66 kDa. Each of these bands underwent autophosphorylation *in vitro*. In addition, a minor set of polypeptides migrating near 95/93 kDa on SDS-PAGE copurified with the S6 kinase activity and also underwent autophosphorylation *in vitro* (61). These 95/93 kDa polypeptides, present at perhaps 5% of the mass of the 70 kDa polypeptide array, precisely coeluted upon gel filtration with the 70/68/66 kDa S6 kinase polypeptides, both sets eluting coincidentally with bovine serum albumin. The 95/93 kDa polypeptides are now known to be identical in sequence with the 70/68/66 kDa polypeptides, except for the presence of a 23 amino acid extension at the polypeptide amino-terminus (38) (see below).

Tryptic and CNBr peptide sequence obtained from the 70 kDa polypeptide led to the isolation of corresponding cDNAs, initially from rat (63, 64), and subsequently from human (38) and *Xenopus* cDNA libraries (51). Two different rat cDNAs were first reported independently; these structures differed only at their 5'-end: one encoded a polypeptide of 502 amino acids (64), whereas the second encoded a polypeptide of 525 residues, whose last 502 amino acids were identical to the entire sequence of the shorter S6 kinase polypeptide (63). Subsequent to the isolation of two distinct rat p70 cDNAs by separate groups, Reinhard and Thomas (65) isolated a second rat cDNA corresponding to that described by Banerjee *et al.* (63), and Grove *et al.* (38) reported cDNAs encoding both p70 S6 kinase polypeptides as isolated from human cDNA libraries. Both p70 S6 kinase isoforms contain a single, centrally located catalytic domain, most closely related (57% identity) to the amino-terminal catalytic domain of the Rsk-1 polypeptide. Immediately carboxy-terminal to the catalytic domain sequence (as defined by Hanks *et al.*, ref. 66), the p70 and amino-terminal Rsk catalytic domains exhibit another homologous segment of ~56 amino acids that is also shared with the protein kinase C family with ~40% mutual identity. Outside of this carboxy-terminally extended catalytic domain, the p70 S6 kinase and the Rsk polypeptides show no significant similarity in amino acid sequence.

We have designated the longer p70 S6 kinase polypeptide as αI, and the shorter polypeptide as αII inasmuch as available evidence indicates that all p70 mRNAs detected by hybridization with these sequences are transcribed from a single p70 S6 kinase gene (designated α), whose expression is diversified at both the mRNA and translational level (38). Northern blots of rat liver mRNA demonstrate four major p70 S6 kinase transcripts (63, 65). A 6 kb mRNA hybridizes with probes from the 5'-UT and coding regions of both rat cDNAs; a 2.5 kb mRNA hybridizes with probes based on the 5'- and 3'-UT sequences of the αI cDNA but not αII cDNA. Finally, 4.0 and 3.2 kb mRNAs are expressed that hybridize with probes corresponding to the 5'- and 3'-UT sequence of αII but not αI cDNA. Genomic Southern blots probed with various segments of the rat cDNA sequences point to the presence of a single, large gene (~25 kb) (65). The mRNA pattern likely arises through alternative splicing and utilization of different polyadenylation sites although the use of an alternative promoter to generate the 2.5 kb mRNA remains a possibility.

In addition to the existence of multiple mRNAs, it is clear that the mRNA encoding the αI polypeptide is translated *in vivo* to both the αI and αII polypeptide. This occurs because of the presence of a very weak translational initiation sequence surrounding the most 5'-ATG in the αI sequence with a C at −3 and an A at +4 (38). In contrast, the sequence surrounding the next ATG in both the rat and human αI polypeptide (encoding αI M24), which corresponds to the initiator methionine in the αII sequence, exhibits a much stronger translational initiation motif with a G at −3 and +4. This situation permits the ribosome to scan past the most 5'-ATG in the αI mRNA at high frequency, and initiate translation at the second ATG. Evidence supporting the operation of such 'leaky' scanning of the αI mRNA is provided by both *in vitro* translation and transient expression in COS cells. In reticulocyte lysates, translation of αI mRNA generates both the αI and αII polypeptides; mutation of αI M24 to Leu or Thr selectively suppresses the expression of the αII polypeptide. In COS cells expression of the wild-type rat or human αI cDNA yields both αI and αII p70 polypeptides, whereas the mutant αI M24-L p70 cDNA generates the longer αI polypeptide only. Significantly, the M24-L mutant expressed in COS cells exhibits 40S S6 kinase activity comparable to that of the αII p70 polypeptide, indicating that the αI p70 polypeptide is catalytically active.

A novel feature of the amino-terminal 23 residue peptide exclusive to the p70 αI polypeptide is the presence of six consecutive arginine residues immediately after the initiator methionine. This polybasic segment confers on the αI polypeptide an anomalous slow mobility on SDS-PAGE, migrating near 90 kDa despite its calculated M_r of ~59 kDa; the αII polypeptide (calculated M_r ~56 kDa) migrates at the expected mobility on SDS-PAGE, and when expressed in mammalian cells, multiple phosphorylation of the p70 αI polypeptide results in a progressive slowing of migration on SDS-PAGE, toward 70 kDa. The αI protein is expressed *in vivo* since, as noted above, highly purified preparations of rat liver p70 S6 kinase contain the 90 kDa forms. The relative abundance of the αI/αII isoform *in situ* is as yet uncertain, inasmuch as differential localization of the two isoforms may require the use of vigorous extraction procedures for accurate assessment. The polybasic sequence near the αI amino-terminus is similar to motifs found in proteins that bind polynucleotides as well as to certain nuclear/nucleolar localization signals. Preliminary experiments by Thomas and colleagues, employing antiserum to the unique 23 amino acid peptide sequence at the αI amino-terminus, support the possibility that the αI isoform is catalytically competent, activated by mitogenic stimulation, and targeted to the nucleus (65). Interestingly, a form of S6 that undergoes phosphorylation on mitogenic stimulation has been identified in the nucleus (67).

As summarized in Table 1, the p70 S6 kinases consist of two isoforms, αI and αII, which differ only by the presence of an additional 23 residues at the amino-terminus of the αI polypeptide (38). The p70 protein arises from mRNA transcribed from a single gene diversified by alternative mRNA splicing (65) and possibly alternate promoter usage, and further diversified through the use of a second translational start site on the mRNA encoding the αI sequence which enables the shorter αII polypeptide to be synthesized also. Both αI and αII polypeptides are expressed *in vivo* (38, 51), and αI may be targeted specifically to the nucleus (65) by virtue of the amino-terminal hexa-arginine sequence.

3.4 Substrate specificity and physiological roles

Initial characterization of the mitogen-activated S6 kinase activity in relatively crude extracts led to the view that the enzyme was highly specific for 40S subunits, inasmuch as mitogen-stimulated kinase activity could not be detected using the conventional basic and acidic polypeptide substrates, i.e. mixed histones, histone subfractions, casein or phosvitin. It is now evident, however, that purified preparations of both the Rsk enzyme, S6K II (42, 49), and p70 S6 kinase (61) can phosphorylate model basic protein substrates such as protamine, certain histones, myelin basic protein, etc. at significant rates. Thus, the inability to detect *in situ* mitogen-activated kinases using these substrates in crude extracts is due in large part to the wide array of kinases that act on these polypeptides including many that are not mitogen-stimulated (e.g. cyclic nucleotide-dependent protein kinases), or whose mitogen-stimulated activity is not retained under the conditions of assay devised for detection of the mitogen-stimulated 40S kinases (e.g. protein kinase C, calcium/calmodulin-dependent kinases). Moreover, assays using microtubule-associated protein-2 (MAP-2) or myelin basic protein enables detection of another broad family of mitogen-stimulated (Ser/Thr) kinases with an entirely different substrate specificity (see Section 4).

3.4.1 Rsk enzymes

Characterization of purified *Xenopus* S6 kinase II showed that this enzyme could phosphorylate a relatively wide number of basic peptide substrates, including the synthetic peptide Kemptide (LRRASLG) and a short S6-derived sequence (RRLSSLRA) (42, 49). *Xenopus* S6 kinase II also catalysed phosphorylation *in vitro* of rabbit skeletal muscle glycogen synthase (Ser 7), troponin I, tyrosine hydroxylase and lamin C, and comparison of the amino acid sequence surrounding the target serine residues in these substrates led to the proposal of RXXSX as the consensus sequence (where R = arginine, S = serine, X = any residue). Although somewhat tentative (in the absence of a systematic evaluation), the putative recognition motif for S6K II is indistinguishable from that of calcium/calmodulin kinase II (68), an enzyme generally considered to be a broad specificity, multifunctional regulator, and overlaps significantly with that of cAMP-dependent protein kinase and protein kinase C. This is perhaps not surprising in view of the presence of two catalytic domains in the Rsk enzymes, each domain homologous with one of those two kinase families. Studies of the individual contributions of each Rsk catalytic domain to substrate specificity would be of considerable interest.

Independent evidence for the operation of S6K II as a multifunctional regulator was provided by Cohen and colleagues from study of the insulin regulation of glycogen synthase (GS) (Fig. 1). As noted earlier, insulin activates the glucose-6-phosphate-independent activity of GS by promoting the dephosphorylation of multiple GS serine residues, primarily a cluster near the carboxy-terminus collectively termed 'site 3' targeted by glycogen synthase kinase-3 (GSK-3, see below; 19). Dephosphorylation of these sites is mediated by the catalytic subunit of protein phosphatase-1 (PP-1), which like GS itself, is bound to the glycogen particle. The 37 kDa catalytic subunit of PP-1 is tethered to glycogen by a 160 kDa polypeptide termed the G-subunit. The association of the phosphatase with glycogen enhances its ability to dephosphorylate and activate GS by ~5- to 8-fold as compared to the rate observed when the catalytic subunit is in solution. The association between the G-subunit and the catalytic domain is regulated; phosphorylation of the G-subunit by cAMP-dependent protein kinase at two serines near the G subunit amino-terminus (termed sites 1 and 2) disrupts the binding of the catalytic subunit

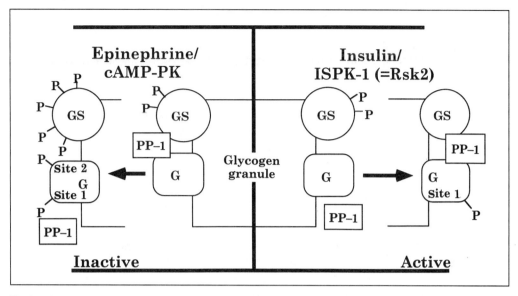

Fig. 1 Regulation of glycogen synthase by type 1 phosphatase in skeletal muscle. Model for the regulation of glycogen synthesis by insulin via ISPK-1 and cAMP-dependent protein kinase via epinephrine. GS, glycogen synthase; PP-1, protein phosphatase-1 catalytic subunit; G, glycogen binding subunit of PP-1. (After 70.)

to the G-subunit and leads to its dissociation from glycogen. Because of the much lower catalytic efficiency of the phosphatase toward GS when in solution, phosphorylation of the G-subunit at sites 1 and 2 serves as one of the mechanisms through which cAMP inhibits GS activity (69). Dent and colleagues (70) observed that selective dephosphorylation of site 2 on the G subunit promotes the reassociation of the PP-1 catalytic subunit; moreover, when this subunit is bound to a G-subunit selectively phosphorylated at site 1, PP-1 catalyses the dephosphorylation of GS (and phosphorylase *b* kinase) 2.5–3-fold faster than when PP-1 is bound to fully dephosphorylated G-subunit. Insulin treatment *in vivo* causes a 2-fold increase in the phosphorylation of the G-subunit at site 1 without affecting site 2. A protein kinase was detected in extracts of rabbit skeletal muscle that targeted the G-subunit specifically at site 1 and the activity of this kinase assayed after partial purification was increased 2–3-fold 15 min after insulin injection *in vivo*. This insulin-stimulated protein kinase active on G-subunit (named ISPK-1) was purified to homogeneity and shown to be a 91 kDa polypeptide, cross-reactive with anti-*Xenopus* S6K II antibodies (55). As noted earlier (Section 3.2), ISPK-1 appears to be identical to Rsk[mo]-2, indicating that ISPK-1 is the rabbit homologue of *Xenopus* S6K II, i.e. rabbit Rsk-2 polypeptide (56).

These observations are important in several respects for understanding the physiological role of the Rsk enzymes in insulin action. First, they identify Rsk-2 and therefore the entire signal transduction pathway upstream as elements that participate in the regulation of a central physiological response to insulin *in vivo*, i.e. activation of glycogen deposition. This does not require that Rsk phosphorylation of G-subunit be the sole or even major mechanism by which insulin regulates GS (and phosphorylase *b* kinase) activity. Clearly cAMP, through cAMP-dependent protein kinase, employs a variety of biochemical strategies in opposing glycogen synthesis and promoting glycogenolysis, suggesting that additional loci are likely to be identified where Rsk or other insulin-responsive signalling intermediates participate in the regulation of glycogen metabolism. A second important aspect of these findings is the effect of Rsk in regulating PP-1 activity. This provides a concrete and plausible explanation for the long-standing conundrum of insulin's ability to concomitantly cause increased Ser/Thr phosphorylation of some polypeptides, and dephosphorylation of others (22, 71). Here again, it is certain that yet other mechanisms will be uncovered through which insulin achieves Ser/Thr dephosphorylation.

There is evidence that Rsk polypeptides translocate to the nucleus upon mitogen stimulation (72), and participate in the phosphorylation and regulation of several nuclear proteins, including lamin C (42) and c-Fos (72). Detailed analyses of the site specificity and functional consequences of these phosphorylations are not yet available but hint at the many additional roles for these enzymes that remain to be investigated.

All of the information described above concerning substrate specificity relates to the Rsk-2 enzymes, i.e. *Xenopus* S6K II and rabbit ISPK-1. In preliminary studies, the properties of purified *Xenopus* S6K I have been compared to S6K II (41). At equal 40S kinase activity, S6K I phosphorylates a synthetic S6 peptide (at 0.1 mM) RRLSSLRA (S6 232–239) at ~ a 2-fold faster rate and Kemptide (LRRASLG) at slightly faster rate than S6K II. However, as additional basic residues are added to the amino-terminus of the S6 peptide (RAKRRRLSSLA), the rate of phosphorylation by S6K II increases 2–3-fold whereas phosphorylation by S6K I falls by 50%. The 32 residue carboxy-terminal S6 CNBr peptide (218–249) is phosphorylated by S6K II at a 4 to 12-fold higher rate than the 232–239 peptide whereas S6K I phosphorylates these two peptides at nearly identical rates (E. Erikson, personal communication). Thus, S6K I and S6K II have distinguishable substrate specificities which may reflect distinct physiological roles.

In summary, the Rsk enzymes are likely to function as broad specificity mitogen-activated Ser/Thr kinases, that phosphorylate an array of targets overlapping with those phosphorylated by calcium/calmodulin kinase II, cAMP-dependent protein kinase, and probably protein kinase C. The strongest candidates for physiological substrates at present are the glycogen binding subunit of PP-1, and, of course, 40S S6. The role of Rsk in the phosphorylation of 40S S6 *in vivo* has been discussed above and will be considered further below (Section 3.4.3).

3.4.2 p70 S6 kinase

In comparison to the Rsk enzymes, the p70 S6 kinase has a narrow substrate specificity. Initial studies employing p70 S6 kinase purified from rat liver (55, 61) indicated that while glycogen synthase is phosphorylated by p70 at ~40% the rate of 40S subunits, comparable to the rate catalysed by cAMP-dependent protein kinase and Rsk-2, p70 phosphorylates the G-subunit of PP-1 at <1% the rate of ISPK-1 (rabbit Rsk-2), does not detectably phosphorylate LRRASLG (Kemptide), and phosphorylates the S6 peptide (232–239) RRLSSLRA only poorly ($K_m = 0.2$ mM). Nevertheless, the synthetic peptide encompassing the carboxy-terminal 31 S6 residues (218–249) is phosphorylated with a K_m of 5–10 μM, indicating that the crucial determinants of p70 specificity are present in this more extended segment of S6 primary sequence. This was confirmed and substantially extended by Flotow and Thomas (40) who showed that p70 S6 kinase phosphorylates a synthetic peptide corresponding to residues 230–249 with kinetic constants nearly identical to those observed for intact 40S subunits. Utilizing variants of this peptide, it was established that Ser236 is the preferred site of phosphorylation, and substitution of Ser235 by Ala actually doubles V_{max}. The crucial determinants for phosphorylation of the peptide by the p70 S6 kinase are Arg231, 232 and 233. Individual substitution by Ala at each of these residues increases K_m by 140X, 5X and 700X, respectively. A lysine at 230 and an arginine at 238 are dispensable without any loss of affinity if arginines 231–233 are intact. The minimal recognition sequence for p70 is thus: R(R)RXXSX. At this time, 40S S6 remains the sole candidate physiological substrate for the p70 S6 kinase, although it is evident from database searches and preliminary experiments that a number of basic nuclear proteins warrant consideration as candidate substrates.

S6 is phosphorylated *in situ* at five sites consequent to mitogen stimulation. These charge variants are well visualized on 2D electrophoretic separations of ribosomal proteins, with the more extensively phosphorylated forms migrating at progressively increasing anodal mobility. Early studies concluded that 5–6 phosphorylations *in situ* occurred in an ordered sequence, based on peptide mapping of the individual charge isoforms (73, 74). It is now known that all 5 sites are clustered within 15 amino acid residues of the S6 protein carboxy-terminus (75).

The sites phosphorylated on rat liver S6 after injection of cycloheximide *in situ* have been identified by amino acid sequence analysis as the five serine residues (Ser236, 235, 240, 244, 247) (75). Moreover, these sites appeared to undergo phosphorylation *in situ* to a diminishing extent in the order 236>235>240>(244, 247). This ordered, multisite phosphorylation could result from the sequential activation of multiple kinases, or a progressive phosphorylation catalysed by a single enzyme. Early studies using partially purified enzyme demonstrated that a chromatographically homogeneous S6 kinase peak from *Xenopus* (probably related to S6K II) could generate the entire array of S6 tryptic ^{32}P-peptides detected during mitogen activation *in situ* (76). By contrast, most kinases capable of phosphorylating 40S subunits on S6 *in vitro*, e.g. cAMP-dependent protein kinase and protein kinase C, phosphorylate only one or two of these sites as detected by peptide mapping. The ability to catalyse 'complete' S6 phosphorylation *in vitro* is probably a reasonable requirement for any 'S6 kinase' to be a candidate for catalysing mitogen-stimulated S6 phosphorylation *in situ*, although no evidence clearly excludes the *in vivo* participation of kinases with less vigorous catalytic activity towards S6.

Employing purified rat liver p70 S6 kinase, Ferrari *et al.* (77) achieved *in vitro* phosphorylation of 3.7 moles P/mole S6, with most of the phosphorylated product migrating on 2D PAGE as the derivative containing 4 moles P/mole S6 with lesser amounts in the 3P and 5P forms. Sequence analysis demonstrated unequivocally the presence of phosphoserine at residues 235, 236, 240, and 244; a peptide fragment containing a fifth site was detected but not identified (presumably Ser247, phosphorylated to a low stoichiometry). Phosphorylation of S6 on *Xenopus* 40S subunits with highly purified *Xenopus* S6K II *in vitro* to an overall stoichiometry of 3 moles P/mole S6, yielded tryptic ^{32}P-peptides that exhibited ^{32}P incorporation into the same four residues as observed for p70 (78). The kinetics of site-specific ^{32}P incorporation indicated that Ser236 was phosphorylated at approximately twice the rate of Ser235, followed by 240 and lastly 244. Thus, both the *Xenopus* S6K II Rsk isoform and p70 αII S6 kinase phosphorylate S6 *in vitro* to a high stoichiometry at four of the five sites phosphorylated *in situ* and both enzymes catalyse a lower level of phosphorylation at a fifth site.

Several lines of evidence support the view that p70 S6 kinase is the dominant, perhaps sole, mitogen-activated 40S kinase operating in mammalian cells. Estimates of the K_m for 40S subunits gives values of 0.5 μM for p70 S6 kinase (40), ~5 μM for *Xenopus* S6K II (49), and 10 μM for S6K I (41). Thus, p70 S6 kinase is the most avid S6 kinase. Selective activation of p70 *in vivo*, as appears to occur after cycloheximide administration, is sufficient to achieve full S6 phosphorylation (31, 59, 75). Moreover, selective inhibition of p70 activation in intact cells without inhibition of Rsk, as observed upon addition of the immunosuppressant rapamycin to mitogenically-stimulated mammalian cells in culture (79–82; see Sections 3.4.3 and 3.5.2), leads to complete loss of mitogen-stimulated S6 phosphorylation. Therefore, the p70 S6 kinase must be the dominant mitogen-activated S6 kinase *in situ*, at least in mammalian cells moving from G_0 toward S. Whether the Rsk proteins function as 40S S6 kinases *in situ* in mitogenically stimulated mammalian cells remains open. This contrasts with the situation in *Xenopus* oocytes, where progesterone stimulates a progressive increase in total S6 kinase activity over 6 h; p70 activity is maximal after 1 h and declines thereafter, and the massive S6 kinase activity evident at later times as well as the 40S phosphorylation that occurs *in situ* appears to be attributable to the Rsk enzymes (51).

3.4.3 Consequences of S6 phosphorylation

Having concluded that p70 and, in some circumstances, Rsk S6 kinases function as 40S kinases *in vivo*, it should be stressed that the functional consequences of S6 phosphorylation are poorly understood. Most reports support the conclusion that S6 phosphorylation facilitates translational initiation, but the evidence is largely correlative and not compelling (for review see 83, 84). Experiments examining the effects of S6 phosphorylation on translational rates in extracts and reconstituted systems are difficult—a fair test requires numerous soluble translational components capable of supporting high rates of translation *in vitro* highly phosphorylated (≥3 mole P/mole S6) translationally competent 40S subunits, and appropriate test mRNAs—conditions rarely achieved.

Experiments addressing the role of S6 and its phosphorylation using mutants and genetic recombination in lower eukaryotes have yielded mixed results. Replacement of S10, the *S. cerevisiae* homologue of mammalian S6, with a mutant S10 polypeptide wherein the two Ser equivalent to S6 Ser235 and 236 are both replaced by Ala, is compatible with survival, although growth is halved (85). This result suggests that S6 phosphorylation is not a crucial regulatory modification. An important caveat, however, is that S10 is shorter at its carboxy-terminus than S6, and does not encode the three more distal Ser phosphorylation sites found in mammalian S6.

A more provocative result derives from deletion of the *air8* gene in *Drosophila*, which encodes the homologue of mammalian S6 (86). Deletion of *air8* results in slowed growth in most tissues of the fly embryo, but leads to tumour development within the haematopoietic system. The proclivity to form haematopoietic tumours is an intrinsic cellular property, manifest on transplantation of $air8^-$ lymph glands into normal host flies. The $air8^-$ embryos survive to a late larval stage, perhaps sustained by maternally derived S6 mRNA transcripts. Although these features suggest that S6 is crucial to growth regulation in a tissue-specific way, the mechanisms underlying this complex phenotype are unknown. The failure of *air8* embryos to develop past late larvae could reflect an overall defect in protein synthesis due simply to a faulty assembly of 40S subunits or to the selective lack of a function provided by S6 that is indispensable to overall protein synthesis. Against such a global defect is the ability of $air8^-$ haematopoietic cells to exhibit a transformed phenotype, which suggests that the translational apparatus is not incompetent, but dysregulated. The absence of S6 may result in the inappropriate expression of certain RNAs, such as impaired expression of tumour suppressers, or over-expression of protooncogene mRNAs, depending on the tissue-specific pattern of gene transcription. It is tempting to speculate that S6 phosphorylation regulates this proposed function of S6 in gating the translation of specific mRNAs. The effects of deletion of the *Drosophila* p70 S6 kinase gene would be of considerable interest.

The immunosuppressant drug rapamycin was recently shown to cause a potent but selective inhibition of p70 S6 kinase and 40S phosphorylation *in situ* in a variety of mammalian cells, without altering mitogen-induced receptor tyrosine phosphorylation, or the activation of Rsk, *erk*-1 or *erk*-2 MAP kinases (79–82; see Section 4). Rapamycin also causes a delay in or partial inhibition of DNA synthesis. Although this correlation is consistent with a role for p70 S6 kinase and/or 40S S6 phosphorylation in growth regulation, a strong conclusion is not warranted. The inhibitory effects of the rapamycin–immunophilin complex are not exerted on p70 itself but on upstream signal transduction elements; these are likely to regulate, in addition to p70, targets whose inhibition may be more directly responsible for slowed growth seen with rapamycin. The selectivity of rapamycin for a pathway crucial to the activation of p70 S6 kinase will nevertheless prove useful in identifying or excluding potential p70 substrates and physiological responses. For example, rapamycin inhibits insulin stimulation of p70 but not glucose transport in 3T3-L1 adipocytes (87).

3.5 Regulation of S6 kinase activity

The major impetus to purify the S6 kinases was the desire to understand the mechanism by which insulin and growth factor receptors activate the catalytic function of these Ser/Thr protein kinases.

The ability of β-glycerophosphate and other protein phosphatase inhibitors to stabilize the activity of S6 kinases during extraction implied that mitogen activation of these enzymes required phosphorylation of Ser/Thr and perhaps Tyr residues on the enzyme polypeptide. When the *Xenopus* S6K II and subsequently the p70 S6 kinases were purified, it was shown that both forms of S6 kinases could be specifically deactivated *in vitro* by treatment with purified Ser/Thr phosphatases, with PP-2A 10-fold more effective that PP-1; tyrosine phosphatases were without effect (88, 89). These results verified that enzyme Ser-P/Thr-P residues were necessary for S6 kinase activity. Nevertheless, the phosphorylation of these crucial Ser/Thr residues was not necessarily the locus of mitogen regulation, and although crucial for enzyme activity, these phosphorylations could have been catalysed as a constitutive co- or post-translational modification. The development of polyclonal antibodies capable of immunoprecipitating the S6 kinase polypeptides, however, established that coincident with mitogen activation, a major increase in ^{32}P-Ser/^{32}P-Thr incorporation occurred. Of equal importance, it was shown by peptide mapping that some (Rsk) (53, 90) or nearly all (p70) (89) of the ^{32}P tryptic peptides that exhibited enhanced ^{32}P incorporation during mitogen stimulation *in situ* were distinct from those that are ^{32}P-labelled by autophosphorylation *in vitro*, indicating that the enhanced phosphate incorporation following mitogen activation is not due entirely to an induced increase in autophosphorylation, but requires the participation of other protein kinases. Nevertheless, the possibility remained that activation of the S6 kinases by a ligand or polypeptide regulator, by altering the conformation of the S6 kinase polypeptide, uncovered previously unavailable phosphorylation sites, which were phosphorylated incidentally by cellular protein kinases. The only direct way to eliminate these caveats was to identify protein kinases capable of phosphorylating and activating the S6 kinase *in vitro* in a fashion that recapitulated the S6 kinase activation by insulin/mitogen *in situ*. This has been accomplished convincingly for the Rsk S6 kinases, but only partially thus far for the p70 S6 kinase.

3.5.1 MAP kinase (*erk*-1 and -2) phosphorylate and activate the Rsk S6 kinases

In 1988, Sturgill *et al.* (90) reported that a 42 kDa MAP-2 kinase, partially purified from insulin-stimulated 3T3-L1 adipocytes, could phosphorylate the *Xenopus* S6K II *in vitro* and increase its S6 kinase activity (see Fig. 2). Phosphorylation and activation of S6K II by MAP kinase occurred whether or not the S6K II substrate, which had been isolated as an active enzyme, was first treated with (Ser/Thr) PP-2A or PP-1 *in vitro*. If S6K II were first dephosphorylated and inactivated by phosphatases, MAP kinase could restore the S6K II to ~30% of its initial activity. The quantitative

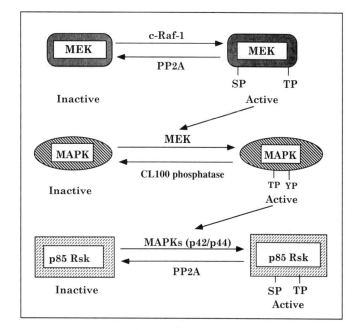

Fig. 2 The signalling cascade by which the Erk subfamily of MAP kinases and the Rsks are immediately regulated.

gain in S6K activity induced by phosphorylation with MAP kinase was similar with or without prior dephosphorylation of the S6K II. These results were rapidly confirmed (91), and nearly identical results were subsequently reported with the rabbit skeletal muscle ISPK-1/Rsk-2 purified by Lavoinne *et al.* (55) and with Rsk immunoprecipitated from serum-deprived 3T3 cells (92). Cobb and colleagues purified a 70 kDa S6 kinase from the livers of insulin-treated rabbits, that was reactive with anti-*Xenopus* Rskα polyclonal antibodies; this enzyme, presumably a slightly proteolysed Rsk isoform, exhibits reactivation by MAP kinase after an initial phosphatase treatment (93). Recombinant rat Rsk-1 expressed transiently in COS cells activated *in situ* by TPA and subsequently deactivated *in vitro* with phosphatase-2A, is also partially reactivated by an insulin-activated MAP kinase (94).

An entirely independent demonstration of the ability of MAP kinases to phosphorylate and reactivate the Rsk S6 kinases was provided in a series of reports cataloguing the EGF-stimulated protein kinases in extracts of NIH 3T3 cells active on the 'short' S6 (232–239) peptide RRLSSLRA (45). Among the seven peaks of EGF stimulated S6 peptide kinase reproducibly separated by MonoQ anion-exchange chromatography (numbered by order of elution), several were ultimately attributed to Rsk polypeptides (MonoQ peaks 1 and 2) and the p70 S6 kinase (MonoQ peak 7). The S6 peptide kinase activity in peaks 3 and 4 each coeluted with a peak of myelin basic protein (MBP) kinase, corresponding to a 42 kDa (Erk-2) and 44 kDa (Erk-1) MBP kinase in MonoQ peaks 3 and 4, respectively. The S6 peptide kinase and MBP kinase activities in MonoQ peaks 3 and 4 could be largely separated by a subsequent gel filtration step. Ahn and Krebs (95) showed that the EGF-activated S6 peptide kinase in peaks 1 and 2 (i.e. Rsks) could be deactivated *in vitro* by phosphatase-2A and reactivated in an ATP-dependent reaction catalysed by the MBP kinases found in either MonoQ peak 3 or peak 4. Moreover, if extracts from serum-starved cells were subjected to MonoQ chromatography, then S6 peptide kinase activity, virtually absent on direct assay of the eluted fractions, could be generated in the MonoQ fractions corresponding to the usual elution position of peaks 1 and 2, by incubation of those fractions with ATP and the EGF-activated MBP kinase activity associated with peaks 3 and 4. No other combination of fractions from basal and EGF-treated extracts generated additional S6 peptide kinase activity in such an ATP-dependent fashion; specifically, the fractions likely to contain the basal, inactivated form of the p70 S6 kinase could not be activated in mixing experiments with fractions from EGF-stimulated cells. These elegant experiments provide additional support for a role for MAP/MBP kinases in the activation of the Rsk kinases (45, 95).

A more difficult question to resolve is whether the *in situ* activation of the Rsk enzymes is attributable entirely to the *erk*-1/*erk*-2 MAP kinases, or whether other elements (i.e. other protein kinases, ligands, or protein cofactors) are also required *in situ*. At this time, the strong preponderance of evidence supports the conclusion that the Erk proteins are the sole immediate upstream activators of the Rsk enzymes. All stimuli that activate Rsk examined thus far also activate *erk*-encoded MAP kinases with similar or greater alacrity (92, 96; comparable studies of the deactivation/down-regulation of the two kinases, however, have not been reported). Moreover, MAP kinases are the major Rsk-1 kinases detected in extracts of unfertilized *Xenopus* eggs (97) or mitogen-stimulated cells (92). Grove *et al.* (94) characterized the alterations in transiently expressed rat Rsk-1 that accompany TPA-induced activation and compared these to the effects of phosphorylation of recombinant Rsk *in vitro* by MAP kinase. In addition to stimulating Rsk 40S and S6 peptide kinase activity, TPA activation of Rsk *in situ* increased the extent of subsequent Rsk autophosphorylation *in vitro* and caused the appearance of new autophosphorylation sites; TPA also slowed the mobility on SDS-PAGE of a subset of Rsk polypeptides, and promoted incorporation of [32]P *in situ* into multiple sites, distributed on 10–12 ^{32}P-tryptic peptides. MAP kinase-catalysed phosphorylation of the PP-2A-treated, inactive recombinant Rsk *in vitro* partially reactivated Rsk catalytic function and reproduced qualitatively nearly all of the other changes seen after TPA activation *in situ*. Prior activation of the recombinant Rsk *in situ* by TPA treatment reduced by over 90% the ability of MAP kinase to phosphorylate the recombinant Rsk-1 *in vitro*, as compared to the phosphatase-treated Rsk; moreover, the TPA-activated Rsk-1 was not further activated by the *in vitro* MAP kinase phosphorylation. The ^{32}P-peptide maps generated for MAP kinase phosphorylation *in vitro* of dephospho, inactive Rsk, exhibited ^{32}P-peptides that comigrated with all of the Rsk ^{32}P-peptides observed after TPA activation *in situ*, plus several Rsk ^{32}P-peptides characteristic of Rsk autophosphorylation *in vitro*. Conversely, MAP kinase phosphorylation of a heat-inactivated dephospho-Rsk generated a much simpler phosphopeptide map. These results indicate the Rsk autophosphorylation is required in addition to direct phosphorylation by *erk*-1/*erk*-2 MAP kinases in order to recapitulate the site-specific pattern of TPA-induced phosphorylation of Rsk *in situ*. Recently, one site of MAP kinase catalysed phosphorylation of ISPK-1 (rabbit muscle Rsk-2) has been identified as a Thr residue in the sequence–TPCY–located just amino-terminal to the APE sequence in subdomain VIII of the carboxy-terminal catalytic domain (56); this Thr residue is homologous to the site of constitutive cotranslational phosphorylation of the cAMP-dependent protein kinase catalytic subunit, a modification known to be crucial to this enzyme's activity, as well as the sites of the post-translational-activating phosphorylation of mammalian cdc2 and Erk-2 (see below).

3.5.2 Regulation of p70 S6 kinase

The compelling indirect evidence indicating that p70 S6 kinase is activated by Ser/Thr phosphorylation catalysed by upstream kinases coupled with the dramatic and easily reproducible activation of Rsk S6 kinase by the *erk*-1/*erk*-2 MAP kinase *in vitro*, pointed to the obvious possibility that MAP kinases might also participate in the activation of p70 S6 kinase. Nevertheless, direct experiments carried out in several labs uniformly observed that the p70 S6 kinase purified from rat liver could not be phosphorylated by a variety of *erk*-encoded MAP kinase preparations, whether or not the p70 had been subjected to prior phosphatase-2A treatment (90, 98). This negative result was congruent with other observations indicating that the p70 and the Rsk S6 kinases, although usually activated coordinately by insulin and mitogens, were regulated by independent mechanisms, at least in part. In some cells (e.g. COS cells), EGF or TPA activates endogenous MAP kinase and Rsk activity but does not alter significantly the constitutive level of p70 S6 kinase activity (38, 82). The reciprocal situation is observed during baculoviral infection of insect cells. Expression of *Xenopus rsk* α in Sf9 cells using baculoviral expression vectors produces a completely inactive Rsk polypeptide; coinfection with baculoviral-encoded v-*src* results in the activation of a portion of the recombinant Rsk (99). In contrast, p70 expressed using baculoviral infection yields activated p70 S6 kinase without the need for v-*src* coinfection. This activation of p70 is a consequence of the viral infection *per se*, inasmuch as infection of Sf9 cells with non-recombinant, wild-type baculovirus produces activation of the endogenous Sf9 cell p70 S6 kinase. Thus, baculoviral infection is capable of fully activating p70 without activating Rsk at all (100).

Although both classes of S6 kinase are activated by growth factor stimulation in most cells, the time course of activation of p70 and Rsk differ with Rsk (and Erk) activation peaking 2–5 min after agonist, whereas the peak of p70 S6 kinase activity is often delayed to 10–30 min (101). This is not due to Rsk regulation of p70 S6 kinase; neither S6 kinase alters the activity of the other (38). Both S6 kinases can exhibit a biphasic pattern of activation during the first hour after mitogen stimulation (101–103). Interpretation of this temporal pattern is complicated by the likelihood that the activation of both proteins is sustained by at least two different mechanisms operating sequentially during the first 60 min; the initial phase of activation is usually resistant to TPA-induced down-regulation of protein kinase C, whereas the later phase usually exhibits a significant (up to 50%) attenuation after TPA down-regulation (101–103). This is a general description; the temporal relationships and the effects of TPA down-regulation differ from cell to cell, and from one agonist to another, both acting on the same cell.

In many systems inhibition of protein synthesis with puromycin or cycloheximide causes a selective increase in the activity of the p70 S6 kinase, evident within 5–30 min, with little or no concomitant activation of Rsk (reviewed in 31, 104). Although the mechanism by which cycloheximide activates p70 S6 kinase in rat liver is unknown, cycloheximide treatment activates, in addition to the p70 S6 kinase, a novel MAP kinase, p54 SAP kinase, that is distinct from and regulated independently of *erk*-encoded MAP kinases (see Section 4.2.3). Moreover, the Erks are not activated in rat liver by cycloheximide (91). The p54 SAP kinase phosphorylates recombinant p70 S6 kinase but not Rsk, and may thus participate in the activation of p70, at least following cycloheximide treatment.

The independent regulation of p70 and Rsk is forcefully illustrated by the ability of the macrolide immunosuppressant, rapamycin, to cause a selective inhibition of the p70 S6 kinase. Rapamycin had been known to inhibit IL-2-induced T cell proliferation at a step in late G_1 without causing inhibition of T cell receptor-activated IL-2 gene expression (105–107). The structurally related macrolide FK506, in contrast, blocks T cell receptor-induced signalling via an inhibition of the calcium-dependent Ser/Thr phosphatase, calcineurin (108, 109). The actions of rapamycin and FK506 are expressed only upon their association with a family of small intracellular polypeptides known as immunophilins (110). Rapamycin and FK506 share sufficient structural homology to bind to the same set of immunophilins (called FK506 binding proteins; FKBPs); nevertheless the functional characteristics of the ligand/FKBP complex differ completely for rapamycin and FK506 so that each agent acts competitively to antagonize the action of the other (106, 107). Addition of rapamycin to serum-stimulated fibroblasts, insulin-stimulated H4 hepatoma cells, or COS cells transfected with the p70, Rsk-1 or Erk-1 cDNAs, inhibits completely the activity of p70 S6 kinase and S6 phosphorylation *in situ*, with no inhibition of endogenous or recombinant Rsk or Erk activity (80–82). FK506 at concentrations 200-fold higher than rapamycin has no effect on p70 S6 kinase but antagonizes the inhibition of p70 caused by rapamycin. The inhibition of p70 S6 kinase is not observed *in vitro* by a direct interaction of p70 with rapamycin/ FKBP-12, -13 or -25 complexes (82). *In situ*, the inhibitory effect of rapamycin is accompanied by a partial dephosphorylation of the p70 polypeptide and complete disappearance of the p70 polypeptides that exhibit slowed migration on SDS-PAGE, indicating that rapamycin blocks or reverses the p70 activation process; moreover, the effects of rapamycin prevail in the presence of okadaic acid (80). Thus, the deactivation of p70 S6 kinase caused by rapamycin is due either to inhibition of an upstream activator or the activation of a novel phosphatase (which is not calcium- or okadaic acid-sensitive) that acts on p70 itself, or on one of its upstream activators. Whatever the target of rapamycin, it is obviously an element that is necessary to the regulation of p70 but irrelevant to the activation of the Rsk or Erk enzymes.

A clue to the molecular mechanisms underlying regulations of p70 S6 kinase was provided by the observation that a segment in the p70 carboxy-terminal tail, starting at K423 of the rat p70 αI sequence just beyond the extended catalytic domain, resembles the region of the S6 polypeptide that encompasses the sites phosphorylated by the p70 S6 kinase. Based on the sequence similarity between the kinase and its substrate (28% identity over 25 amino acids), Banerjee et al. (63) proposed a regulatory model wherein this segment of the p70 S6 kinase carboxy-terminal tail acts as an autoinhibitory pseudosubstrate domain (see Chapter 2). In the basal state, the pseudosubstrate segment would occupy the substrate binding site of the p70 S6 kinase preventing access to the S6 polypeptide substrate. Activation of the p70 S6 kinase requires that the pseudosubstrate segment be displaced from the substrate binding site; this was proposed to occur through the phosphorylation of multiple Ser/Thr residues in this p70 segment, catalysed by insulin/mitogen-activated protein kinases situated upstream. The amino acid sequence in this pseudosubstrate segment contains six Ser/Thr residues, five of which are immediately followed by a proline residue, including two cdc2 consensus phosphorylation sites: (R/K) S/T X R/K (68). This feature suggested that phosphorylation of this pseudosubstrate segment was catalysed by mitogen-activated, proline-directed kinases, such as the MAP/Erk kinases and the cyclin-dependent kinases.

Several experiments support this model. A 37 residue synthetic peptide encompassing the putative autoinhibitory sequences of p70 (termed **S**6 kinase **a**utoinhibitory **p**seudosubstrate (SKAIPS) peptide) (Fig. 3) and a 32 mer peptide encompassing the homologous sequences in S6 (S6 peptide) were used to test several predictions of the pseudosubstrate hypothesis (111). If it were to function within the p70 polypeptide as a 'pseudo' substrate, the endogenous segment corresponding to the SKAIPS peptide should not be susceptible to phosphorylation by the p70 S6 kinase; otherwise, the endogenous segment would simply undergo autophosphorylation and be dislodged. Consistent with this prediction, the SKAIPS peptide, at concentrations up to 1 mM, is not detectably phosphorylated by p70, nor by cAMP-dependent protein kinase or protein kinase C, whereas the homologous 32 mer S6 peptide is a high affinity (K_m ~10μM) substrate for p70. Nevertheless, the SKAIPS peptide, despite its inability to serve as a substrate, does act as a high affinity inhibitor of the p70 S6 kinase, antagonizing the p70 S6 kinase-catalysed phosphorylation of 40S subunit with a potency equal to that of the 32 mer S6 peptide (IC_{50} ~20 μM). The SKAIPS peptide also competitively inhibits p70-catalysed phosphorylation of the 32 mer S6 peptide. This provides strong evidence that the SKAIPS peptide binds to the same site as the 32 mer S6 peptide substrate, indicating that the SKAIPS sequences are capable of serving as an endogenous autoinhibitory pseudosubstrate domain. The model also predicts that mitogen activation of p70 requires multiple phosphorylation events within the endogenous pseudosubstrate segment. This prediction was fulfilled with the demonstration

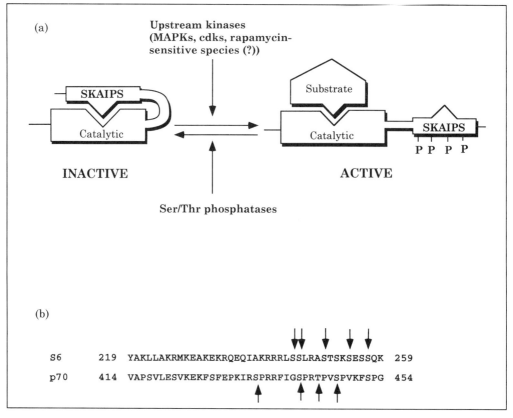

Fig. 3 (a) Mechanism of autoinhibition of basal p70 S6 kinase by the SKAIPS domain, and disinhibition of p70 by upstream activating kinases. (b) Comparison of SKAIPS and S6 sequences and phosphorylation sites. Phosphorylation sites are denoted with arrows. (After 111 and 113.)

that activated p70 polypeptides isolated from cycloheximide-treated rat liver and mitogen-activated 3T3 cells are phosphorylated at residues Ser434, 441 447, and Thr444 (using αI sequence numbering); each of these residues is situated within the putative pseudosubstrate site and is followed immediately by a proline (112, see Fig. 3). A further prediction is that the SKAIPS peptide will serve as a substrate for the mitogen-activated protein kinases that catalyse phosphorylation of p70 on this segment *in situ*. Insulin rapidly activates SKAIPS peptide kinase activity in H4 hepatoma cells; three peaks of insulin-stimulated SKAIPS peptide kinase are resolved by MonoQ anion-exchange chromatography of extracts of H4 hepatoma cells (113). Subsequent analysis confirmed that each of these peaks contained proline-directed kinases. MonoQ peak 1 was identified as Erk-2; MonoQ peak 2 was further resolved by gel filtration as a mixture of Erk-1 (eluting near 50 kDa) and a cyclin/cdc2 complex (eluting near 150 kDa). The enzymes comprising MonoQ peak 3 of SKAIPS peptide kinase activity were not definitively identified but appeared to include a third 40 kDa immunoreactive *erk* isoform, among a mixture of several kinases. The identification of cdc2 as an insulin/mitogen-stimulated protein kinase is a novel observation, but consistent with the *in situ* phosphorylation of the consensus cdc2 sites observed by Ferrari *et al.* (112). *Xenopus* MPF readily phosphorylates both SKAIPS peptide and intact rat liver p70 S6 kinase, and complete tryptic digestion of these two substrates yields identical 2D ^{32}P-peptide maps. Thus cdc2 or a closely related cyclin-dependent kinase are strong candidates for participation in the phosphorylation of p70 *in situ* during mitogen activation. The cdc2 catalytic subunit responsible for p70 phosphorylation *in situ* is likely to be complexed with one of the 'G$_1$' cyclins; little is known, however, concerning the specific identity of the cyclin in this complex or the mechanism by which mitogens regulate the activity of the G$_1$ cyclin/cdc2 complex.

The identification of Erks as insulin-activated SKAIPS peptide kinases was surprising, inasmuch as p42 MAP kinase/Erk-2 is incapable of phosphorylating purified rat liver p70 S6 kinase, even after dephosphorylation of the p70 polypeptide by phosphatase-2A (90, 98). The inability of p42 MAP kinase to phosphorylate the intact rat liver p70 despite rapid phosphorylation of the SKAIPS peptide cannot be simply attributed to inaccessibility of the p70 autoinhibitory domain, in view of its rapid phosphorylation *in vitro* by MPF. A potential explanation was suggested by the properties of recombinant p70 S6 kinase expressed in COS cells (38); although the majority of recombinant p70 polypeptides underwent phosphorylation *in situ*, only the most extensively phosphorylated p70 polypeptides (corresponding to those with the slowest mobility on SDS-PAGE) copurified on MonoQ chromatography with 40S kinase activity. We reasoned that perhaps a set of preliminary phosphorylations of p70 were required in order for MAP kinase to act on this substrate, and these phosphorylations were reversed during the PP-2A deactivation of rat liver p70 S6 kinase. By this reasoning, some portion of the recombinant p70, expressed in COS cells, which is predominantly in a partially phosphorylated but as yet inactive state, might serve as a substrate for Erk-2. Indeed both Erk-2 and p54 SAP kinase were shown to phosphorylate COS cell-expressed p70 S6 kinase yielding equivalent peptide maps (113). Unlike MPF, which phosphorylated p70 only within the SKAIPS domain, the MAP kinases phosphorylated p70 both within the SKAIPS sequences (although with a different preference from MPF), as well as outside the SKAIPS sequences; the identity of the latter sites and their relevance to sites phosphorylated during mitogen stimulation *in situ* is not known.

Despite evidence implicating both cdc2 and some isoforms of MAP kinase in the phosphorylation of p70 *in situ*, activation of COS cell-expressed p70 by phosphorylation *in vitro* with MPF, Erk-2, or p54 SAP kinase, alone or in any combination, has not been observed. Comparison of the peptide maps of p70 phosphorylated by these enzymes *in vitro* to maps obtained from p70 isolated from ^{32}P-labelled mitogen-activated cells indicates that these proline-directed kinases phosphorylate only a subset of the sites that undergo mitogen-stimulated phosphorylation *in situ* (113). Consequently, it is likely that one or more as yet uncharacterized kinases are necessary, in addition to cdc2 and one of the MAP kinases, to recapitulate the multisite phosphorylation and activation of p70 induced by mitogen *in situ*. It is possible that the unknown p70 S6 kinase-kinase is situated on the signal transduction pathway selectively inhibited by rapamycin.

In summary, the regulatory elements immediately upstream of the p70 S6 kinase appear to be distinct from, largely independent of and much more complex than those required for activation of Rsk S6 kinase. Activation of p70 S6 kinase requires multisite phosphorylation of the enzyme in a pseudosubstrate segment, catalysed by proline-directed kinases including cdc2 and probably one or more MAP kinases, either p54 or Erks-1/2. In addition, a rapamycin-sensitive phosphorylation, catalysed by unknown kinase(s), also appears to be indispensable for p70 activation. The tight regulation of p70 S6 kinase activation and the likely participation of cdc2 strongly suggest that the multiple phosphorylation of 40S S6 will prove crucial to the regulation of cell division.

4. The MAP kinases

4.1 Early observations

The MAP (**M**itogen-**A**ctivated **P**rotein) kinases were discovered by Sturgill and Ray (114), who were exploring the possibility that an insulin-induced (Ser/Thr) phosphorylation of the protein phosphatase-1 regulatory protein called Inhibitor-2 (I-2) might provide a mechanism by which insulin promotes (Ser/Thr) phosphorylation of some polypeptides while causing dephosphorylation of others. Extracts from insulin-treated 3T3-L1 cells did not catalyse I-2 phosphorylation at an increased rate; however, the (Ser/Thr) phosphorylation of a high molecular weight polypeptide contaminant in the I-2 isolate, identified as microtubule-associated protein-2 (MAP-2), was substantially insulin-stimulated. Authentic purified MAP-2 also served as a substrate for this insulin-stimulated (Ser/Thr) kinase activity. In characterizing this novel activity, they showed that the insulin-stimulated MAP-2 kinase activity in 3T3-L1 cells was activated more rapidly and was chromatographically distinct from the insulin-stimulated S6 kinase (now known to be p70 S6 kinase) also present in these extracts (115). Thus the MAP-2 kinase eluted upon gel filtration near 35 kDa (116). In addition, the MAP-2 kinase, like the S6 kinase, required phosphatase inhibitors for reliable recovery. Sturgill and Ray provided two observations that brought this insulin-stimulated MAP-2 kinase to wide attention. As discussed earlier, they demonstrated in collaboration with Erikson and Maller, that the insulin-stimulated MAP-2 kinase could phosphorylate and activate the *Xenopus* Rsk enzyme S6 kinase II (90). In addition, they observed that activation of the MAP-2 kinase by insulin was accompanied by both tyrosine and threonine phosphorylation of the MAP-2 kinase polypeptide (117) and that both of these phosphorylations were necessary for activation of the MAP kinase (118). Treatment of the partially purified insulin- or PMA-activated p42 MAP kinase with either Ser/Thr-specific phosphatase 2A or a Tyr-specific phosphatase (CD45) led to a complete deactivation of the kinase (118). This represented one of the first demonstrations of an insulin/mitogen-stimulated tyrosine phosphorylation that was crucial to the regulation of the function of the protein substrate. This discovery raised the exciting possibility that the MAP-2 kinase was itself a substrate for receptor protein tyrosine kinases and served as a direct link between the membrane-bound tyrosine kinases and multiple potential intracellular targets regulated by Ser/Thr phosphorylation, i.e. the substrates of the MAP-2 kinase itself.

The likelihood that the MAP-2 kinase was a ubiquitous effector in mitogen action was strengthened by the demonstration that the MAP-2 kinase polypeptide was apparently identical to a set of 40–45 kDa polypeptides previously detected as a substrate for tyrosine phosphorylation in response to a wide variety of mitogens and transforming agents (119–123). In recognition of these possibilities, the name based on a serendipitous substrate (MAP-2) was swapped for **Mitogen-Activated Protein (MAP) kinase** (124). Concomitant to this work, a number of groups (45, 95, 125–129) were describing apparently novel mitogen-activated kinases which ultimately proved to be very closely related or identical to the insulin-stimulated 42 kDa MAP-2 kinase initially detected by Sturgill and colleagues (see Cobb *et al.* (130)). Although subsequent work has shown that the recruitment of the MAP kinases after receptor activation is less direct and more complex than initially anticipated, the promise that this novel family of Ser/Thr protein kinases would provide important clues to the nature of the early steps in signalling, and come to occupy a wide and crucial role in cell regulation has been amply fulfilled.

4.2 Structural features of the MAP kinases: a multigene family

4.2.1 Similarities with yeast signalling kinases

Boulton *et al.* (131) purified a 44 kDa, insulin-activated MAP kinase from a CHO line overexpressing recombinant insulin receptors, and cloned a partial cDNA encoding a polypeptide which they named extracellular signal-regulated kinase-1 (*erk*-1) (132). The sequence encoded a Ser/Thr protein kinase that showed a striking homology (56%) to *FUS3* (133) and *KSS1* (134), two gene products previously identified as participants in the mating pathway of *Saccharomyces cerevisiae*. These yeast genes encode protein kinases which function to arrest the yeast cell cycle at G1 in response to mating factor binding at the cell surface. A more recently identified member of this protein kinase subfamily is *spk1+*, isolated from *Schizosaccharomyces pombe* (135); *spk1+* encodes a protein kinase that when overexpressed from multicopy plasmids, confers resistance to the ability of staurosporine to inhibit the growth of *S. pombe*. Thus, Spk1 is probably involved in the growth control of the fission yeast. More recently, two new subfamilies of yeast MAP kinases have been identified; HOG1 appears involved in osmo-sensing, whereas MPK1 is necessary for cell wall synthesis and is downstream of yeast protein kinase C (reviewed in 136).

The structural similarities between the mammalian MAP kinase, Erk-1 and yeast kinases involved in growth control (133–136), coupled with the demonstration that MAP kinases are also present in the oocytes of marine invertebrates and amphibians (130, 137, 138) where they are regulated in a cell cycle dependent manner (discussed below), indicates that the MAP kinase subfamily represents a set of ubiquitous, multifunctional and highly conserved elements regulating cellular growth and differentiation.

4.2.2 Cloning of additional mammalian Erks

Using low stringency screening, Boulton *et al.* (139) isolated two additional cDNAs encoding rat MAP kinase family members: *erk*-2 and *erk*-3. *erk*-2, 83% identical to *erk*-1, corresponds to the 42 kDa MAP kinase originally characterized biochemically by Sturgill and colleagues and subsequently cloned by that group from a murine cDNA library (140). In addition, a second Erk-2 isoform probably reflecting an mRNA splice variant, has been observed (141). *erk*-3, a 62-kDa species not previously detected at the biochemical level, is only ~43% identical to *erk*-1 and -2 overall, but contains segments in the catalytic domain that show more extensive identity and establish *erk*-3 as a legitimate *erk* polypeptide. Studies employing antipeptide antisera point to the existence of a 45 kDa Erk-like protein kinase with chromatographic behaviour distinct from Erk-1 and Erk-2. A cDNA corresponding to this isoform, tentatively named Erk-4, has not yet been reported.

All of the Erks described thus far are highly expressed in brain where they show distinct spatial and developmental patterns of expression (139, 142). *erk*-2 and -3 show reciprocal patterns of expression, with *erk*-3 expression highest in hindbrain and *erk*-2 expression highest in forebrain; by contrast, *erk*-1 expression is relatively uniform throughout the brain. P19 embryocarcinoma cells, which can differentiate into neuronal lineages upon stimulation with retinoic acid, or muscle lineages upon stimulation with DMSO, show distinct patterns of *erk* expression depending on the mode of differentiation elicited. *erk*-1 expression decreases, while *erk*-3 increases with either developmental programme. By contrast, *erk*-2 expression increases with neuronal differentiation, and is unchanged during myogenic conversion.

4.2.3 The p54 SAP kinase subfamily

p54 SAP kinase was originally purified from rat liver as the dominant MAP-2 kinase activated by *in vivo* administration of cycloheximide (91). This kinase displayed overlapping but distinct functional properties as compared to the 42- and 44-kDa MAP kinases (Erks-2 and -1). Degenerate oligonucleotides based on p54 tryptic peptide sequences were used to generate a partial cDNA by PCR (143). Five closely related, but distinct p54 cDNAs were isolated from a rat brain library (p54α_I, p54α_{II}, p54β, p46β and p46γ) which share greater than 95% identity in amino acid sequence. p54α, p54β, and p54γ are likely encoded on distinct genes; p54α_{II} is identical to α_I except for the presence of 15 alternate amino acid residues in subdomain IX of the catalytic domain, probably arising by differential splicing. p54β and p46β differ only by a differential splice which affects their C-terminal region generating 54 and 46kDa proteins respectively. The p54 SAP kinases, like the *erks*, are most abundantly expressed in brain, where all three p54 gene products are detected.

The p54 SAP kinases, like the Erks, are strikingly homologous to *FUS3* and

Table 2 Comparison between the catalytic domains of MAP kinase relatives in yeast and mammals. Figures are percentage identity

	p54α	p54β	p54γ	Slt-2	p44 *mapk*	Hog-1	Kss-1	Fus-3	Spk-1
p54α	—								
p54β	90	—							
p54γ	88	95	—						
Slt-2	45	44	44	—					
p44 *mapk*	44	43	43	49	—				
Hog-1	43	42	44	46	49	—			
Kss-1	42	41	42	52	52	50	—		
Fus-3	42	42	42	51	52	51	61	—	
Spk-1	43	42	43	51	56	52	57	58	—

KSS1 (42% identity). The p54 SAP kinases are also about 42% identical to both Erk-1 and -2 but only 29% identical to Erk-3. Based on sequence comparisons, it is likely that Erk-1 and -2 form one branch of the MAP kinase family, the p54 SAP kinases a second branch, and Erk-3 a third branch. As shown by the recent cloning of new MAP kinase isoforms in yeast, more members for each branch and perhaps yet more branches are likely to emerge (see Table 2). As discussed below, although there are common features in the substrate specificity and regulatory properties of the MAP kinases that probably reflect the structural homologies, the two branches of the MAP family characterized thus far have distinctive functional properties.

4.3 Substrate specificity of the MAP kinases

4.3.1 MAP kinases are proline-directed

The elucidation of the amino acid sequence surrounding the sites phosphorylated by MAP kinase on MBP: VIPRTPPP (144) and EGFR: VEPLTPSG (145–147), raised the possibility that the MAP kinases required a proline immediately carboxy-terminal to the site of phosphorylation, a property exhibited by the cyclin-dependent kinases (see Chapter 5), the subfamily next most closely related in catalytic domain sequence (40% identity between Erk-1 and -2 and cdc2, for example). The proline-directed substrate specificity of the MAP kinases was estab-lished using peptide substrates corresponding to the sequences surrounding Thr669 of the EGF receptor (ELVEPLTPSGEAPNQALLR) and the site on myelin basic protein (APRTPGGR) phosphorylated *in vitro* by the 42 kDa MAP kinase (113, 147, 148). Systematic variation in the sequences surrounding the single Thr phos-phorylation site in the EGFR peptide and the single Thr in the MBP peptide verified the essential role of the proline at +1 carboxy-terminal to the site of phosphorylation. Phosphorylation of the EGFR-derived peptides by the 42- and 44-kDa MAP kinases also diminished substantially if the proline at −2 in EGFR peptide sequence was altered or displaced. From these studies, a 'consensus' phosphorylation sequence was proposed for the Erk-1 and -2 subfamily; Pro–X–Ser/Thr–Pro, where X is neutral or basic. It is unlikely, however, that the proline at

−2 is a crucial requirement for substrate recognition inasmuch as the phosphorylation sites of several candidate physiological substrates of the Erks lack a P at −2. By example, Rsk-1 and Rsk-2 (54), major Erk-1 and Erk-2 substrates, contain no P–X–S/T–P sites in their primary structure; c-Jun residues Ser63 and 73 (149), major sites of MAP kinase phosphorylation *in vitro* and probably *in situ*, lack a proline at −2. Furthermore, several observations suggest that Erk-1 and Erk-2 specificity determinants of greater importance than the Pro at −2 of the substrate remain to be discovered. The K_m of EGFR and MBP synthetic peptides for Erks is ≥0.4 mM, despite the presence of the canonical P–X–S/T–P motif (146, 148); this low affinity indicates that crucial determinants are lacking in these peptides; such addition of determinants may be less related to primary sequence than to prior phosphorylations nearby, or to the ability of the region surrounding the phosphorylation site to adopt a specific secondary or tertiary structure. Interestingly, although c-Jun is an excellent substrate *in vitro* v-Jun is not phosphorylated by Erk-2; v-Jun encodes Ser63 and 73, but has undergone deletion of a segment ending several residues amino-terminal to these sites (the delta deletion) which likely acts as a distal recognition motif. The puzzling discrepancy in the ability of Erk and p54 SAP kinase to phosphorylate recombinant COS p70 S6 kinase versus the inability of these enzymes to phosphorylate rat liver p70 S6 kinase was discussed in Section 3.5.2.

The specificity of p54 SAP kinase, although not fully elucidated, is easily distinguished from that of the Erks by use of MBP and EGFR-derived synthetic peptides (113). Although equally dependent on a proline at +1, p54 phosphorylation of the EGFR T669 peptide is undiminished by omission of the proline at −2. In addition, p54 exhibits virtually no phosphorylation of the (relatively basic) MBP peptides, which are avidly phosphorylated by both the Erks and *Xenopus* MPF. The Erk-1 and -2 and p54 branches of the MAP kinase family display clearly distinct substrate specificities with protein substrates as well, as was first observed when p54 SAP kinase and p42 MAP kinase were compared for their ability to phosphorylate myelin basic protein; at equivalent MAP-2 phosphorylating activity, p42 MAP kinase phosphorylates MBP ~12 fold more rapidly than does p54 SAP kinase. A more physiologically important difference is in the ability to phosphorylate *Xenopus* S6K II (Rsk-2): p42 MAP kinase catalyses substantial phosphorylation and partial reactivation of the *Xenopus* S6K II, whereas p54, at an equal MAP-2 kinase activity, is entirely unable to phosphorylate *Xenopus* S6K II (91). Conversely, p54 is 5–10-fold more active as a c-Jun kinase (149) and ATF-2 kinase (150) than is p42 MAP kinase (Erk-2). The basis for these differences remains to be elucidated.

4.4 Physiological functions of MAP kinases

As for other protein kinases, the evidence pointing to the physiological roles of the MAP kinases are of several kinds: (i) similarities between the effects of mitogens on target protein function *in situ* and the effect of phosphorylation by MAP kinase *in vitro*; (ii) peptide mapping studies comparing the sites of mitogen-stimulated phosphorylation *in situ* to those phosphorylated by MAP kinases *in vitro*; and (iii) identification of MAP kinases by conventional purification as major mitogen-activated protein kinases active on mitogen-regulated polypeptide targets (e.g. Rsk, p70 S6 kinase or c-Jun). The strength of such data, as support for the participation of MAP kinase *in situ*, is cumulative rather than compelling, when compared to, e.g. data provided by the use of a highly selective kinase activator (e.g. cAMP) or inhibitor (e.g. PKI) or by specific gene knockout, none of which has been reported for the MAP kinases. Based on available evidence, important roles for MAP kinases have been proposed in several aspects of the mitogen-regulated cellular response, which can be arranged into three broad categories: (i) regulation of the activity of other Ser/Thr kinases; (ii) regulation of gene expression; and (iii) regulation of cytoskeletal function.

4.4.1 MAP kinases regulate other Ser/Thr protein kinases

The central roles of the Erk isoforms in the regulation of Rsk-1 and -2 is clearly established. The ability of both p54 SAP kinases and the Erks to phosphorylate p70 S6 kinase, and the likelihood that one or both of these MAP kinase subfamilies participates in the mitogen activation of p70 S6 kinase has been discussed above in Section 3.5.2. At present, two further Ser/Thr kinases appear to be likely substrates for MAP kinases *in situ*.

(i) c-Raf-1

This ubiquitous mitogen-activated Ser/Thr kinase protooncogene was introduced in Chapter 3. Mitogens may activate c-Raf-1 through both protein kinase C-dependent and -independent pathways; however the nature of the proximal biochemical steps that mediate c-Raf activation remain uncertain (151, 152). In response to PDGF (153, 154) or insulin (154–156), c-Raf kinase and autophosphorylating activities are increased *in situ* in parallel with an increase in ^{32}P incorporation into the c-Raf-1 polypeptide, exclusively onto Ser/Thr residues; concomitantly, there is a retardation of c-Raf-1 mobility on SDS-PAGE. Treatment of ^{32}P–c-Raf-1 isolated from insulin-stimulated cells with the (Ser/Thr) protein phosphatase-1 leads to a decrease in c-Raf-1 peptide kinase activity (155); thus c-Raf-1 Ser/Thr phosphorylation is required to maintain c-Raf-1 kinase activity. Recombinant c-Raf-1 whose ATP binding site has been mutagenically inactivated exhibits the same site-specific pattern of mitogen-stimulated phosphorylation *in situ* as seen with wild-type, active c-Raf-1, indicating that mitogen-stimulated c-Raf-1 Ser/Thr phosphorylation *in situ* is catalysed by mitogen-activated protein kinases other than c-Raf-1 itself (154).

c-Raf-1 is a 74 kDa polypeptide that contains an amino-terminal regulatory domain and a carboxy-terminal kinase catalytic domain (151, 152). Deletion of progressively longer segments in from the c-Raf amino-terminus yields an abrupt increase in raf transforming potency (and presumably, kinase activity) as truncation proceeds past residue 273 to encompass a non-catalytic, but highly conserved, Ser/Thr-rich domain located between residues 273 and 305 (157); this corresponds closely to the site of truncation that occurs spontaneously in the oncogene v-*raf* (158). Such a truncation also removes all sites for mitogen-stimulated c-Raf-1

Ser/Thr phosphorylation (159). Many of the Ser/Thr residues in this segment are immediately followed by a proline. Taken together, these data are compatible with the hypothesis that mitogen-activated, proline-directed protein (Ser/Thr) kinases may phosphorylate c-Raf-1 within the (Ser/Thr)-rich segment of its amino-terminal inhibitory domain, and thereby contribute to the mitogen activation of c-Raf-1 kinase. In fact, Erk-1 and -2 MAP kinases, but not p54 SAP kinase, can phosphorylate native and recombinant c-Raf-1, whereas v-Raf-1 is not a substrate for the Erks (160–163). Lee *et al.* (161, 162) observed that extracts from insulin-treated cells exhibit several peaks of c-Raf-1 kinase kinase activity upon MonoQ chromatography; three of the major peaks correspond to Erk-1, Erk-2, and probably a third Erk-like enzyme. Comparison of tryptic peptide maps of ^{32}P–c-Raf-1, phosphorylated *in situ* in PDGF, insulin or EGF-stimulated ^{32}P-labelled cells, or phosphorylated *in vitro* by p42 MAP kinase, indicates that MAP kinase phosphorylates *in vitro* sites on c-Raf-1 that correspond to many, but not all of the sites phosphorylated *in situ* during insulin/growth factor stimulation (160, 163).

Despite considerable evidence supporting a role for *erk* enzymes in the phosphorylation of c-Raf-1 *in situ*, phosphorylation *in vitro* of baculoviral, recombinant c-Raf-1 or c-Raf-1 immunoprecipitated from 3T3 cells with MAP kinase does not alter c-Raf-1 kinase activity toward several substrates (160, 163); c-Raf-1 autophosphorylating activity and mobility on SDS-PAGE are also unaffected by MAP kinase-catalysed c-Raf-1 phosphorylation *in vitro*. Thus, no evidence directly supports a role for Erks in c-Raf-1 activation. In many respects, the findings in regard to Erk and c-Raf-1 are similar to those seen with Erk and p70 S6 kinase (113). An important difference, however, is the recent discovery that c-Raf-1 is an activator of the MAP kinase-kinase (163), i.e. c-Raf-1 is actually situated immediately upstream of the Erk activator in the mitogen-activated signal transduction pathway (see Section 4.5.3). This finding greatly diminishes the likelihood that MAP kinase plays a primary role in the activation of c-Raf-1. Several formulations concerning the functional significance of Erk-catalysed c-Raf-1 phosphorylation in the mitogen regulation of c-Raf-1 activity remain consistent with available data: (i) Factors other than Erk, serve as the primary activators of c-Raf-1; such 'primary' activators might include another protein kinase, or ligand/polypeptide Raf-1 activators that operate through non-covalent mechanisms. Nevertheless, Erks may still participate in a secondary phase of c-Raf-1 regulation, perhaps serving to 'lock in' an activated state by phosphorylating the c-Raf Ser/Thr-rich domain. (ii) Erk-catalysed Raf-1 phosphorylation could reflect a down-regulation of c-Raf-1. (iii) Finally, Erk-catalysed c-Raf phosphorylation may be functionally irrelevant to c-Raf-1 activation, and only reflect increased exposure of the conserved Ser/Thr segment secondary to activation-induced changes in c-Raf-1 conformation. These alternatives will be distinguished by future experiments.

(ii) (Mitogen-activated protein kinase)-activated protein kinase-2

In the course of purifying protein kinases from skeletal muscle of insulin-treated rabbits, Cohen and colleagues isolated a novel protein kinase that phosphorylates the N-terminal CNBr fragment of glycogen synthase (46). This kinase, initially called ISPK-2, was purified as a mixture of 60- and 53-kDa polypeptides, that yielded tryptic peptides of identical sequence. The kinase activity was completely inactivated by phosphatase-2A *in vitro*, and restored by phosphorylation *in vitro* with the p42 MAP kinase; the 60 and 53 kDa polypeptides each incorporated ~1 mole P/subunit. Although presumed to be activated *in vivo* in response to insulin, direct demonstration of insulin/growth factor regulation of the 60/53 kDa kinase is not yet available, and the enzyme was therefore renamed as **MAP kinase-activated protein kinase-2** (MAPKAP kinase-2) on the basis of its regulation by MAP kinase.

The substrate specificity and physiological targets of MAPKAP kinase-2 are not known; the enzyme phosphorylates glycogen synthase rapidly, predominantly at serine 7 near the N-terminus, and a synthetic peptide corresponding to this sequence (KKPLNRTLSVASLPGLamide) is phosphorylated with favourable kinetics. Recently, Stokoe *et al.* (165) have observed that the small mammalian heat-shock proteins, murine hsp25 and human hsp27, previously shown to be phosphorylated *in situ* in response to growth factors and phorbol esters at two serine residues that reside in a sequence context similar to glycogen synthase serine 7 (hsp25 LLRQLSSG) (166); hsp27 LSRQLSSG (167), are rapidly phosphorylated *in vitro* by MAPKAP kinase-2 at these sites. Although the RXXS/T motif exhibited by these MAPKAP kinase-2 sites is similar to sites phosphorylated by ISPK-1/Rsk-2 (=MAP-KAP kinase 1), MAPKAP kinase-2 does not phosphorylate 40S S6, S6 peptides or the G subunit of protein phosphatase-1. The identification of the hsp proteins as avid substrates *in vitro* for MAPKAP KINASE-2 should enable a detailed examination of the specificity determinants of MAPKAP kinase-2 as compared to Rsk-2.

(iii) The EGF receptor

The EGF receptor is phosphorylated at Thr 669 (within context PLTP) by an EGF-stimulated Ser/Thr kinase. Davis and colleagues have purified this enzyme extensively and identified it as an Erk-2 homologue (168); similar evidence has been observed by Takishima *et al.* (147). The functional effects of Erk phosphorylation of the EGFR at T669 are disputed, but mutagenesis of Thr 669 to Ala has little effect on EGF-stimulated receptor tyrosine kinase activity.

4.4.2 MAP kinases phosphorylate numerous transcription factors and may mediate agonist regulation of gene expression

Alterations in gene expression at the transcriptional level are a central component of the cellular response to mitogenic stimuli. A growing number of mitogen-regulated transcriptional regulatory proteins are strong candidates to be among the physiological substrates of the Erk and p54 SAP kinases.

(i) c-Jun

An important class of mitogen-regulated genes are those whose regulatory DNA sequences contains the heptameric motif TGA C/G TCA, defined by work from several groups as the phorbol ester (TPA) response element, or TRE. Mitogens and active phorbol ester action at this *cis*-acting element is mediated by a transcription factor complex known as AP-1, which typically consists of a heterodimer of the c-*jun* and c-*fos* protooncogene products (169). The c-Jun polypeptide contains several functionally independent domains: the carboxy-terminal half of the protein encodes a basically charged DNA binding domain followed by a dimerization domain of the leucine zipper-type near the protein carboxy-terminus (170, 171). The amino-terminal half of the protein contains a domain crucial for the transactivation (172, 173). The DNA binding and transactivating function are each regulated by phosphorylation in a coordinate but independent manner. Phosphorylation of c-Jun at four sites (Thr231, Thr239, Ser243, Ser249) in a segment immediately amino-terminal to the DNA binding domain inhibits DNA binding (174). Phorbol ester or mitogen stimulation of cells produces an increase in c-Jun binding to DNA, coincident with the dephosphorylation of these Ser/Thr residues. Phosphorylation of these sites *in vitro* can be catalysed efficiently by glycogen synthase kinase-3 (GSK-3) (174) or casein kinase-2 (175). The p42 MAP kinase can also phosphorylate Ser243 *in vitro* (146), but is unlikely to participate *in situ*, inasmuch as treatments that activate MAP kinases (e.g. phorbol ester, growth factors) result in dephosphorylation of Ser243 (174). The mechanism underlying the dephosphorylation of the c-Jun C-terminal residues after mitogens treatment is unclear. Protein kinase C has been shown to phosphorylate and inhibit GSK-3β *in vitro* (176), and provides one plausible mechanism; activation of a phosphatase acting on these carboxy-terminal c-Jun sites has been proposed, but as yet no direct evidence supports this idea.

In several cell types, phorbol esters, activated Ras and polypeptide growth factors stimulate phosphorylation of the c-Jun amino-terminal transactivation domain concomitant with increased transcription (177). Two amino-terminal mitogen-stimulated phosphorylation sites, designated 'X' and 'Y,' have been detected on tryptic phosphopeptide maps and correspond to Ser73 and 63, respectively (149). The importance of these residues to c-Jun transactivation has been established by the demonstration that mutagenesis of one or both serines inhibits or abolishes TPA-stimulated expression from cotransfected TRE-CAT reporters. Recombinant c-Jun is phosphorylated *in vitro* by the MAP kinases, with p54 SAP kinase exhibiting 5–10-fold greater potency than p42 (Erk-2); tryptic phosphopeptide maps show that ^{32}P incorporation is predominantly (Erk-2) or exclusively (p54) into 'X' and 'Y'. c-Jun phosphorylation by p54 SAP kinase *in vitro* is associated with a stoichiometric shift of the c-Jun polypeptide to a slower mobility on SDS-PAGE, as occurs after TPA treatment *in situ*; at equal MAP-2 phosphorylating activity, Erk-2 induces a lesser shift. Interestingly, GSK-3 phosphorylation of the carboxy-terminal sites to a similar stoichiometry does not induce such a shift, indicating that altered c-Jun conformation rather than charge is the basis for slowed electrophoretic mobility. c-Jun molecules mutated at Ser63 and 73 show reduced or absent phosphorylation by MAP kinases *in vitro*.

The major c-Jun kinases detected in extracts of TPA-stimulated U937 cells co-purify with Erk-2 and Erk-1 (178). Alder and colleagues (179) have also recovered a

45 kDa c-Jun kinase active towards sites X and Y; the relation of this enzyme to the MAP kinases is not yet known. v-Jun, which is rendered constitutively active in transcriptional regulation, in part by a deletion of the segment corresponding to c-Jun residues 31 to 57, (the delta deletion) (180) does not undergo mitogen-stimulated amino-terminal phosphorylation *in situ*, nor MAP kinase-mediated phosphorylation *in vitro*, despite the presence of the serine analogous to Ser63 and 73 of c-Jun (see Section 4.3.1). Thus, mitogen regulation of c-Jun function is mediated through coordinate changes in the phosphorylation of functionally distinct domains: dephosphorylation of multiple sites near the DNA binding domain enhances DNA binding activity, whereas phosphorylation at Ser63 and 73 in the amino-terminal region, catalysed by one or more MAP kinases, is necessary for transcriptional activation.

(ii) c-Myc

c-Myc is a short-lived polypeptide protooncogene whose synthesis is increased within minutes of mitogen stimulation, i.e. an 'early growth response' gene product (181). c-Myc is a helix-loop-helix polypeptide, that forms a heterodimer with Max, another helix-loop-containing polypeptide, through their respective leucine zipper domains (182); the complex binds DNA in a sequence-specific fashion and apparently functions as transcriptional regulatory factor (183, 184). Myc (but not Max (185)) contains an amino-terminal transactivating domain, distinct from the DNA binding, helix-loop-helix and leucine zipper domains (186). Phosphorylation of c-Myc occurs within this transactivating domain, at Ser62, consequent to growth factor treatment of cells (187–189), and a 41 kDa MAP kinase (homologous to Erk-2) can specifically phosphorylate Myc Ser62 *in vitro* (146). Fusion of c-Myc residues 1–103 onto the Gal 4 DNA binding domain confers serum stimulated chloramphenicol acetyl transferase (CAT) activity from a cotransfected Gal 4 CAT reporter plasmid. Mutation of Myc Ser62 to Ala in the Gal4–Myc fusion protein virtually abolishes the mitogen-stimulated CAT expression. Cotransfection with p41 MAP kinase cDNA with the Gal4–Myc plasmid further stimulates CAT expression from Gal CAT, both in the absence and presence of serum, but is without effect when introduced with Gal4/Myc Ser62 to Ala (188). Thus, as with c-Jun, considerable evidence indicates that phosphorylation within the c-Myc transactivation domain is crucial to mitogen regulation of c-Myc function, and is catalysed *in situ* by MAP kinase(s).

(iii) ATF-2

ATF-2/(CRE-BP1) (190) is a member of the large polypeptide family (191, 192) that binds to the cAMP response element (CRE) octameric motif, TGACGTCA. Unlike the cAMP-responsive transactivating factor CREB, which is a high affinity substrate *in vitro* for cAMP-dependent protein kinase, ATF-2 is a very poor substrate for this kinase. Like c-Jun, ATF-2 contains a carboxy-terminal DNA binding domain followed by a leucine zipper, and an amino-terminal regulatory domain. N-terminally-truncated ATF-2 expressed in bacteria can bind to a consensus cAMP-response element (CRE)-containing oligonucleotide (193); however, the full-length prokaryotic recombinant ATF-2 polypeptide is unable to bind DNA (150). In contrast, when expressed in Sf9 cells using a baculoviral vector, the full-length ATF-2 binds DNA avidly. Treatment of the baculoviral recombinant ATF-2 with PP-2A abolishes its ability to bind DNA, whereas phosphorylation of the inactive ATF-2 with MAP kinase restores DNA binding. p54 SAP kinase is 5–10-fold more active in phosphorylating ATF-2 than p42/44 MAP kinases; cAMP-dependent protein kinase is less than 5% as active in ATF-2 phosphorylation as p42 MAP kinase. Moreover, in contrast to the ability of MAP kinase to restore ATF-2 DNA binding, a similar extent of ATF-2 phosphorylation catalysed by cAMP-dependent protein kinase has no effect. MAP kinase phosphorylates ATF-2 in an amino-terminal domain, distal to the DNA binding and leucine zipper domains that are situated in the carboxy-terminal region of the ATF-2 polypeptide. Abdel-Hafiz *et al.* (150) have proposed that in the unphosphorylated ATF-2 polypeptide, as produced in *E. coli*, the amino-terminal region is folded over and bound to the carboxy-terminal domain so as to inhibit the DNA binding function. MAP kinase-catalysed phosphorylation of the ATF-2 amino-terminal segment is proposed to disrupt this intramolecular interaction and free the ATF-2 DNA binding domain.

(iv) p62^{TCF}

Transcription of the protooncogene c-*fos*, another 'early growth response' gene is stimulated by many serum growth factors (194). Essential for this process is a *cis* DNA sequence, the serum response element (SRE). *In vivo*, transcriptional activation through the SRE is mediated by a protein complex containing the serum response factor p67^{SRF} and an accessory protein, the ternary complex factor p62^{TCF} (elk-1), which together bind to the SRE DNA; disruption of this complex impairs c-*fos* induction by extracellular stimuli (195). p62^{TCF} is phosphorylated by the p42/44 MAP kinases *in vitro* and MAP kinase phosphorylation of p62^{TCF} enhances ternary complex formation (196). This phenomenon appears to be indispensable to the mitogen stimulation of c-*fos* transcription. p62^{TCF}/elk-1 is a member of the Ets gene family (197); other members, e.g. Ets-1 and Ets-2 (which cooperate with the AP-1 transcription factor) are also phosphorylated *in situ*, and are candidate MAP kinase substrates.

4.4.3 Regulation of cytoskeletal dynamics by MAP kinases

A role for MAP kinases in the interphase–metaphase rearrangement of microtubules was demonstrated by Gotoh *et al.* (198) who were able to induce this characteristic cytoskeletal rearrangement in cell-free extracts of interphase *Xenopus* oocytes, by adding purified, active p42 MAP kinase from either *Xenopus* or mammalian sources. This interphase–metaphase transition in the microtubule arrays is characterized by a decrease in the growing speed of the microtubules as well as a shortening of the steady state length of the microtubules to one reminiscent of metaphase II extracts. The MAP kinase substrates that underlie this microtubule rearrangement are not known, although microtubule-associated proteins related to MAP-2 (which is expressed predominantly in brain) are attractive candidates (199).

A provocative observation was provided by Drewes *et al.* (200) who showed that MAP kinases can phosphorylate the tau protein at sites phosphorylated in the tau polypeptides isolated from the neurofibrillary tangles found in Alzheimer's disease. A characteristic of the unique structure of tau polypeptides isolated from the paired helical filaments of Alzheimer brains is their reactivity with a class of specific α tau antibodies. Recognition of tau by these antibodies is lost if the tau polypeptide is treated with phosphatase; reactivity can be restored upon phosphorylation of tau *in vitro* with the 42 kDa MAP kinase (201). These workers suggest that the neurofibrillary tangles emblematic of Alzheimer's disease may arise from an inappropriate hyperphosphorylation of tau, attributable to MAP kinases. This could reflect a constitutive activation of the MAP kinase by signals generated upstream, or a deficiency in the phosphatases that act on tau, on the MAP kinases, or on their upstream regulators. Similar findings were reported by others and other protein kinases have also been implicated (202–205).

4.4.4 Phosphorylation of phospholipase A$_2$

Several reports (206, 207) have shown that a 110 kDa form of phospholipase A$_2$ (PLA$_2$), a Ca-dependent enzyme known to be regulated by numerous extracellular agonists, can be phosphorylated by p42 MAP kinase *in vitro* in a site-specific fashion (but not by the p54 SAP kinase), and activated ~2–3-fold. Recombinant PLA$_2$ undergoes phosphorylation at Ser505 *in situ* in agonist-stimulated cells concomitant with a 5–10-fold increase in PLA$_2$ activity. If this reflects a mode of regulation operative on endogenous PLA$_2$, it places the p42 MAP kinase upstream of a wide array of intracellular effectors, including various isoforms of protein kinase C.

4.5 Regulation of MAP kinases

4.5.1 Requirement for tyrosine and threonine phosphorylation at two sites in subdomain VIII of the catalytic domain

The ability of CD45, a Tyr-specific phosphatase, and phosphatase-2A, a Ser/Thr-specific phosphatase, to each independently deactivate the p42 MAP kinase, established that both Ser/Thr and Tyr phosphorylation are required to maintain the active state of p42/p44 MAP kinase (118; see Fig. 2); a similar situation is observed with p54 SAP kinase (208). A requirement for both Ser/Thr and Tyr phosphorylation typifies the activated state of the MAP kinases, regardless of the nature of the initiating stimulus, e.g. whether via activation of receptor tyrosine kinases or heterotrimeric G proteins (209, 210) through Ca-dependent or Ca-independent pathways (211) or protein kinase C-dependent or protein kinase C-independent pathways (96). The only other kinases regulated by both Ser/Thr and Tyr phosphorylation are p34^{cdc2} and GSK-3/shaggy, as discussed in Chapters 5 and 8. A protein phosphatase encoded by the CL100 gene dephosphorylates both the phosphotyrosine and phosphothreonine on p42 MAP kinase and is able to suppress activation of MAP kinase by Ras in *Xenopus* extracts (212; see Section 4.5.4). The CL100 gene is transcriptionally induced in response to growth factors and stress, perhaps ensuring MAP kinase activation is transient.

Payne *et al.* (209) identified the sites of phosphorylation on murine p42 MAP kinase as Thr183 and Tyr185 (numbering of Erk-2; ref. 141) in subdomain VIII of the catalytic domain. The sequence surrounding these sites is highly conserved among the *erk* family, KSS1, and FUS3, and corresponds to the rough consensus GFX**TEY**VATR (X is a hydrophobic amino acid). The p54 SAP kinases differ slightly, containing the sequence -TPY- in the homologous region (143). Thr 183 is homologous to a Thr in the cAMP-dependent protein kinase catalytic subunit whose phosphorylation occurs cotranslationally and is crucial to proper folding of the domain into an active conformation (213); the Thr phosphorylation in cAMP-dependent protein kinase appears to be autocatalysed inasmuch as it is present when this kinase is expressed as a prokaryotic recombinant. Erk-1 or Erk-2, by contrast, are recovered in a very low activity state when expressed as prokaryotic recombinant polypeptides, but incubation *in vitro* with [γ-^{32}P]ATP leads to the slow autophosphorylation of the Thr and Tyr residues corresponding to 183/185 with a parallel slow gain in protein kinase activity toward exogenous substrates, such as MBP (214–216). MAP kinase purified in a mitogen-activated state, and then de-activated with tyrosine phosphatase or phosphatase-2A *in vitro* also shows a slow Tyr/Thr autophosphorylation on incubation with MgATP with some regain in kinase activity (217).

4.5.2 MAP kinase-kinase: a novel dual specificity kinase phosphorylates and activates MAP kinases

Ahn *et al.* (218) were first to identify a MAP kinase 'activator' in extracts of EGF-treated Swiss 3T3 cells. This activator of Erk-1 and -2 could reactivate the phosphatase-inactivated 42- and 44-kDa MAP kinases, back to the level observed prior to phosphatase treatment. The MAP kinase activator was a cytosolic activity, readily resolved from the endogenous MAP kinase on MonoQ anion-exchange columns. The dual specificity reactivation function coeluted from (mostly flowed through) MonoQ anion-exchange chromatography appearing as two peaks, and the dual specificity of reactivation exhibited by each peak continued to copurify on Mono-S and gel filtration chromatography, suggesting that a single 'activator' protein(s) could evoke both Tyr and Thr phosphorylation of MAP kinases. Although also active on recombinant Erks, partially purified preparations of the MAP kinase activator showed little or no ability to catalyse autophosphorylation or phosphorylation of common polypeptide substrates. Moreover, although most protein kinases will continue to phosphorylate substrates that have been de-natured, mild heat inactivation of MAP kinase abolished completely its ability to undergo phosphorylation in the presence of the MAP kinase activator. Consequently, it was not immediately clear whether the MAP kinase activator was, in fact, a protein kinase, or a polypeptide activator that stimulated Erk autophosphorylation via a non-covalent interaction. Similar conclusions were reached by

Gomez and Cohen (219) examining MAP kinase activation in NGF-treated PC12 cells. They observed in addition that the MAP kinase activating element itself was regulated by Ser/Thr phosphorylation, as its activator function could be abolished by Ser/Thr phosphatases but not Tyr phosphatases.

Several potential mechanisms can be envisioned for MAP kinase activation through dual phosphorylation of the MAP kinase polypeptide. Phosphorylation of Erk Tyr_{185} could be catalysed by a (receptor or cellular) Tyr kinase, concomitantly with phosphorylation of the Thr_{183} by a distinct Ser/Thr kinase. Alternatively the action of only one of these kinases (Tyr or Ser/Thr) may be sufficient to initiate activation followed by Erk autophosphorylation at the other residue. A single protein kinase that phosphorylates both the Tyr and Thr residues on Erk could be operative. Finally, a ligand or a polypeptide lacking any kinase activity might be made available by mitogen action, and serve to activate Erk by accelerating greatly the rate of Erk Tyr/Thr autophosphorylation. Although the major 'activator' of Erk-1 and Erk-2 seen after mitogen stimulation has now been identified definitively as a protein kinase of dual specificity, i.e. one that phosphorylates both the Erk Thr and Tyr residues, the existence of the alternative pathways for activation of other MAP kinases (i.e. Erk-3 or p54), and even under some circumstances, Erk-1 and -2 activation, remains plausible. Thus, a 44 kDa MBP kinase isolated from sea star inactivated *in vitro* with the CD45 tyrosine phosphatase can be phosphorylated by the non-receptor tyrosine kinase $p56^{lck}$ with a partial regain in MAP kinase activity (220). Nevertheless, the likelihood that this phosphorylation represents a major mode of MAP kinase activation is quite low, inasmuch as the extent of MAP kinase activation by $p56^{lck}$ *in vitro* was far less than that achieved by extracellular agonists, including those known to activate $p56^{lck}$ *in situ*.

Convincing evidence that the MAP kinase activator is a protein kinase came from studies examining the phosphorylation of kinase-inactive but non-denatured MAP kinase polypeptides. Adams and Parker (221) used FSBA-inactivated p42 MAP kinase to demonstrate that partially purified MAP kinase activator could phosphorylate *in vitro* a phosphatase-treated MAP kinase, at both Tyr and Thr. Several laboratories (222–225) using a recombinant Erk-1 or Erk-2 substrate whose ATP binding site had been mutagenically inactivated, observed that partially purified preparations of MAP kinase activator catalysed the phosphorylation *in vitro* of both Tyr and Thr on the MAP kinase polypeptide. Posada and Cooper (226) mutated Thr188 or Tyr190 of the recombinant *Xenopus* p42 MAP kinase to Ala and Phe, respectively; although either mutant was inactive when expressed in *Xenopus* oocytes, the recombinant protein underwent stimulated phosphorylation at the remaining native residue, indicating strongly that phosphorylation of MAP kinase by other protein kinases was the major route of MAP kinase phosphorylation *in situ*. Extracts from mitogen-stimulated mammalian cells catalyse a similar dual phosphorylation of the inactive MAP kinase polypeptide, although phosphorylation of the tyrosine proceeds at a considerably faster rate than threonine (227). Several groups achieved partial (228) or essentially complete (229–231) purification of activator, coincident with the isolation of 45- and/or 46-kDa polypeptides (or both;

229, 231). Similar patterns of selective dual phosphorylation are seen with highly purified isolates of MAP kinase activator, i.e. phosphorylation of the Tyr in the Thr→Ala MAP kinase mutant and the Thr in the Tyr→Phe mutants, with Tyr phosphorylation occurring much more rapidly; prior Tyr phosphorylation of the MAP kinase appears to decrease the K_m for MAP kinase substrate (232). A faster rate of MAP kinase tyrosine phosphorylation after agonist stimulation also occurs *in situ* (233) and mono (Tyr) phosphorylated, but inactive MAP kinase polypeptides can be detected in unstimulated PC12 cells (139). These results provide strong evidence that MAP kinase activator is indeed a protein kinase with dual (Thr/Tyr) specificity toward exogenous substrates.

It is important to note that this Erk activator, although capable of activating both the p42 and p44 MAP kinases, will not reactivate p54 SAP kinase whether the substrate used is phosphatase-inactivated purified rat liver enzyme, or an inactive prokaryotic recombinant p54 enzyme (143). Thus, the p54 SAP kinase activator is an entirely distinct element yet to be identified.

Several groups isolated cDNAs that encode tryptic peptide sequences derived from the Erk-specific MAP kinase kinase (234–237). Crews *et al.* (234) named this gene *mek* (**M**AP/**E**rk **k**inase): when expressed in bacteria, MEK is capable of phosphorylating Erk-1 *in vitro* and activating its MBP kinase activity, as expected for the Erk activator. Interestingly *mek* is homologous to two yeast genes: *STE7* from *S. cerevisiae* and *byr1* from *S. pombe*. Genetic studies place *STE7* upstream of the MAP kinase homologues *FUS3/KSS1*, and *byr1* upstream of *spk1* (see 238 and Chapter 8). Inspection of MEK amino acid sequence places it squarely among the Ser/Thr protein kinases, but does not point to the basis of its novel dual specificity.

4.5.3   Regulation of MEK

MEK purified from mitogen-treated cells is phosphorylated on Thr (230, 239) or Ser and Thr residues (240) and can be inactivated completely with Ser/Thr phosphatases (Fig. 2; 164, 219, 227, 229, 230). In attempting to identify the Ser/Thr protein kinase upstream of MEK, we observed that cells transformed with the v-*raf* oncogene display constitutively active Erk-1, Erk-2 and MEK. This observation led us to investigate whether phosphatase-inactivated MEK could be directly reactivated *in vitro* by active Raf-1. An amino-terminally truncated, constitutively active form of c-Raf-1 was shown to catalyse the Ser/Thr phosphorylation of a PP-2A inactivated MEK, concomitant with the complete reactivation of MEK activity, in an ATP-dependent fashion (164). Similar observations have been subsequently reported from other laboratories (241, 242). More recently, we observed that mitogen stimulation of NIH 3T3 cells or H4 hepatoma cells increases the kinase activity of endogenous c-Raf-1 as assayed by phosphorylation of MEK *in vitro*. The endogenous c-Raf-1 kinase is activated more rapidly than the endogenous MEK activity consistent with the sequential activation of these kinases *in situ*, as part of the usual programme of mitogen action (163).

The association of Raf and MEK does not appear as transient as usually observed for kinase/substrate interactions. When co-expressed in baculovirus infected insect cells, immunoprecipitates of MEK contain Raf (243). Likewise, purified MEK tightly binds immunoprecipitated Raf in an interaction resistant to high salt or non-ionic detergent. These findings not only identify c-Raf-1 as a likely immediate upstream activator of the MEK (i.e. MEKK), but in addition establishes MEK as the first known physiological substrate for the c-Raf-1 kinase protooncogenes.

c-Raf-1 is not the only protein kinase capable of activating MEK. The existence of MEK activators other than c-Raf-1 was suggested by the observation that expression of v-Raf in PC-12 cells fails to activate MAP kinase, whereas Ras expression does so (244). Similarly, whereas v-*raf* expression in NIH 3T3 cells leads to constitutive activation of the Erk pathway, v-*raf* expression in Rat-1a cells does not lead to constitutive MAP kinase activation. Conversely, chronic Erk activation in Rat-1a cells is produced by expression of the *gip2* oncogene, a constitutively activated heterotrimeric GTP-binding protein (G-protein) α subunit of the $G_{\alpha i2}$ type; this suggests the presence of a novel MEK or MEK activator (other than c-Raf-1) in Rat-1a cells (245, 246). Genetic studies in *S. cerevisiae* place *STE-11* upstream of the *mek* homologue *STE-7* (247, 248) and in *S. pombe* place *byr2* upstream of *byr1* (238). Although their overall domain topology is similar, the Ste-11 and Byr2 protein kinases bear little resemblance to c-Raf-1 in amino acid sequence.

Using oligonucleotides to the conserved sequences between Ste-11 and Byr2, a mouse protein kinase gene was identified exhibiting 35% identity (75% similarity) with the catalytic domains of the two yeast kinases (249). This protein, termed MEK kinase (MEKK), is widely expressed and phosphorylates and activates MEK *in vitro*. Why might two distinct protein kinases activate the MAP kinase pathway? The response to yeast mating factor is regulated through heterotrimeric G proteins. It is possible that MEKK specifically conveys mitogenic signals originating from this signal generator[2]. Since receptor protein tyrosine kinases do not exist in yeast, the Raf arm of the input may have evolved in multicellular organisms to wire in these signals. Thus, distinct pathways may partially converge at MEK.

4.5.4 Upstream of c-Raf-1 and MEKK

There is considerable evidence that the product of the protooncogene c-*ras* lies upstream of c-Raf-1 in the MAP kinase pathway. Ras is a small (21 kDa) G-protein that is mutated to oncogenically activated forms in ~30% of all human cancers (250). The *ras* proteins are thought to be active as signal transducers in the GTP-bound state, and inactive in the GDP-bound state. Ras proteins have a weak auto-inactivating GTPase activity which is greatly enhanced by GTPase-activating proteins (GAPs). Accordingly, GAPs, including $p120^{GAP}$ and neurofibromin, the product of the von Recklinghausen neurofibromatosis gene, are considered tumour suppressors (251). Ras proteins are also capable of a slow exchange of bound nucleotide and free nucleotide that can be greatly accelerated by guanine nucleotide exchange or release factors, proteins that promote the dissociation of bound

---

[2] Nevertheless thrombin, endothelin and other ligands that act through heterotrimeric G protein-linked receptors also activate c-Raf-1 kinase activity (164).

guanine nucleotides and exchange with free guanine nucleotides. In the cell, GTP is present in large excess over GDP, and therefore the availability of active nucleotide release factors will favour the formation of the GTP bound state of Ras, other factors remaining constant. Mutant Ras proteins with impaired basal and GAP-activatable GTPase activity are powerfully transforming when expressed in cultured cells, as are Ras mutants that have an intrinsically high nucleotide exchange. Numerous studies have shown that activation of tyrosine kinase receptors, or expression of oncogenic non-receptor tyrosine kinases results in the loading of Ras with GTP, and the consequent activation of Ras-dependent pathways (252, 253).

Implication of Ras in the activation of mitogen-stimulated Ser/Thr kinases came initially from the demonstration that Erk activation is seen within 5 min of scrape-loading activated Ras polypeptides into cultured cells (254) and 90 min after microinjection into *Xenopus* oocytes (255). Conversely, overexpression of Ras-GAP-blocked phorbol ester stimulation of Erk-2, but did not interfere with the Erk activation induced by AlF_4, a selective stimulator of heterotrimeric G proteins, which perhaps reflects the independent action of MEKK (256). Additional compelling evidence is provided by the effects of dominant inhibitory mutants of *ras*, such as N17-*ras*. These functionally impaired *ras* mutants when overexpressed are believed to act as competitive antagonists of endogenous Ras through the binding of nucleotide release factors, and can inhibit almost completely NGF, PDGF and insulin-induced activation of MAP kinases, Rsk and c-Raf-1 in PC12 cells and 3T3 cells (257–260). Two groups have shown that addition of Ras polypeptide to a cell-free extract prepared from *Xenopus* oocytes can yield activation of MAP kinase. Carboxy-terminally processed Ras polypeptide (i.e. farnesylated and cleaved, carboxymethylated and palmitoylated) appears to be necessary to elicit this activation (261, 262). The availability of a cell-free system wherein Ras can activate MAP kinase suggests that the polypeptide elements which couple Ras to Erk, in addition to MEK and c-Raf-1 (or another MEK activator), are entirely cytosolic.

An overwhelming weight of evidence places Ras upstream of Raf on the same pathway. Thus, like v-src and activated receptor-tyrosine kinases, v-Ras induces hyperphosphorylation (presumably reflecting activation) of c-Raf-1 (263). Anti-sense Raf-1 RNA or kinase inactive Raf-1 polypeptides introduced into 3T3 cells inhibit DNA synthesis (264) and the activation of gene expression induced by serum and PMA, as well as transformation caused by v-Raf, and by the Ki-Ras and Ha-Ras oncogenes (265). In addition, dominant-negative mutants of Ras block the activation of c-Raf-1 (257) caused by receptor tyrosine kinases. Genetic evidence from the *Drosophila seitenless* tyrosine kinase pathway also places *Drosophila* Ras upstream of the *Drosophila* D-Raf gene product (*polehole*) (266). Another gene, *son of seitenless (SOS)*, encodes a guanine nucleotide exchange release factor and is situated downstream of the receptor tyrosine kinase and upstream of *Drosophila Ras* (267), whereas a RasGAP acts as a negative regulator (268). A similar order is derived from analysis of the *Drosophila Torso* genes, which encode a receptor tyrosine kinase, and lie upstream of D-raf (269). Genetic studies in *C. elegans*, reviewed elsewhere (270; see Chapter 8), strongly implicate Ras as an obligatory intermediate between the receptor tyrosine kinase and c-Raf-1. Such studies, however, cannot identify the specific biochemical steps, nor the number of elements/reactions necessary for Ras activation of Raf-1.

How does Ras activate the MAP kinase cascade? Immobilized Ras has been used to fish for interacting proteins (271). Using Ras locked into different states of activation, it was shown that Raf and MEK and MAP kinase bind to columns containing activated but not inactive Ras. Furthermore, the yeast two-hybrid system which assays for interacting proteins *in vivo* has revealed that the N-terminal domain of Raf binds to Ras. Using the two hybrid (272–274) approach in which only fragments of cDNAs are expressed, a Ras bait commonly selected portions of the Raf N-terminal domain from a library (~50% of clones) (274) whereas a Raf N-terminal bait selected the small GTPase Rap-1, a Ras antagonist with an 'effector' domain identical to Ras (272). The domain of Raf required for interaction with Ras is contained within amino acids 51–131, in so far as this minimal domain is sufficient for binding to Ras *in vitro*. In budding yeast, the two-hybrid assay has also demonstrated interaction of Ras with the N-terminal domain of the protein kinase, Byr2 (273).

Raf binding to Ras does not activate Raf kinase but probably positions Raf at the membrane so as to enable its activation by another, as yet unidentified, element. Nevertheless, although questions remain, the recent advances in the MAP kinase signalling pathway document most, if not all, of the components of a transduction system from Ras to the nucleus (Fig. 4).

While little is currently known biochemically about upstream regulation of MEKK, by analogy with the yeast pathways, it is likely that heterotrimeric G proteins play a role. Like Raf, MEKK, Byr2 and Ste-11 have extended amino-terminal domains of ~400 residues, although these exhibit few similarities at the amino acid level, even within the domains within Raf and Byr2 required for binding Ras (271, 273).

Although it is possible that stimulation of both Raf and MEKK is required for full activation of the MAP kinase pathway it is more likely that the two pathways are each capable of full MAP kinase activation and operate *in situ* in a stimulus-specific or temporally distinct fashion. MEKK agonists may include heterotrimeric G-proteins (258), protein kinase C agonists (259), and possibly cyclin-dependent kinases. Thus in cell-free extracts from *Xenopus* oocyte. MAP kinase can be activated not only by oncogenic Ras but also by addition of MPF (cdc2/cyclin B) (262). These results suggest that a cdc2-mediated pathway of MAP kinase activation operates in parallel to the Ras pathway, perhaps via MEKK. This is not yet established; although Ras-induced activation of MAP kinase proceeds without intervening cdc2 activation, the possibility that cdc2 activation of MAP kinase proceeds by way of Ras activation is not eliminated by present information.

What lies upstream of the G-proteins? In the case of Ras, recent reports show that activation and autophosphorylation of the EGF receptor lead to association with SHC (275), a protein which consists solely of an SH2 domain and a glycine/proline-rich region (276). This association, presumably through the SHC SH2

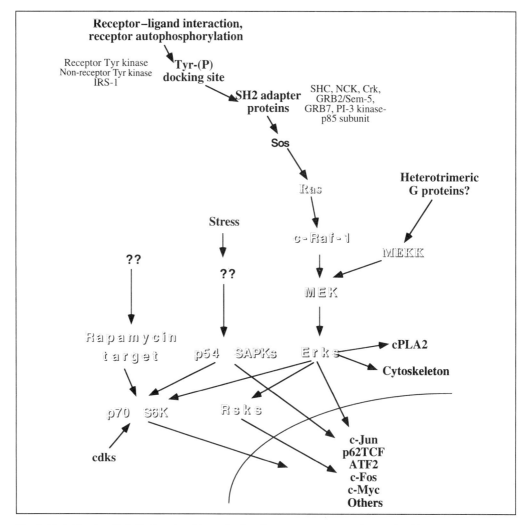

Fig. 4 Model for insulin/mitogen signal transduction from a receptor protein-tyrosine kinase to the nucleus. Elements discussed in this chapter are highlighted.

domain, results in the Tyr phosphorylation of the SHC polypeptide. As a consequence of SHC Tyr phosphorylation, a complex is formed between SHC and GRB2 (277), a protein with the domain structure SH3–SH2–SH3. GRB2 is homologous to the product of the *C. elegans* gene, *sem-5*. Genetic studies implicate *sem-5* as an upstream activator of *let-60* gene product (i.e. Ras) and a downstream target of *let-23* (i.e. receptor tyrosine kinase) (278; see Chapter 8). There is evidence that SH3 domains such as those found in GRB2/sem-5 can regulate the members of the *ras* family (279).

The elements linking the heterotrimeric G proteins to the MAP kinases have yet to be identified.

4.5.5 Regulation of the p54 SAP kinases

Although the p54 SAP kinase family is regulated by tyrosine and threonine phosphorylation, similar to the p42/44 MAP kinases, Erk-1 and -2, their physiological mechanisms of activation appear to be quite distinct. As was stated in Section 4.5.2, the p54 enzymes are neither activated nor phosphorylated by the MEK protein kinases. Antibodies raised to bacterially expressed p54β specifically immunoprecipitate a c-Jun kinase activity that phosphorylates serines 63 and 73 of that protein (143). Extracts from cells treated with cycloheximide contain 10-fold more precipitable kinase activity which must result from activation of the enzyme (since there is no ongoing protein synthesis). Using this as an assay for potential p54 SAP kinase agonists, a number of agents were found to stimulate the enzyme (Fig. 5). Many of these 'agonists', such as heat-shock, TNF-α and tunicamycin, poorly activated the Erk enzymes (143). In contrast, many of the well characterized Erk agonists, particularly growth factors and phorbol esters, poorly stimulated p54 activity (Fig. 5) (some cell specificity was observed, for example, PMA activates p54 in HepG2 cells). All of the p54 agonists induced c-Jun phosphorylation at serines 63 and 73, strongly implicating this kinase as a physiological c-Jun kinase (143).

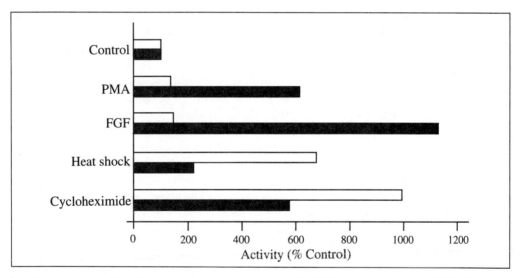

Fig. 5. Comparison of the effect of various treatments on p54 SAPK (open boxes) and p42/44 Erk (closed boxes) activities in NIH 3T3 cells.

These data clearly indicate that the Erk and p54 kinases lie on distinct pathways. The p54 agonists share a common theme in so far as they are all linked to cellular insult or stress. Accordingly, this family has been termed Stress-Activated Protein Kinase or SAPK. The family presumably plays a role in transducing stress signals throughout the cell. Identification of the components upstream of the kinase should reveal a novel cascade which may be analogous to the sensory pathway identified in yeast (see Chapter 8).

5. Conclusions and perspectives

This review has attempted to summarize current knowledge concerning two sub-families of insulin/mitogen-activated protein Ser/Thr kinases, the S6 kinases and the MAP kinases. Information in this area seems to be expanding geometrically, especially with regard to the MAP kinases. The list of candidate physiological substrates for the MAP kinase is rapidly growing, and suggests a ubiquitous role for this kinase subfamily in the regulation of other multipotential proteins, such as other protein (Ser/Thr) kinases (especially Rsk and MAPKAP kinase-2) and transcriptional regulatory proteins (c-Jun, c-Myc, ATF-2, $p62^{TCF}$) and most recently, phospholipase A_2. The p42/44 MAP kinases are the sole immediate upstream regulator of Rsk, which itself appears to be a protein kinase whose specificity is broad, defined in part by a requirement for basic residues amino-terminal to the site of phosphorylation. This is radically different from the specificity of the MAP kinases which require a proline immediately carboxy-terminal to the site of phosphorylation. The ability of MAP kinases to regulate other protein kinases of differing specificity greatly broadens the range of substrates subordinate to MAP kinase regulation; a similar reasoning applies to MAP kinase regulation of transcriptional factor function, although the time scale of the physiological responses governed by this group of substrates will be somewhat more delayed. One very familiar response controlled in part by the p42/44 MAP kinase is the insulin regulation of glycogen synthesis in skeletal muscle, mediated by Rsk phosphorylation of the glycogen-binding subunit of protein phosphatase-1; the resultant increase in phosphatase catalytic efficiency leads to the dephosphorylation and activation of glycogen synthase. The control of protein phosphatase activity by Rsk further diversifies the range of substrates susceptible to regulation by MAP kinase.

The ability of a 'mitogen'-activated kinase to promote glycogen storage in mature skeletal muscle in response to insulin raises the question as to whether the function of MAP kinases in cell physiology is primarily to promote the mitogenic response, or to execute the specialized, tissue-specific programmes that characterize differentiated cells. A definitive answer is not yet available for several reasons: little study of MAP kinases has been carried out *in vivo* or with postmitotic tissues; most studies have focused on oocytes or cultured cells; the latter, even when exhibiting a differentiated phenotype, are usually tilted toward a mitogenic programme as compared to their *in vivo* tissue homologue. Moreover, constitutively active mutants as well as specific stimulators or inhibitors of the MAP kinases applicable to intact cells have not yet been identified. Although the putative role of the MAP kinases in the activation of the protooncogenes, c-Jun and c-Myc, points to a positive role for MAP kinases in promoting cell division, no compelling direct evidence is as yet available to indicate whether MAP kinase activation is in itself mitogenic or is even a necessary event for mitogen-induced cell division. Conversely, several observations suggest that the dominant cellular role of the MAP kinases may be in the control of differentiated responses, and that the enzymes are perhaps even antimitotic in some circumstances. Thus, in yeast, the MAP kinase homologues Kss1 and Fus3 are crucial to the response to mating factors; the latter direct the cell to interrupt its cell division cycle, and initiate a programme of gene expression, cytoskeletal and nuclear reorganization aimed toward meiosis and mating. Although the specific biochemical functions of Fus3/Kss1 in this programme are not known, the overall sense of the programme is directed toward terminating cell division and adopting a specific differentiated state. The recently described yeast MAP kinases, Mpk1 and Hog1, each appear to be crucial to some aspect of yeast cell structural integrity. In mammalian tissues, the highest abundance of the p42/44 and p54 subfamilies of MAP kinase are observed in the central nervous system, further pointing to an important role of the MAP kinases in postmitotic cells. Although a definitive conclusion concerning the physiological roles of the MAP kinases is not yet warranted, the cAMP-dependent protein kinase subfamily provides an instructive example. In general, continuous elevation of cAMP levels is inhibitory to cell division, but transient elevations occur during the cell cycle, which may be crucial to the traversal of specific steps. Moreover in some tissues, particularly of endocrine origin, cAMP-dependent protein kinase, in addition to its acute regulator role, appears to be a dominant determinant of cellular phenotype and may become a protagonist of cell division.

Based on these considerations, we infer that the major physiological roles of the MAP kinases will be in the functional regulation of differentiated cells, including specifically the maintenance of the differentiated state itself, through their actions in regulating gene expression. Activation of p42/44 MAP kinases both during the G_1–S and G_2·M transitions suggests that a positive role in promoting those transitions will be uncovered; however, it may prove that, as with cAMP-dependent protein kinase, continuous elevations of MAP kinase activity in most cell backgrounds will prove inhibitory to cell division.

The understanding of the identity and order of the components in the signal transduction pathway between the receptor tyrosine kinases and the p42/44 MAP kinases has advanced rapidly. MEK, the novel dual specificity (Ser/Thr and Tyr) kinases (two isoforms thus far) appear to be the primary (perhaps sole) regulator immediately upstream of the p42/44 MAP kinases. In contrast, there are at least two protein kinases, c-Raf-1 and MEKK directly upstream of MEK. Preliminary evidence indicates that MEK is very restricted in its range of substrates, being unable, for example, to phosphorylate the p54 MAP kinase, an enzyme that, like Erk, is also regulated by 'dual' specificity phosphorylation. Nevertheless, much further work on the substrate specificity of the MEKs is necessary, especially with regard to the ability of MEK to act on other Erk family members such as Erk-4. MEK is the first physiological substrate of c-Raf-1 identified thus far; it is not yet clear whether c-Raf-1 is itself a kinase of broad or narrow specificity. Despite the unusually stable interactions between Raf and MEK, it seems likely that c-Raf-1 will have at least a small set of cellular substrates in addition to MEK, and will therefore have a broader array of subordinate substrates than even the large set defined by the p42/44 MAP kinases.

The understanding of the cellular role and regulatory mechanisms controlling the p70 S6 kinase continues to lag behind the Rsk enzymes. The p70 clearly has a more narrow substrate specificity, but the nuclear localization of the α_I isoform suggests that interesting p70 targets, in addition to 40S ribosomal subunits, are yet to be uncovered. The much more complex activation mechanism of p70 S6 kinase appears to require p70 phosphorylation by a cyclin-dependent kinase (cdk), a MAP kinase (p54 and/or p42/44) and a third, as yet unidentified, protein kinase whose activity/effect is reversed selectively by the macrolide immunosuppressant, rapamycin. The availability of rapamycin as a selective inhibitor for a signal transduction crucial to the activation off p70 will enable a survey of the aspects of the cellular response that require p70 activity as well as the operative of the rapamycin in sensitive pathway. A major goal is to identify the rapamycin-sensitive element, so as to reconstruct *in vitro* the activation of the p70; once this is achieved, it will be necessary to define the insulin/mitogen signal transduction pathways upstream of the cdk, the p54 MAP kinase and the 'rapamycin-sensitive' element, and whether these elements are also subordinate to Ras and/or Raf, as recently elucidated for the MAP kinases. A major theme that has emerged in comparing the p70 to the Rsk S6K kinases, as well as the p42/44 MAP kinases to the p54 MAP kinases, is the existence of parallel signal transduction pathways. Some of these originate immediately at the receptor, and their existence is graphically exemplified by the effects of point mutations at specific receptor tyrosine phosphorylation sites. Other branch points likely occur downstream, although specific examples are as yet lacking. It seems likely that multiple effector pathways emanate from Ras, of which Raf is on one avenue; similarly, the existence of Raf-1 substrates other than MEK is also to be expected. Pathways also converge, as shown by MEKK and Raf activation of MEK. Clearly, substantial challenges remain before the insulin/mitogen signal transduction pathways are fully elucidated.

References

1. Rosen, O. M. (1987) After insulin binds. *Science*, **237**, 1452.
2. Ullrich, A. and Schlessinger, J. (1990) Signal transduction by receptors with tyrosine kinase activity. *Cell*, **61**, 203.
3. Samuelson, L. E. and Klausner, R. D. (1992) Tyrosine kinase and tyrosine-based activation motifs. *J. Biol. Chem.*, **267**, 24913.
4. Avruch, J., Alexander, M. C., Palmer J. L., Pierce, M. W., Nemenoff, R. A., Blackshear, P. J. *et al*. (1982) The role of insulin-stimulated protein phosphorylation in insulin action. *Fed. Proc.*, **41**, 2629.
5. Avruch, J., Nemenoff, R. A., Pierce, M., Kwok, Y. C., and Blackshear, P. J. (1985) Protein phosphorylations as a mode of insulin action. In *Molecular basis for insulin action* (ed M. P. Czech), p. 263. Plenum Press, New York.
6. Czech, M. P., Klarlund, J. K., Yagaloff, K. A., Bradford, A. P., and Lewis, R. E. (1988) Insulin receptor signalling: activation of multiple serine kinases. *J. Biol. Chem.*, **263**, 11017.
7. Tornqvist, H. E. and Avruch, J. (1988) Relation of site-specific beta subunit tyrosine autophosphorylation to insulin activation of the insulin receptor (tyrosine) protein kinase activity. *J. Biol. Chem.*, **263**, 4593.

8. Jove, R. and Hanafusa, H. (1987) Cell transformation by the viral src oncogene. *Ann. Rev. Cell Biol.*, **3**, 31.

9. Cooper, J. A. (1990) In *Peptide and protein phosphorylation*, (ed. G. Kemp), pp. 85–113. CRC, Boca Raton.

10. Koch, C. A., Anderson, D., Moran, M. F., Ellis, C., and Pawson, T. (1991) SH2 and SH3 domains: elements that control interactions of cytoplasmic signalling proteins. *Science*, **252**, 668.

11. Myers, M. G. Jr., Backer, J. M., Su, X. J., Shoelson, S., Hu, P., Schlessinger, J. *et al.* (1992) IRS-1 activates phosphatidylinositol 3'-kinase by associating with *src* homology 2 domains of p85. *Proc. Natl. Acad. Sci. USA*, **89**, 10350.

12. Rhee, S. G. and Choi, K. D. (1992) Regulation of inositol phospholipid-specific phospholipase C isoenzymes. *J. Biol. Chem.*, **267**, 12393.

13. Schindler, C., Shuai, K., Prezioso, V. R., and Darnell, J. E. Jr. (1992) Interferon-dependent tyrosine phosphorylation of a latent cytoplasmic transcription factor. *Science*, **257**, 809.

14. Krebs, E. G. (1972) Protein kinases. *Current Topics in Cell Reg.*, **5**, 99.

15. Schulman, H. (1988) Calcium-calmodulin kinase-2. *Advances in Second Messenger & Protein Phosphorylation Res.*, **22**, 39.

16. Nishizuka, Y. (1992) Intracellular signalling by hydrolysis of phospholipids and activation of protein kinase C. *Science*, **238**, 607.

17. Berridge, M. J. (1993) Inositol trisphosphate and calcium signalling. *Nature*, **361**, 315.

18. Larner, J. (1988) Insulin signalling mechanisms — lessons from the old testament of glycogen metabolism and from the new testament of molecular biology. *Diabetes*, **37**, 262.

19. Cohen, P. (1988) Muscle glycogen synthase. In *The enzymes* (ed. P. Boyer and E. G. Krebs), p. 461. Academic Press, New York.

20. Larner, J., Takeda, Y., Brewer, H. B., Huang, L. C., Hazen, R., Brooker G. *et al.* (1976) In *Metabolic interconversion of enzymes* (ed. S. Shaltiel), pp. 71–85. Springer Verlag, Berlin.

21. Benjamin, W. B. and Singer, I. (1975) Actions of insulin, epinephrine, and dibutyryl cyclic adenosine 5'-monophosphate on fat cell protein phosphorylations. Cyclic adenosine 5'-monophosphate dependent and independent mechanisms. *Biochemistry*, **14**, 3301.

22. Avruch, J., Leone, G. R., and Martin, D. B. (1976) The effect of insulin and epinephrine on phosphopeptide metabolism in adipocytes. *J. Biol. Chem.*, **251**, 1511.

23. Forn, J. and Greengard, P. (1976) Regulation of lipolytic and antilipolytic compounds of the phosphorylation of specific proteins in isolated intact fat cells. *Arch. Biochem. Biophys.*, **176**, 721.

24. Avruch, J., Witters, L. A., Alexander, M. C., and Bush, M. A. (1978) The effect of insulin and glucagon on the phosphorylation of hepatic cytoplasmic peptides. *J. Biol. Chem.*, **253**, 4754.

25. Alexander, M. C., Kowaloff, E. M., Witters, L. A., Dennihy, D. T., and Avruch, J. (1979) Purification of a hepatic 123,000-dalton hormone-stimulated ^{32}P-peptide and its identification as ATP-citrate lyase. *J. Biol. Chem.*, **254**, 8052.

26. Ramakrishna, S. and Benjamin, W. B. (1979) Fat cell protein phosphorylation: identification of phosphoprotein-2 as ATP-citrate lyase. *J. Biol. Chem.*, **254**, 9232.

27. Witters, L. A. (1981) Insulin stimulates the phosphorylation of acetyl-CoA carboxylase. *Biochem. Biophys. Res. Commun.*, **100**, 872.

28. Brownsey, R. W. and Denton, R. M. (1982) Evidence that insulin activates fat-cell acetyl-CoA carboxylase by increased phosphorylation at a specific site. *Biochem. J.*, **202**, 77.

29. Haselbacher, G. K., Humbel, R. E., and Thomas, G. (1979) Insulin-like growth factors: insulin or serum increase phosphorylation of ribosomal S6 during transition of stationary chick embryo fibroblasts in early G1 phase of the cell cycle. *FEBS Lett.*, **100**, 185.

30. Nielsen, P. J. and Thomas, G. (1982) Increased phosphorylation of ribosomal protein S6 during meiotic maturation of *Xenopus* oocytes. *Proc. Natl. Acad. Sci. USA*, **79**, 2937.

31. Erikson, R. L. (1991) Structure, expression and regulation of protein kinases involved in phosphorylation of ribosomal protein S6. *J. Biol. Chem.*, **266**, 6007.

32. Smith, C. J., Rubin, C. S., and Rosen, O. M. (1980) Insulin-treated 3T3-L1 adipocytes and cell-free extracts derived from them incorporated ^{32}P into ribosomal protein S6. *Proc. Natl. Acad. Sci. USA*, **77**, 2641.

33. Cobb, M. H. and Rosen, O. M. (1983) Description of a protein kinase from insulin-treated 3T3-L1 cells that catalyzes the phosphorylation of ribosomal S6 and casein. *J. Biol. Chem.*, **258**, 12472.

34. Novak-Hofer, I. and Thomas, G. (1984) An activated S6 kinase in extracts from serum- and epidermal growth factor-stimulated Swiss 3T3 cells. *J. Biol. Chem.*, **259**, 5995.

35. Wu, M. and Gerhart, J. C. (1980) Partial purification and characterization of the maturation-promoting factors from eggs of *Xenopus laevis*. *Dev. Biol.*, **79**, 465.

36. Nemenoff, R. A., Price, D. J., Mendelsohn, J., Carter E. A., and Avruch, J. (1988) An S6 kinase activated during liver regeneration is related to the insulin-stimulated S6 kinase in H4 hepatoma cells. *J. Biol. Chem.*, **263**, 19455.

37. Pelech, S. L. and Krebs, E. G. (1987) Mitogen-activated S6 kinase is stimulated via protein kinase-C dependent and independent pathways in Swiss 3T3 cells. *J. Biol. Chem.*, **262**, 11598.

38. Grove, J. R., Banerjee, P., Balasubramanyam, A., Coffer, P. J., Price, D. J., Avruch, J., and Woodgett, J. R. (1991) Cloning and expression of two human p70 S6 kinase polypeptides differing only at their amino termini. *Mol. Cell. Biol.*, **11**, 5541.

39. Sweet, L. J., Alcorta, D. A., and Erikson, R. L. (1990) Two distinct enzymes contribute to biphasic S6 phosphorylation in serum-stimulated chicken embryo fibroblasts. *Mol. Cell. Biol.*, **10**, 2787.

40. Flotow, H. and Thomas, G. (1992) Substrate recognition determinants of the mitogen-activated S6 kinase from rat liver. *J. Biol. Chem.*, **267**, 3074.

41. Erikson, E. and Maller, J. L. (1991) Purification and characterization of ribosomal protein S6 kinase I from *Xenopus* eggs. *J. Biol. Chem.*, **266**, 5249.

42. Erikson, E. and Maller, J. L. (1988) Substrate specificity of ribosomal protein S6 kinase II from Xenopus eggs. *Second Messengers and Phosphoproteins*, **12**, 135.

43. Pelech, S. L., Olwins, B. B., and Krebs, E. G. (1986) Fibroblast growth factor treatment of Swiss 3T3 cells activates a subunit S6 kinase that phosphorylates a synthetic peptide substrate. *Proc. Natl. Acad. Sci. USA*, **83**, 5968.

44. Cicerelli, M. F., Pelech, S. L., and Krebs, E. G. (1988) Activation of multiple protein kinases during the burst in protein phosphorylation that precedes the first meiotic cell division in *Xenopus* oocytes. *J. Biol. Chem.*, **263**, 2009.

45. Ahn, N. G., Weiel, J. E., Chan, C. P., and Krebs, E. G. (1990) Identification of multiple epidermal growth factor-stimulated protein serine/threonine kinase from Swiss 3T3 cells. *J. Biol. Chem.*, **265**, 11487.

46. Stokoe, D., Campbell, D. G., Nakielny, S., Hidaka, H., Leevers, S. J., Marshall, C.,

and Cohen, P. (1992) MAPKAP kinase-2; a novel protein kinase activated by mitogen-activated protein kinase. *EMBO J.*, **11**, 3985.

47. Wolfe, L., Bradford, A. P., Klarlund, J. K., and Czech, M. P. (1992) Purification and characterization of a cytosolic insulin-stimulated serine kinase from rat liver. *J. Biol. Chem.*, **267**, 9749.

48. Maller, J. L. (1990) *Xenopus* oocytes and the biochemistry of cell division. *Biochemistry*, **29**, 3157.

49. Erikson, E. and Maller, J. L. (1986) Purification and characterization of a protein kinase from Xenopus eggs highly specific for ribosomal protein S6. *J. Biol. Chem.*, **261**, 350.

50. Erikson, E., Stefanovic, D., Blenis, J., Erikson, R. L., and Maller, J. L. (1987) Antibodies to *Xenopus* egg S6 kinase II recognizes S6 kinase from progesterone- and insulin-stimulated *Xenopus* oocytes and from proliferating chick embryo fibroblasts. *Mol. Cell. Biol.*, **7**, 3147.

51. Lane, H. A., Morley, S. J., Doree, M., Kozma, S. C., and Thomas, G. (1992) Identification and early activation of a *Xenopus laevis* p70^{S6K} following progesterone-induced meiotic maturation. *EMBO J.*, **11**, 1743.

52. Jones, S. W., Erikson, E., Blenis, J., Maller, J. L., and Erikson, R. L. (1988) A *Xenopus* ribosomal protein S6 kinase has two apparent kinase domains that are each similar to distinct protein kinases. *Proc. Natl. Acad. Sci. USA*, **85**, 3377.

53. Erikson, E. and Maller, J. L. (1989) *In vivo* phosphorylation and activation of ribosomal protein S6 kinase during *Xenopus* oocyte maturation. *J. Biol. Chem.*, **264**, 13711.

54. Alcorta, D. A., Crews, C. M., Sweet, L. J., Bankston, L., Jones, S. W., and Erikson, R. L. (1989) Sequence and expression of chicken and mouse rsk: homologs of *Xenopus laevis* ribosomal S6 kinase. *Mol. Cell. Biol.*, **9**, 3850.

55. Lavoinne, A., Erikson, E., Maller, J. L., Price, D. J., Avruch, J., and Cohen, P. (1991) Purification and characterization of the insulin-stimulated protein kinase from rabbit skeletal muscle; close similarity to S6 kinase II. *Eur. J. Biochem.*, **199**, 723.

56. Sutherland, C., Campbell, D. G., and Cohen, P. (1993) Identification of insulin-stimulated protein kinase-1 as the rabbit equivalent of rsk^{mo}-2; identification of two threonines phosphorylated during activation by MAP kinases. *Eur. J. Biochem.*, **212**, 581.

57. Gressner, A. M. and Wool, I. G. (1974) The phosphorylation of liver ribosomal proteins *in vivo*. Evidence that only a single small subunit protein (S6) is phosphorylated. *J. Biol. Chem.*, **249**, 6917.

58. Gressner, A. M. and Wool, I. G. (1974) The stimulation of the phosphorylation of ribosomal protein S6 by cycloheximide and puromycin. *Biochem. Biophys. Res. Commun.*, **60**, 1482.

59. Nielsen, P. J., Manchester, K. L., Towbin, H., Gordon, J., and Thomas, G. (1982) The phosphorylation of ribosomal protein S6 in rat tissues following cycloheximide injection, in diabetes, and after denervation of diaphragm. Simple immunological determination of the extent of the S6 phosphorylation on protein blots. *J. Biol. Chem.*, **257**, 12316.

60. Price, D. J., Grove, J. R., and Avruch, J. (unpublished observations).

61. Price, D. J., Nemenoff, R. A., and Avruch, J. (1989) Purification of a hepatic S6 kinase from cycloheximide-treated rats. *J. Biol. Chem.*, **264**, 13825.

62. Jeno, P., Jaggi, N., Luther, H., Siegemann, M., and Thomas, G. (1989) Purification and characterization of a 40S ribosomal S6 kinase from vanadate-stimulated Swiss 3T3 cells. *J. Biol. Chem.*, **264**, 1293.

63. Banerjee P., Ahmad, M. F., Grove, J. R., Kozlosky, C., Price, D. J., and Avruch, J.

(1990) Molecular structure of a major insulin/mitogen-activated 70-kDa S6 protein kinase. *Proc. Natl. Acad. Sci. USA*, **87**, 9550.

64. Kozma, S. C., Ferrari, S., Bassand, P., Siegemann, M., Totty, N., and Thomas, G. (1990) Cloning of the mitogen-activated S6 kinase from rat liver reveals an enzyme of the second messenger subfamily. *Prac. Natl. Acad. Sci. USA*, **87**, 7365.

65. Reinhard, C. and Thomas, G. (1992) A single gene encodes two isoforms of the p70 S6 kinase: activation upon mitogenic stimulation. *Proc. Natl. Acad. Sci. USA*, **89**, 4052.

66. Hanks, S. K., Quinn, M. A., and Hunter, T. (1988) The protein kinase family: conserved features and deduced phylogeny of the catalytic domains. *Science*, **241**, 42.

67. Franco, R. and Rosenfeld, M. G. (1990) Hormonally inducible phosphorylation of a nuclear pool of ribosomal protein S6. *J. Biol. Chem.*, **265**, 4321.

68. Kennelly, P. J. and Krebs, E. G. (1991) Consensus sequences as substrate specificity determinants for protein kinases and protein phosphatases. *J. Biol. Chem.*, **266**, 15555.

69. Cohen, P. (1989) The structure and regulation of protein phosphatases. *Ann. Rev. Biochem.*, **58**, 453.

70. Dent, P., Lavoinne, A., Nakielny, S., Caudwell, F. B., Watt, P., and Cohen, P. (1990) The molecular mechanism by which insulin stimulates glycogen synthesis in mammalian skeletal muscle. *Nature*, **348**, 302.

71. Avruch, J., Witters, L. A., Alexander, M. C., Bush, M. A., and Crapo, L. (1979) Insulin and the phosphorylation of intracellular proteins. *Proc. ICN/UCLA Symp. Transmembrane Signalling*, **31**, 621.

72. Chen, R., Chung, J., and Blenis, J. (1991) Nuclear localization and regulation of erk- and rsk-encodes protein kinases. *Mol. Cell. Biol.*, **12**, 915.

73. Martin-Perez, J. and Thomas, G. (1983) Ordered phosphorylation of 40S ribosomal protein S6 after serum stimulation in quiescent 3T3 cells. *Proc. Natl. Acad. Sci. USA*, **80**, 926.

74. Martin-Perez, J., Siegmann, M., and Thomas, G. (1984) EGF, PDF$_2$, and insulin induces the phosphorylation of identical S6 peptides in Swiss mouse 3T3 cells: effects of cAMP on early sites of phosphorylation. *Cell*, **36**, 287.

75. Kreig, J., Hofsteenge, J., and Thomas, G. (1988) Identification of the 40S ribosomal protein S6 phosphorylation sites induced by cycloheximide. *J. Biol. Chem.*, **263**, 11473.

76. Wettenhall, R. E. H. and Morgan, F. J. (1984) Phosphorylation of hepatic ribosomal protein S6 on 80 and 40S ribosomes. Primary structure of S6 in the region of the major phosphorylation sites for cAMP-dependent protein kinases. *J. Biol. Chem.*, **259**, 2084.

77. Ferrari, S., Bandi, H. R., Hofsteenge, J., Bussian, B. M., and Thomas, G. (1991) Mitogen-activated 70K S6 kinase. Identification of *in vitro* 40S ribosomal phosphorylation sites *J. Biol. Chem.*, **268**, 22770.

78. Wettenhall, R. E. H., Erikson, E., and Maller, J. L. (1992) Ordered multisite phosphorylation of *Xenopus* ribosomal protein S6 by S6 kinase II. *J. Biol. Chem.*, **267**, 9021.

79. Chung, J., Juo, C. J., Crabtree, G. R., and Blenis, J. (1992) Rapamycin-FKBP specificity blocks growth-dependent activation of signalling by the 70 kd S6 protein kinase. *Cell*, **69**, 1227.

80. Kuo, C. J., Chung, J., Fiorentino, D. F., Flanagan, W. M., Blenis, J., and Crabtree, G. R. (1992) Rapamycin selectively inhibits interleukin-2 activation of p70 S6 kinase. *Nature*, **358**, 70.

81. Price, D. J., Grove, J. R., Calvo, V., Avruch, J., and Bierer, B. E. (1992) Rapamycin-induced inhibition of the 70-kilodalton S6 protein kinase. *Science*, **257**, 973.

82. Calvo, V., Crews, C. M., Vik, T. A., and Bierer, B. E. (1992) Interleukin 2 stimulation of p70 S6 kinase activity is inhibited by the immunosuppresent rapamycin. *Proc. Natl. Acad. Sci. USA*, **89**, 7571.

83. Traugh, J. A. and Pendergast A. M. (1986) Regulation of protein synthesis by phosphorylation of ribosomal protein S6 and aminoacyl-tRNA synthetases. *Prog. Nucl. Acid. Res.*, **33**, 195.

84. Hershey, J. W. B. (1989) Protein phosphorylation controls translation rates. *J. Biol. Chem.*, **264**, 20823.

85. Krase, C., Johnson, S. P., and Warner, J. R. (1985) Phosphorylation of the yeast equivalent of ribosomal protein S6 is not essential for growth. *Proc. Natl. Acad. Sci. USA*, **82**, 7515.

86. Watson, K. L., Konrad, K. D., Woods, D. F., and Bryant, P. J. (1992) *Drosophila* homolog of the human S6 ribosomal protein is required for tumor suppression in the hematopoietic system. *Proc. Natl. Acad. Sci. USA*, **89**, 11302.

87. Fingar, D. C., Hausdorff, S. F., Blenis, J., and Birnbaum, M. J. (1993) Dissociation of pp70 ribosomal protein S6 kinase from insulin-stimulated glucose transport in 3T3-L1 adipocytes. *J. Biol. Chem.*, **268**, 3005.

88. Ballou, L. M., Jeno, P., and Thomas, G. (1988) Protein phosphatase 2A inactivates the mitogen-stimulated S6 kinase from Swiss mouse 3T3 cells. *J. Biol. Chem.*, **263**, 1188.

89. Price, D. J., Gunsalus, J. R., and Avruch, J. (1990) Insulin activates a 70,000 dalton S6 kinase through serine/threonine-specific phosphorylation of the enzyme polypeptide. *Proc. Natl. Acad. Sci. USA*, **87**, 7944.

90. Sturgill, T. W., Ray, L. B., Erikson, E., and Maller, J. L. (1988) Insulin-stimulated MAP-2 kinase phosphorylates and activates ribosomal protein S6 kinase II. *Nature*, **334**, 715.

91. Kyriakis, J. M. and Avruch, J. (1990) pp54 MAP-2 kinase. A novel serine/threonine protein kinase regulated by phosphorylation and stimulated by poly-L-lysine. *J. Biol. Chem.*, **265**, 17355.

92. Chung, J., Pelech, S. L., and Blenis, J. (1991) Mitogen-activated Swiss mouse 3T3 RSK kinase I and II are related to pp44[mpf] from sea star oocytes and participate in the regulation of pp90[rsk] activity. *Proc. Natl. Acad. Sci. USA*, **88**, 4981.

93. Gregory, J. S., Boulton, T. G., Sang, B., and Cobb, M. H. (1989) An insulin-stimulated ribosomal protein S6 kinase from rabbit liver. *J. Biol. Chem.*, **264**, 18397.

94. Grove, J. R., Price, D. J., Banerjee, P., Balasubramanyam, A., Ahmad, M. F., and Avruch, J. (1993) Regulation of an epitope-tagged recombinant Rsk-1 S6 kinase by phorbol ester and erk/MAP kinase. *Biochemistry*, **32**, 7727.

95. Ahn, N. G. and Krebs, E. G. (1990) Evidence for an epidermal growth factor-stimulated protein kinase cascade in Swiss 3T3 cells. Activation of serine peptide kinase activity by myelin basic protein kinases *in vitro*. *J. Biol. Chem.*, **265**, 11495.

96. Chung, J., Chen, R. H., and Blenis, J. (1991) Coordinate regulation of pp90[rsk] and a distinct protein-serine/threonine kinase activity that phosphorylates recombinant pp90[rsk] *in vitro*. *Mol. Cell. Biol.*, **11**, 1868.

97. Barrett, C. B., Erikson, E., and Maller, J. L. (1992) A purified S6 kinase kinase from *Xenopus* eggs activates S6 kinase II and autophosphorylation on serine, threonine, and tyrosine residues. *J. Biol. Chem.*, **267**, 4008.

98. Ballou, L. M., Luther, H., and Thomas, G. (1991) MAP2 kinase and 70K S6 kinase lie on distinct signalling pathways. *Nature*, **349**, 348.

99. Vik, T. A., Sweet, L. J., and Erikson, R. L. (1991) Coinfection of insect cells with recombinant baculovirus expression pp60$^{v\text{-}src}$ results in the activation of a serine-specific protein kinase pp90^{rsk}. *Proc. Natl. Acad. Sci. USA*, **87**, 2685.

100. Kozma, S. C., McGlynn, E., Siegmann, M., Reinhard, C., Ferrari, S., and Thomas, G. (1993) Active baculovirus recombinant p70^{s6k} and p85^{s6k} produced as a function of the infectious response. *J. Biol. Chem.*, **268**, 7134.

101. Chen, R., Sarnecki, C., and Blenis, J. (1991) Regulation of pp90^{rsk} phosphorylation and S6 phosphotransferase activity in Swiss 3T3 cells by growth factor-, phorbol ester-, and cyclic AMP-mediated signal transduction. *Mol. Cell. Biol.*, **11**, 1861.

102. Susa, M., Oliver, A. R., Fabbro, D., and Thomas, G. (1989) EGF induces biphasic S6 kinase activation: late phase is protein kinase C-dependent and contributes to mitogenicity. *Cell*, **57**, 814.

103. Susa, M., Vulevic, D., Lane, H. A., and Thomas, G. (1992) Inhibition of down-regulation of protein kinase C attenuates late phase p70^{s6k} activation induced by epidermal growth factor but not platelet-derived growth factor or insulin. *J. Biol. Chem.*, **267**, 6905.

104. Blenis, J. (1991) Growth-regulated signal transduction by MAP kinases and Rsks. *Cancer Cells*, **3**, 445.

105. Dumont, F. J., Staruch, M. J., Koprak, S. L., Melino, M. R., and Sigal, N. H. (1990) Distinct mechanisms of suppression of murine T cell activation by the related macrolides FK-506 and rapamycin. *J. Immunol.*, **144**, 251.

106. Bierer, B. E., Mattila, P. S., Standaert, R. F., Herzenberg, L. R., Burakoff, S. J., Crabtree, G., and Schreiber, S. L. (1990) Two distinct signal transmission pathways in T lymphocytes are inhibited by complexes formed between an immunophilin and either FK506 or rapamycin. *Proc. Natl. Acad. Sci. USA*, **87**, 9231.

107. Dumont, P. J., Melino, M. R., Staruch, M. J., Koprak, S. L., Fischer, P. A., and Sigal, N. H. (1990) The immunosuppressive macrolides FK-506 and rapamycin act as reciprocal antagonists in murine T cells. *J. Immunol.*, **144**, 1418.

108. Liu, F., Farmer, J. D. Jr., Lane, W. S., Friedman, J., Weissman, I., and Schreiber, S. L. (1991) Calcineurin is a common target of cyclophilin–cyclosporin A and FKBP–FK506 complexes. *Cell*, **66**, 807.

109. Fruman, D. A., Klee, C. B., Bierer, B. E., and Burakoff, S. J. (1992) Calcineurin phosphatase activity in T lymphocytes is inhibited by FK506 and cyclosporin A. *Proc. Natl. Acad. Sci. USA*, **89**, 3686.

110. Schreiber, L. (1991) Chemistry and biology of the immunophilins and their immunosuppressive ligands. *Science*, **251**, 283.

111. Price, D. J., Mukhopadhyay, N. K., and Avruch, J. (1991) Insulin-activated kinases phosphorylate a pseudosubstrate synthetic inhibitor of the p70 S6 kinase. *J. Biol. Chem.*, **266**, 16281.

112. Ferrari, S., Bannwarth, W., Morley, S. J., Totty, N. F., and Thomas, G. (1992) Activation of p70^{s6k} is associated with phosphorylation of four clustered sites displaying Ser/Thr–Pro motifs. *Proc. Natl. Acad. Sci. USA*, **89**, 7282.

113. Mukhopadhyay, N. K., Price, D. J., Kyriakis, J. M., Pelech, S. J., Sanghera, J. and Avruch, J. (1992). An array of insulin-activated proline-directed (Ser/Thr) protein kinases phosphorylate the p70 S6 kinase. *J. Biol. Chem.*, **267**, 3325.

114. Sturgill, T. W. and Ray, L. B. (1986) Muscle proteins related to microtubule associated protein-2 are substrates for an insulin-stimulatable kinase. *Biochem. Biophys. Res. Commun.*, **134**, 565.

115. Ray, L. B. and Sturgill, T. W. (1987) Rapid stimulation by insulin of a serine/threonine kinase in 3T3-L1 adipocytes that phosphorylate microtubule-associated protein 2 *in vitro*. *Proc. Natl. Acad. Sci. USA*, **84**, 1502.

116. Ray, L. B. and Sturgill, T. W. (1988) Insulin-stimulated microtubule-associated protein kinase is detectable by analytic gel chromatography as a 35-kDa protein in myocytes, adipocytes, and hepatocytes. *Arch. Biochem. Biophys.*, **262**, 307.

117. Ray, L. B. and Sturgill, T. W. (1988) Insulin-stimulated microtubule-associated protein kinase is phosphorylated on tyrosine and threonine *in vivo*. *Proc. Natl. Acad. Sci. USA*, **85**, 3753.

118. Anderson, N. G., Maller, J. L., Tonks, N. K., and Sturgill, T. W. (1990) Requirement for integration of signals from two distinct phosphorylation pathways for activation of MAP kinase. *Nature*, **343**, 651.

119. Gilmore, T. and Martin, G. S. (1983) Phorbol ester and diacylglycerol induces protein phosphorylation at tyrosine. *Nature*, **306**, 47–87.

120. Bishop, R., Martinez, R., Nakamura, K. D., and Weber, M. J. (1983) A tumor promoter stimulates phosphorylation on tyrosine. *Biochem. Biophys. Res. Commun.*, **115**, 536.

121. Cooper, J. A., Sefton, M. B., and Hunter, T. (1984) Diverse mitogenic agents induce the phosphorylation of two related 42,000-dalton proteins on tyrosine in quiescent chick cells. *Mol. Cell. Biol.*, **4**, 30.

122. Cooper, J. A. and Hunter, T. (1985) Major substrate for growth factor-activated protein-tyrosine kinases is a low-abundance protein. *Mol. Cell. Biol.*, **5**, 3304.

123. Kazlaukas, A. and Cooper, J. A. (1988) Protein kinase C mediates platelet-derived growth factor-induced tyrosine phosphorylation of p42. *J. Cell. Biol.*, **106**, 1395.

124. Rossomando, A. J., Payne, D. M., Weber, M. J., and Sturgill, T. W. (1988) Evidence that pp42, a major tyrosine kinase target, is a mitogen-activated serine/threonine protein kinase. *Proc. Natl. Acad. Sci. USA*, **86**, 6940.

125. Hoshi, M., Nishida, E., and Sakai, H. (1988) Activation of a Ca^{2+}-inhibitable protein kinase that phosphorylates microtubule-associated protein 2 *in vitro* by growth factors, phorbol esters, and serum in quiescent cultured human fibroblasts. *J. Biol. Chem.*, **263**, 5396.

126. Gotoh, H., Nishida, E., Yamashita, T., Hoshi, M., Kawakami, M., and Sakai, H. (1990) Microtubule-associated-protein (MAP) kinase activated by nerve growth factor and epidermal growth factor in PC12 cells. Identity with the mitogen-activated MAP kinase of fibroblastic cells. *Eur. J. Biochem.*, **193**, 661.

127. Hoshi, M., Nishida, E., and Sakai, H. (1989) Characterization of a mitogen-activated, Ca^{2+}-sensitive microtubule-associated protein-2 kinase. *Eur. J. Biochem.*, **184**, 477.

128. Pelech, S. J., Tombes, R. M., Meijer, L., and Krebs, E. G. (1988) Activation of myelin protein kinases during Echinoderm oocyte maturation and egg fertilization. *Dev. Biol.*, **130**, 28.

129. Sanghera, J. S., Paddon, H. B., Bader, S. A., and Pelech, S. L. (1990) Purification and characterization of a maturation-activation of a myelin basic protein kinase from sea star oocytes. *J. Biol. Chem.*, **265**, 52.

130. Cobb, M. H., Boulton, T. G., and Robbins, D. J. (1991) Extracellular signal-regulated kinases: ERKs in progress. *Cell Reg.*, **2**, 965.

131. Boulton, T. G., Gregory, J. S., and Cobb, M. H. (1991) Purification and properties of an extracellular signal-regulated kinase 1, an insulin-stimulated microtubule-associated protein 2 kinase. *Biochemistry*, **30**, 278.

132. Boulton, T. G., Yancopoulos, G. D., Gregory, J. S., Slaughter, C., Moomaw, C., Hsu,

J., and Cobb, M. H. (1990) An insulin-stimulated protein kinase homologous to yeast kinases involved in cell cycle control. *Science*, **249**, 64.

133. Elion, E. A., Grisafi, P. L., and Fink, G. R. (1990) *FUS3* encodes a cdc^{2+}/CDC28-related kinase required for the transition from mitosis into conjugation. *Cell*, **60**, 649.

134. Courchesne, W. E., Kunisawa, S., and Thorner, J. (1989) A putative kinase overcomes pheromone-induced arrest of cell cyling in *S. cerevisiae*. *Cell*, **58**, 1107.

135. Toda, T., Shimanuki, T., and Yanagida, M. (1991) Fission yeast genes that confer resistance to staurosporine encode an AP-1-like transcription factor and a protein kinase related to the mammalian ERK1/MAP2 and budding yeast *FUS3* and *KSS1* kinases. *Genes Dev.*, **5**, 60.

136. Errede, B. and Levin, D. E. (1993) A conserved kinase cascade for MAP kinase activation in yeast. *Curr. Opinions in Cell Biol.*, **5**, 254.

137. Posada, J., Sanghera, J., Pelech, S., Aebersold, R., and Cooper, J. A. (1991) Tyrosine phosphorylation and activation of homologous protein kinases during oocyte maturation and mitogenic activation of fibroblasts. *Mol. Cell. Biol.*, **11**, 2517.

138. Gotoh, Y., Moriyama, K., Matsuda, S., Kawasaki, H., Suzuki, K., Yahara, I. *et al.* (1991) *Xenopus* M phase MAP kinase: isolation of its cDNA and activation by MPF. *EMBO J.*, **10**, 2661.

139. Boulton, T. G., Nye, S. H., Robbins, D. J., Ip, N. Y., Radziejewska, E., Morgenbesser S. D. *et al.* (1991) ERKs: a family of protein-serine/threonine kinases that are activated and tyrosine phosphorylated in response to insulin and NGF. *Cell*, **65**, 663.

140. Her, J. H., Wu, J., Rall, T. B., and Sturgill, T. W. (1991) Sequence of pp42/MAP kinase, a serine/threonine kinase regulated by tyrosine phosphorylation. *Nucleic Acid Res.*, **19**, 3743.

141. Gonzalez, F. A., Raden, D. L., Rigby, M. R., and Davis, R. J. (1992) Heterogeneous expression of four MAP kinase isoforms in human tissues. *FEBS Lett.*, **304**, 170.

142. Boulton, T. G. and Cobb, M. H. (1991) Identification of multiple extracellular signal-regulated kinases (ERKS) with antipeptide antibodies *Cell Reg.*, **2**, 357.

143. Kyriakis, J. M., Banerjee, P., Nikolakaki, E., Dai, T., Rubie, E. A., Ahmad, M. F. *et al.* (1994) The stress-activated protein kinase subfamily of c-Jun kinases. *Nature*, **369**, 156.

144. Sanghera, J. S., Aebersold, R., Morrison, H. D., Bures, E. I., and Pelech, S. J. (1990) Identification of the sites in myelin basic protein that are phosphorylated by meiosis-activated protein kinase pp44^{mpk}. *FEBS Lett.*, **273**, 223.

145. Countaway, J. L., Northwood, I. C., and Davis, R. J. (1989) Mechanism of phosphorylation of the epidermal growth factor receptor on threonine 669. *J. Biol. Chem.*, **264**, 10828.

146. Alvarez, E., Northwood, I. C., Gonzalez, F. A., Latour, D. A., Seth, A., Abate, C. *et al.* (1991) Pro–Leu–Ser/Thr–Pro is a consensus primary sequence for substrate protein phosphorylation. Characterization of the phosphorylation of c-myc and c-jun proteins by an epidermal growth factor receptor threonine 669 protein kinase. *J. Biol. Chem.*, **266**, 15277.

147. Takishima, K., Griswold-Prenner, I., Ingebritsen, T., and Rosner, M. R. (1991) Epidermal growth factor (EGF) receptor T669 peptide kinase from 3T3-L1 cells in an EGF-stimulated 'MAP' kinase. *Proc. Natl. Acad. Sci. USA*, **88**, 2520.

148. Gonzalez, F. A., Raden, D. L., and Davis, R. J. (1991) Identification of substrate recognition determinants for human ERK1 and ERK2 protein kinases. *J. Biol. Chem.*, **266**, 22159.

149. Pulverer, B. J., Kyriakis, J. M., Avruch, J. R., Nikolakaki, E., and Woodgett, J. R. (1991) Phosphorylation of c-jun mediated by MAP kinases. *Nature*, **353**, 670.

150. Abdel-Hafiz, H. A.-M., Heasley, L. E., Kyriakis, J. M., Avruch, J., Kroll, J. M., Johnson, G. L., and Hoeffler, J. P. (1992) Activating transcription factor-2 DNA-Binding activity is stimulated by phosphorylation catalyzed by p42 and p54 microtubule-associated protein kinases. *Mol. Endocrinol.*, **6**, 2079.

151. Rapp, U. R. (1991) Role of Raf-1 serine/threonine protein kinase in growth factor transduction. *Oncogene*, **6**, 495.

152. Li, P., Wood, K., Mamon, H., Haser, W., and Roberts, T. (1991) RAF-1: A kinase currently without a cause but not lacking in effects. *Cell*, **64**, 479.

153. Morrison, D. K., Kaplan, D. R., Escobedo, J. A., Rapp, U. R., Roberts, T. M., and Williams, L. T. (1989) Direct activation of the serine/threonine kinase activity in Raf-1 through tyrosine phosphorylation by PDGFβ receptor. *Cell*, **58**, 641.

154. Izumi, T., Tamemoto, H., Nagao, M., Kadowaki, T., Takaku, F., and Kasuya, M. (1991) Insulin and platelet-derived growth factor stimulate phosphorylation of the c-*raf* product at serine and threonine residues in intact cells. *J. Biol. Chem.*, **266**, 7933.

155. Kovacina, K. S., Yonezawa, K., Brautigan, D. L., Tonks, N. K., Rapp, U. R., and Roth, R. (1990) Insulin activates the kinase activity of the Raf-1 proto-oncogene by increasing its serine phosphorylation. *J. Biol. Chem.*, **265**, 12115.

156. Blackshear, P. J., Haupt, D. M., App, H., and Rapp, U. R. (1990) Insulin activates the Raf-1 protein kinase. *J. Biol. Chem.*, **265**, 12131.

157. Stanton, V. P. Jr., Nichols, D. W., Laudano, A.P., and Cooper, G. M. (1989) Definition of the human *raf* amino-terminal regulatory region by deletion mutagenesis. *Mol. Cell. Biol.*, **9**, 636.

158. Heidecker, G., Huleihel, M., Cleveland, L., Kolch, W., Beck, T. W., Lloyd, P. *et al.* (1990) Mutational activation of c-*raf*-1 and definition of the minimal transforming sequence. *Mol. Cell. Biol.*, **10**, 2503.

159. McGrew, B. R., Nichols, D. W., Stanton, V. P. Jr., Cai, H., Wholf, R. C., Patel, V. *et al.* (1992) Phosphorylation occurs in the amino terminus of the Raf-1 protein. *Oncogene*, **7**, 33.

160. Anderson, N. G., Li, P., Marsden, L. A., Williams, N., Roberts, T., and Sturgill, T. W. (1991) Raf-1 is a potential substrate for mitogen-activated protein kinase *in vivo*. *Biochem. J.*, **277**, 573.

161. Lee, R. M., Rapp, U. R., and Blackshear, P. J. (1991) Evidence for one or more Raf-1 kinase kinase(s) activated by insulin and polypeptide growth factors. *J. Biol. Chem.*, **266**, 10351.

162. Lee, R. M., Cobb, M. H., and Blackshear, P. J. (1992) Evidence that extracellular signal-regulated kinases are the insulin-activated Raf-1 kinase kinases. *J. Biol. Chem.*, **267**, 1088.

163. Kyriakis, J. M., Force, T. L., Rapp, U. R., Bonventre, J. V., and Avruch, J. (1993) Mitogen regulation of c-Raf-1 protein kinase activity toward mitogen-activated protein kinase-kinase. *J. Biol. Chem.*, **268**, 16009.

164. Kyriakis, J. M., App, H., Zhang, X.-F., Banerjee, P., Brautigan, D., Rapp, U. R. and Avruch, J. (1992) Raf-1 activates MAP kinase-kinase. *Nature*, **358**, 417.

165. Stokoe, D., Engel, K., Campbell, D. G., Cohen, P., and Gaestel, M. (1992) Identification of MAPKAP kinase 2 as a major enzyme responsible for the phosphorylation of the small mammalian heat shock proteins. *FEBS Lett.*, **313**, 307.

166. Gaestel, M., Schroder, W., Benndorf, R., Lippmann, C., Buchner, K., Hucho, F. *et al.*

(1991) Identification of the phosphorylation sites of the murine small heat shock protein hsp25. *J. Biol. Chem.*, **266**, 14721.

167. Landry, J., Lambert, H., Zhou, M., Lavoie, J. N., Hickey, E., Weber, L. A., and Anderson, C. W. (1992) Human HSP27 is phosphorylated at serines 78 and 82 by heat shock and mitogen-activated kinases that recognize the same amino acid motif at S6 kinase II. *J. Biol. Chem.*, **267**, 794.

168. Northwood, I. C., Gonzalez, F., Wartmann, M., Raden, D. L., and Davis, R. J. (1991) Isolation and characterization of two growth factor-stimulated protein kinases that phosphorylate the epidermal growth factor-stimulated receptor at threonine 669. *J. Biol. Chem.*, **266**, 15266.

169. Curran, T. and Franza, B. R. Jr. (1988) Fos and jun: the AP-1 connection. *Cell*, **55**, 395.

170. Bohmann, D., Bos, T. J., Admon, A. Nishimura, T., Vogt, P. K., and Tjian, R. (1987) Human proto-oncogene c-jun encodes a DNA binding protein with structural and functional properties of transcriptional factor AP-1. *Science*, **238**, 1386.

171. Angel, P., Allegretto, E. A., Okino, S. T., Hattori, K., Boyle, W. J., and Hunter, T. (1987) Oncogene jun encodes a sequence-specific trans-activator similar to AP-1. *Nature*, **332**, 166.

172. Angel, P., Smeal, T., Meek, J., and Karin, M. (1989) Jun and v-jun contain multiple regions that participate in transcriptional activation in an interdependent manner. *New Biologist*, **1**, 35.

173. Bohmann, D. and Tjian, R. (1989) Biochemical analysis of transcriptional activation by Jun: differential activity of c- and v-Jun. *Cell*, **59**, 709.

174. Boyle, W. J., Smeal, T., Defize, L. H. K., Angel, P., Woodgett, J. R., Karin, M., and Hunter, T. (1991) Activation of protein kinase C decreases phosphorylation of c-Jun at sites that negatively regulate its DNA binding activity. *Cell*, **64**, 573.

175. Lin, A., Frost, J., Deng, T., Smeal, T., Al-Alawi, N., Kikkawa, U. *et al.* (1992) Casein kinase II is a negative regulator of c-Jun DNA binding and AP-1 activity. *Cell*, **70**, 777.

176. Goode, N., Hughes, K., Woodgett, J. R., and Parker, P. J. (1992) Differential regulation of glycogen synthase kinase-3β by protein kinase C isotypes. *J. Biol. Chem.*, **267**, 16878.

177. Binetruy, B., Smeal, T., and Karin, M. (1991) Ha-Ras augments c-Jun and stimulates phosphorylation of its activation domain. *Nature*, **351**, 122.

178. Pulverer, B. J., Hughes, K., Franklin, C. C., Kraft, A. S., Leevers, S. J., and Woodgett, J. R. (1993) Co-purification of mitogen-activated protein kinases with phorbol ester-induced c-Jun kinase activity in U937 leukaemic cells. *Oncogene*, **8**, 407.

179. Alder, V., Polotskaya, A., Wagner, F., and Kraft, A. S. (1992) Affinity-purified c-Jun amino-terminal protein kinase requires serine/threonine phosphorylation for activity. *J. Biol. Chem.*, **267**, 17001.

180. Bos, T. J., Monteclaro, F. S., Mitsunobu, F., Ball, A. P. Jr., Chang, C. H. W., Nishimura, F., and Vogt, P. K. (1990) Efficient transformation of chicken embryo fibroblasts by c-Jun requires structural modification in coding and noncoding sequences. *Genes Dev.*, **4**, 1677.

181. Luscher, B., Christenson, E., Litchfield, D. W., Krebs, E. G., and Eisenmann, R. N. (1990) Myb DNA binding inhibited by phosphorylation at a site deleted during oncogenic activation. *Nature*, **344**, 517.

182. Blackwood, E. M. and Eisenmann, R. M. (1991) Max: a helix-loop zipper protein that forms a sequence-specific DNA-binding complex with Myc. *Science*, **251**, 1211.

183. Blackwell, K., Kretiner, L., Blackwood, E. M., Eisenmann, R. M., and Weintraub, H. (1990) Sequence-specific DNA binding by the c-Myc protein. *Science*, **250**, 1149.

184. Blackwood, E. M., Luscher, B., and Eisenmann, R. M. (1992) Myc and Max associate *in vivo*. *Genes Dev.*, **6**, 71.

185. Kato, G. J., Lee, W. M. F., Chen, L., and Dang, C. V. (1992) Max: functional domains and interaction with c-Myc. *Genes Dev.*, **6**, 81.

186. Kato, G. J., Barrett, J., VillaGarcia, M., and Dang, C. V. (1990) An amino-terminal c-Myc domain required for neoplastic transformation activates transcription. *Mol. Cell. Biol.*, **10**, 5914.

187. Seth, A., Alvarez, E., Gupta, S., and Davis, R. J. (1991) A phosphorylation site located in the NH$_2$-terminal domain of c-Myc increases transactivation of gene expression. *J. Biol. Chem.*, **266**, 13521.

188. Seth, A., Gonzalez, F. A., Gupta, S., Raden, D. L., and Davis, R. J. (1992) Signal transduction within the nucleus by mitogen-activated protein kinase. *J. Biol. Chem.*, **267**, 24796.

189. Saksela, K., Makela, T. P., Hughes, K., Woodgett, J. R., and Alitalo, K. (1991) Activation of protein kinase C increases phosphorylation in the L-*myc* trans-activator domain at a GSK-3 target site. *Oncogene*, **7**, 123.

190. Maekawa, T., Sakura, H., Kanei-ishii, C., Sudo, T., Yoshimura, T., Fujisawa, J. *et al.* (1989) Leucine zipper structure of the protein CRE-BP1 binding to the cyclic AMP response element in brain. *EMBO J.*, **8**, 2023.

191. Ziff, E. (1990) Transcription factors: a new family gathers at the cAMP response site. *TIGS*, **6**, 69.

192. Habener, J. F. (1990) Cyclic AMP response element binding proteins: a cornucopia of transcription factors. *Mol. Endocrinol.*, **4**, 1087.

193. Hoeffer, J. P., Lustbader, J. W., and Chen, C.-Y. (1991) Identification of multiple nuclear factors that interact with cyclic adenosine 3',5'-monophosphate response element-binding protein and activating transcription factor-2 by protein-protein interactions. *Mol. Endocrinol.*, **5**, 256.

194. Herschman, H. R. (1991) Primary response genes induced by growth factors and tumor promoters. *Ann. Rev. Biochem.*, **60**, 281.

195. Herrera, R. E., Shaw, P. E., and Nordheim, A. (1989) Occupation of the *c-fos* serum response element *in vivo* by a multi-protein complex is unaltered by growth factor induction. *Nature*, **340**, 68.

196. Gille, H., Sharrocks, A. D., and Shaw, P. E. (1992) Phosphorylation of transcription factor p62^{TCF} by MAP kinase stimulates ternary complex formation at *c-fos* promoter. *Nature*, **258**, 414.

197. Macleod, K., Leprince, D., and Stehelin, D. (1992) The *ets* gene family. *TIBS*, **17**, 251.

198. Gotoh, Y., Nishida, E., Matsuda, S., Shiinga, N., Kosako, H., Shiokawa, K. *et al.* (1991)
In vitro effects on microtubule dynamics of purified *Xenopus* M phase-activated MAP kinase. *Nature*, **349**, 251.

199. Shiinga, N., Moriguchi, T., Ohta, K., Gotoh, Y., and Nishida, E. (1992) Regulation of a major microtubule-associated protein by MPF and MAP kinase. *EMBO J.*, **11**, 3977.

200. Drewes, G., Lichtenberg-Kraag, B., Doring, F., Mandelkow, E.-M., Biernat, J., Goris, J. *et al.* (1992) Mitogen activated protein (MAP) kinase transforms tau protein into an Alzheimer-like state. *EMBO J.*, **11**, 2131.

201. Lichtenberg-Kraag, B., Mandelkow, E.-M., Biernat, J., Steiner, B., Schroter, C., Gustke,

N. *et al.* (1992) Phosphorylation-dependent epitopes of neurofilament antibodies on tau protein and relationship with Alzheimer tau. *Proc. Natl. Acad. Sci. USA*, **89**, 5384.

202. Ishiguro, K., Takamatsu, M., Tomizawa, K., Omori, A., Takkahashi, M., Arioka, M. *et al.* (1992) Tau protein kinase I converts normal tau protein into A68-like component of paired helical filaments. *J. Biol. Chem.*, **267**, 10897.

203. Goedert, M., Cohen, F. S., Jakes, R., and Cohen, P. (1992) p42 map kinase phosphorylation sites in microtubule-associated protein tau are dephosphorylated by protein phosphatase 2A$_1$. *FEBS Lett.*, **312**, 95.

204. Vulliet, R., Halloran, S. M., Braun, R. K., Smith, A. J., and Lee, G. (1992) Proline-directed phosphorylation of human tau protein. *J. Biol. Chem.*, **267**, 22570.

205. Hanger, D. P., Hughes, K., Woodgett, J. R., Brion, J.-P., and Anderton, B. H. (1992). Glycogen synthase kinase-3 induces Alzheimer's disease-like phosphorylation of tau: generation of paired helical filament epitopes and neuronal localisation of the kinase. *Neurosci. Letts.*, **147**, 58.

206. Nemenoff, R. A., Winitz, S., Qian, N.-X., Van Putten, V., Johnson, G. L., and Heasley, L. E. (1993) Phosphorylation and activation of a high molecular weight form of phospholipase A$_2$ by p42 microtubule-associated protein 2 kinase and protein kinase C. *J. Biol. Chem.*, **268**, 1960.

207. Lin, L.-L., Wartmann, M., Lin, A. Y., Knopf, J. L., Seth, A., and Davis, R. J. (1993) cPLA$_2$ is phosphorylated and activated by MAP kinase. *Cell*, **72**, 269.

208. Kyriakis, J. M., Brautigan, D. L., Ingebritsen, T. S., and Avruch, J. (1991) pp54 microtubule-associated protein-2 kinase requires both tyrosine and serine/threonine phosphorylation for activity. *J. Biol. Chem.*, **266**, 10043.

209. Payne, D. M., Rossomando, A. J., Martino, P., Erickson, A. K., Her, J., Shabanowitz, J. *et al.* (1991) Identification of the regulatory phosphorylation sites in pp42/mitogen-activated protein kinase (MAP kinase). *EMBO J.*, **10**, 885.

210. Anderson, N. G., Kilgour, E., and Sturgill, T. W. (1991) Activation of mitogen-activated protein kinase in BC$_3$H1 myocytes by fluoroaluminate. *J. Biol. Chem.*, **266**, 10131.

211. Chao, T. O., Byron, K. L., Lee, K., Villereal, M., and Rosner, M. R. (1992) Activation of MAP kinases by calcium-dependent and calcium-independent pathways. Stimulation by thapsigargin and epidermal growth factor. *J. Biol. Chem.*, **267**, 19876.

212. Alessi, D. R., Smythe, C., and Keyse, S. M. (1993) The human CL100 gene encodes a Tyr/Thr-protein phosphatase which potently and specifically inactivates MAP kinase and suppresses its activation by oncogenic ras in *Xenopus* oocyte extracts. *Oncogene*, **8**, 2015.

213. Knighton, D. R., Zheng, J., Ten Eyck, L. F., Ashford, V. A., Xuong, N.-H., Taylor, S. S., and Sowadski, J. M. (1991) Crystal structure of the catalytic subunit of cyclic adenosine monophosphate-dependent protein kinase. *Science*, **253**, 407.

214. Seger, R., Ahn, N. G., Boulton, T. G., Yancopoulos, G. D., Panayotatos, N., Radziejewska, E. *et al.* (1991) Microtubule-associated protein 2 kinases, ERK1 and ERK2, undergo autophosphorylation on both tyrosine and threonine residues: implications for their mechanism of activation. *Proc. Natl. Acad. Sci. USA*, **88**, 6142.

215. Crews, C. M., Alessandrini, A. A., and Erikson, R. L. (1991) Mouse *erk1* gene product is a serine/threonine protein kinase that has the potential to phosphorylate tyrosine. *Proc. Natl. Acad. Sci. USA*, **88**, 8845.

216. Wu, J., Rossomando, A. J., Her, J.-H., Del Vecchio, R., Weber, M. J., and Sturgill,

T. W. (1991) Autophosphorylation *in vitro* of recombinant 42-kilodalton mitogen-activated protein kinase in tyrosine. *Proc. Natl. Acad. Sci. USA*, **88**, 9508.

217. Scimeca, J.-C., Ballotti, R., Nguyen, T. T., Filloux, C., and Van Obberghen, E. (1991) Tyrosine and threonine phosphorylation of an immunoaffinity-purified 44-kDa MAP kinase. *Biochemistry*, **30**, 9313.

218. Ahn, N. G., Seger, R., Bratlien, R. L., Diltz, C. D., Tonks, N. K., and Krebs, E. G. (1991) Multiple components in an epidermal growth factor-stimulated protein kinase cascade. *In vitro* activation of a myelin basic protein/microtubule-associated protein 2 kinase. *J. Biol. Chem.*, **266**, 4220.

219. Gomez, N. and Cohen, P. (1991) Dissection of the protein kinase cascade by which nerve growth factor activates MAP kinases. *Nature*, **353**, 170.

220. Ettehadieh, E., Sanghera, J. S., Pelech, S. J., Hess-Bienz, D., Watts, J., Shastri, N., and Aebersold, R. (1991) Tyrosyl phosphorylation and activation of MAP kinase by p56[lck]. *Science*, **265**, 863.

221. Adams, P. D. and Parker, P. J. (1992) Activation of mitogen-activated protein (MAP) kinase by a MAP kinase-kinase. *J. Biol. Chem.*, **267**, 13135.

222. Nakielny, S., Cohen, P., Wu, J., and Sturgill, T. (1992) MAP kinase activator from insulin-stimulated skeletal muscle is a protein threonine-tyrosine kinase. *EMBO J.*, **11**, 2123.

223. Rossomando, A., Wu, J., Weber, M. J., and Sturgill, T. W. (1992) The phorbol ester-dependent activator of the mitogen-activated protein kinase p42[mapk] is a kinase with specificity for the threonine and tyrosine regulatory sites. *Proc. Natl. Acad. Sci. USA*, **89**, 5221.

224. Alessandrini, A. A., Crews, C. M., and Erikson, R. L. (1992) Phorbol ester stimulates a protein-tyrosine/threonine kinase that phosphorylates and activates the *Erk-1* gene product. *Proc. Natl. Acad. Sci. USA*, **89**, 8200.

225. Crews, C. M. and Erikson, R. L. (1992) Purification of a murine protein-tyrosine/threonine kinase that phosphorylates and activates the *ERK-1* gene product: relationship to the fission yeast *bry1* gene product. *Proc. Natl. Acad. Sci. USA*, **89**, 8205.

226. Posada, J. and Cooper, J. A. (1992) Requirements for phosphorylation of MAP kinase during meiosis in *Xenopus* oocytes. *Science*, **255**, 212.

227. L'Allemain, G., Her, J.-H., Wu, J., Sturgill, T. W., and Weber, M. J. (1992) Growth factor-induced activation of a kinase activity which causes regulatory phosphorylation of p42/microtubule-associated protein kinase. *Mol. Cell. Biol.*, **12**, 2223.

228. Shirakabe, K., Gotoh, Y., and Nishida, E. (1992) A mitogen-activated protein (MAP) kinase activating factor in mammalian mitogen-stimulated cells is homologous to *Xenopus* M phase MAP kinase activator. *J. Biol. Chem.*, **267**, 16685.

229. Seger, R., Ahn, N. G., Posada, J., Munar, A. S., Jensen, A., Cooper, J. A. *et al.* (1992) Purification and characterization of mitogen-activated protein kinase activator(s) from epidermal growth factor-stimulated A431 cells. *J. Biol. Chem.*, **267**, 14373.

230. Matsuda, S., Kosako, H., Takenaka, K., Moriyama, K., Sakai, H., Akyama, T. (1992) *Xenopus* MAP kinase activator: identification and function of a key intermediate in the phosphorylation cascade. *EMBO J.*, **11**, 973.

231. Nakielny, S., Campbell, D. G., and Cohen, P. (1992) MAP kinase kinase from rabbit skeletal muscle. *FEBS Lett.*, **310**, 41.

232. Haystead, T. A. J., Dent, P., Wu, J., Haystead, C. M. M., and Sturgill, T. W. (1992) Ordered phosphorylation of p42[mapk] by MAP kinase kinase. *FEBS Lett.*, **306**, 17.

233. Robbins, D. J. and Cobb, M. H. (1992) Extracellular signal-regulated kinases autophos-

phorylate on a subset of peptides phosphorylated in intact cells in response to insulin and nerve growth factor: analysis by peptide mapping. *Mol. Biol. Cell*, **3**, 299.

234. Crews, C., Alessandrini, A. A., and Erikson, R. L. (1992) The primary structure of MEK, a protein kinase that phosphorylates the *ERK* gene product. *Science*, **258**, 478.

235. Seger, R., Seger, D., Lozeman, F. J., Ahn, N. G., Graves, L. M., Campbell, J. S. *et al.* (1992) Human T-cell MAP kinase-kinases that are related to yeast signal transduction kinases. *J. Biol. Chem.*, **267**, 25628.

236. Wu, J., Harrison, J. K., Vincent, L. A., Haystead, C., Haystead, T. A. J., Michel, H. *et al.* (1993) Molecular structure of a protein-tyrosine/threonine kinase activating p42 mitogen-activated protein (MAP) kinase: MAP kinase kinase. *Proc. Natl. Acad. Sci. USA*, **90**, 173.

237. Ashworth, A., Nakielny, S., Cohen, P., and Marshall, C. (1992) The amino acid sequence of a mammalian MAP kinase kinase. *Oncogene*, **7**, 2555.

238. Neiman, A. M., Stevenson, B. J., Xu, H.-P., Sprague, G. F., Herskowitz, I., Wigler, M., and Marcus, S. (1993) Functional homology of protein kinases required for sexual differentiation in *Schizosaccharomyces pombe* and *Saccharomyces cerevisiae* suggests a conserved signal transduction module in eukaryotic organisms. *Mol. Biol. Cell*, **4**, 107.

239. Kosako, H., Gotoh, Y., Matsuda, S., Ishikawa, M., and Nishida, E. (1992) *Xenopus* MAP kinase activator is a serine/threonine/tyrosine kinase activated by threonine phosphorylation. *EMBO J.*, **11**, 2903.

240. Ahn, N. G., Seger, R., and Krebs, E. G. (1992) The mitogen-activated protein kinase activator. *Current Opinions in Cell Biol.*, **4**, 992.

241. Dent, P., Haser, W., Haystead, T. A. J., Vincent, L. A., Roberts, T. M., and Sturgill, T. W. (1992) Activation of mitogen-activated protein kinase kinase by v-Raf in NIH 3T3 cells and *in vitro*. *Science*, **257**, 1404.

242. Howe, L. R., Leevers, S. J., Gomez, N., Nakielny, S., Cohen, P., and Marshall, C. J. (1992) Activation of the MAP kinase pathway by the protein kinase raf. *Cell*, **71**, 335.

243. Huang, W., Alessandrini, A., Crews, C. M., and Erikson, R. L. (1993) Raf-1 forms a stable complex with MEK1 and activates MEK1 by serine phosphorylation. (1993) *Proc. Natl. Acad. Sci. USA*, **90**, 10947.

244. Wood, K. W., Sarnecki, C., Roberts, T. M., and Blenis, J. (1992) *ras* mediates nerve growth factor receptor modulation of three signal-transducing protein kinases: MAP kinase, Raf-1 and RSK. *Cell*, **68**, 1041.

245. Gupta, S., Gallego, C., Johnson, G. L., and Heasley, L. E. (1992) MAP kinase is constitutively activated in *gip2* and *src* transformed rat 1a fibroblasts. *J. Biol. Chem.*, **267**, 7987.

246. Gallego, C., Gupta, S. K., Heasley, L. E., Quian, N. X., and Johnson, G. L. (1992) Mitogen-activated protein kinase activation resulting from selective oncogenic expression in NIH3T3 and rat 1a cells. *Proc. Natl. Acad. Sci. USA*, **89**, 7355.

247. Stevenson, B. J., Rhodes, N., Errede, B., and Sprague, G. F. Jr. (1992) Constitutive mutants of the protein kinase STE11 activates the yeast pheromone response pathway in the absence of the G protein. *Genes Dev.*, **6**, 1293.

248. Cairns, B. R., Ramer, S. W., and Kornberg, R. D. (1992) Order of action of components in the yeast pheromone response pathway revealed with a dominant allele of the STE11 kinase and multiple phosphorylation of the STE7 kinase. *Genes Dev.*, **6**, 1305.

249. Lange-Carter, C. A., Pleiman, C. M., Gardner, A. M., Blumer, K. J., and Johnson, G.

L. (1993) A divergence in the MAP kinase regulatory network defined by MEK kinase and Raf. *Science*, **260**, 315.

250. Barbacid, M. (1987) *ras* genes. *Ann. Rev. Biochem.*, **56**, 779.

251. Hall, A. (1992) Signal transduction through small GTPases — a tale of two GAPs. *Cell*, **69**, 389.

252. Kaziro, Y., Itoh, H., Kozasa, T., Nakafuka, M., and Satoh, T. (1991) Structure and function of signal-transducing GTP-binding proteins. *Ann. Rev. Biochem.*, **60**, 349.

253. Satoh, T., Nakafuku, M., and Kaziro, Y. (1992) Function of Ras as a molecular switch in signal transduction. *J. Biol. Chem.*, **267**, 24149.

254. Leevers, S. and Marshall, C. J. (1992) Activation of extracellular signal-regulated kinase, ERK2, by p21^{ras} oncoprotein. *EMBO J.*, **11**, 569.

255. Pomerance, M., Schweighoffer, F., Tocque, B., and Pierre, M. (1992) Stimulation of mitogen-activated protein kinase by oncogenic Ras p21 in *Xenopus* oocytes. *J. Biol. Chem.*, **251**, 16155.

256. Nori, M., L'Allemain, G., and Weber, M. J. (1992) Regulation of tetradecanoyl phorbol acetate-induced responses in NIH 3T3 cells by GAP, the GTPase-activating protein associated with p21$^{c\text{-}ras}$. *Mol. Cell. Biol.*, **12**, 936.

257. Thomas, S. M., DeMarco, M., D'Arcangelo, G., Halegoua, S., and Brugge, J. S. (1992) Ras is essential for nerve growth factor- and phorbol ester-induced tyrosine phosphorylation of MAP kinases. *Cell*, **68**, 1031.

258. Wood, K. W., Sarnecki, C., Roberts, T. M., and Blenis, J. (1992) *ras* mediates nerve growth factor receptor modulation of three signal-transducing protein kinases: MAP kinase, Raf-1, and RSK. *Cell*, **68**, 1041.

259. de Vries-Smits, A. M. M., Th. Burgering, B. M., Leevers, S. J., Marshall, C. J., and Bos, J. L. (1992) Involvement of p21^{ras} in activation of extracellular signal-regulated kinase 2. *Nature*, **357**, 602.

260. Robbins, D. J., Cheng, M., Zhen, E., Vanderbilt, C. A., Feig, L. A., and Cobb, M. H. (1992) Evidence for a Ras-dependent extracellular signal-regulated protein kinase (ERK) cascade. *Proc. Natl. Acad. Sci. USA*, **89**, 6924.

261. Hattori, S., Fukuda, M., Yamashita, T., Nakamura, S., Gotoh, Y., and Nishida, E. (1992) Activation of mitogen-activated protein kinase and its activator by *ras* in intact cells and in a cell-free system. *J. Biol. Chem.*, **267**, 20346.

262. Shibuya, E. K., Polverino A. J., Chang, E., Wigler, M., and Ruderman, J. V. (1992) Oncogenic Ras triggers the activation of 42-kDa mitogen-activated protein kinase in extracts of quiescent *Xenopus* oocytes. *Proc. Natl. Acad. Sci. USA*, **89**, 9831.

263. Morrison, D. K., Kaplan, D. R., Rapp, U., and Roberts, T. M. (1988) Signal transduction from membrane to cytoplasm: growth factors and membrane-bound oncogene products increase Raf-1 phosphorylation and associated protein kinase activity. *Proc. Natl. Acad. Sci. USA*, **85**, 8855.

264. Kolch, W., Heidecker, G., Lloyd, P., and Rapp, U. R. (1991) Raf-1 protein is required for growth of induced NIH/3T3 cells. *Nature*, **349**, 426.

265. Bruder, J. T., Heidecker, G., and Rapp, U. R. (1992) Serum-, TPA-, and Ras-induced expression from Ap-1/Ets-driven promoters requires Raf-1 kinase. *Genes & Dev.*, **6**, 545.

266. Dickson, B., Sprenger, F., Morrison, D., and Hafen, E. (1992) Raf functions downstream in the Ras1 Sevenless signal transduction pathway. *Nature*, **360**, 600.

267. Simon, M. A., Bowtell, D. D. L., Dodson, G. S., Laverty, T. R., and Rubin, G. M. (1992) Ras1 and a putative guanine nucleotide exchange factor perform crucial steps in signalling by the sevenless protein tyrosine kinase. *Cell*, **67**, 701.

268. Gaul, U., Mardon, G., and Rubin, G. M. (1992) A putative Ras GTPase activating protein acts as a negative regulator of signalling by the sevenless receptor tyrosine kinase. *Cell*, **68**, 1007.

269. Ambrosio, L., Mahowald, A. P., and Perrimon, N. (1992) Requirement of the *Drosophila raf* homologue for *torso* function. *Nature*, **342**, 288.

270. Steinberg, P. W. and Horvitz, H. R. (1991) Signal transduction during *C. elegans* vulval induction. *TIG*, **7**, 366.

271. Moodie, S. A., Willumsen, B. M., Weber, M. J., and Wolfman, A. (1993) Complexes of Ras·GTP with Raf-1 and mitogen-activated protein kinase kinase. *Science*, **260**, 1658.

272. Zhang, X.-F., Settleman, J., Kyriakis, J. M., Takeuchi-Suzuki, E., Elledge, S. J., Marshall, M. S. *et al.* (1993) Normal and oncogenic p21^{ras} proteins bind to the amino-terminal regulatory domain of c-Raf-1. *Nature*, **364**, 308.

273. Van Elst, L. V., Barr, M., Marcus, S., Polverino, A., and Wigler, M. (1993) Complex formation between RAS and RAF and other protein kinases. *Proc. Natl. Acad. Sci. USA*, **90**, 6213.

274. Vojtek, A. B., Hollenberg, S. M., and Cooper, J. A. (1993) Mammalian Ras interacts directly with the serine/threonine kinase Raf. *Cell*, **74**, 205.

275. Rozakis-Adcock, M., McGlade, J., Mbamalu, G., Pelicci, D., Daly, G., Li, W. *et al.* (1993) Association of the Shc and Grb2/Sem5-containing proteins is implicated in activation of the Ras pathway by tyrosine kinases. *Nature*, **360**, 689.

276. Pelicci, G., Lanfrancone, L., Grignani, F., McGlade, J., Cavallo, F., Forni, G. *et al.* (1992) A novel transforming protein (SHC) with an SH2 domain is implicated in mitogenic signal transduction. *Cell*, **70**, 93.

277. Lowenstein, E. J., Daly, R. J., Batzer, A. G., Li, W., Margolis, B., Lammers, R. *et al.* (1992) The SH2 and SH3 domain-containing protein GRB2 links receptor tyrosine kinases to ras signaling. *Cell*, **70**, 431.

278. Clark, S. G., Stern, M. J., and Horvitz, R. (1992) *C. elegans* cell-signalling gene *sem-5* encodes a protein with SH2 and SH3 domains. *Nature*, **356**, 340.

279. Duchesne, M., Schweighoffer, F., Parker, F., Clerc, F., Frobert, Y., Thang, M. N., and Tocque, B. (1992) Identification of the SH3 domain of GAP as an essential sequence for Ras-GAP-mediated signaling. *Science*, **259**, 525.

5 | Cyclin-dependent protein kinases

KATHLEEN L. GOULD

1. Introduction

Cyclin-dependent protein kinases (Cdks) are regulators of the eukaryotic cell cycle. This role was first established through genetic studies in yeast. Subsequently, it was recognized that Cdks regulate cell division in all eukaryotic cells. Cdks are the catalytic subunits of a heterodimeric complex. They are approximately the size of a minimal protein kinase catalytic domain, 34 kDa, and alone, they are inactive. Activation requires association with their partner, a member of the cyclin family of proteins. Cyclins were so named because their abundance varies acutely during the cell cycle. The cyclin subunits are more diverse in size and structure than are the catalytic subunits, and cyclins have a variety of distinct regulatory roles.

Each Cdk is activated transiently at a particular stage in the cell cycle, and this activation is thought to trigger the next events in the cell cycle. To ensure the coordination of these events, activation of Cdks is strictly regulated. The regulatory mechanisms governing Cdk activation are intricate and not fully understood, but clearly involve an elaborate cascade of protein phosphorylation and dephosphorylation events.

By far, the best understood of the Cdk catalytic subunits is Cdc2, and general aspects of Cdc2 structure, regulation and substrate selection will be the focus of this chapter. To illustrate differences and similarities, other Cdks will be briefly discussed. Due to the wealth of knowledge concerning Cdks and the limitations of space, it is not possible to include all the detailed information about Cdks in this chapter or to cite all of the original references. Numerous recent reviews emphasize various aspects of Cdk function and readers are referred to these where appropriate.

2. Discovery of Cdks

Catalytic subunits of Cdks were first identified in the budding yeast, *Saccharomyces cerevisiae*, and the fission yeast, *Schizosaccharomyces pombe*, using genetic screens for conditional lethal mutants unable to progress through the cell cycle at a high temperature (reviewed in 1, 2). Such mutants are called CDC, for c̲ell d̲ivision cycle. CDC mutants arrest at the particular point in the cell cycle where the mutant gene product normally functions. Cell growth is not impaired in these mutants and they can become quite large. There are nearly 50 CDC mutants from each yeast whose arrest points are scattered through every stage of the cell cycle (G1, S, G2 or M).

2.1 Nomenclatures

Several CDC mutants identify homologous gene functions in the two yeasts. However, since the CDC mutants were numbered in the order in which they were isolated in each genetic screen, genes with the same number in both yeasts (i.e. *S. cerevisiae CDC2* and *S. pombe cdc2*) are not related and, conversely, related genes have different numbers (i.e. *CDC28* and *cdc2*). A further confusion in the terminology of cell cycle genes arises because of the different conventions for naming genes in the two yeast systems (discussed in 1). Wild type genes in *S. pombe* are written in lower case italicized letters (i.e. *cdc2*) whereas wild type genes in *S. cerevisiae* are written in upper case italicized letters (i.e. *CDC28*). The protein products of the genes are indicated by non-italicized letters, often with the first letter capitalized for *S. pombe* proteins (i.e. Cdc2 and CDC28).

2.2 A brief history of Cdc2

The study of CDC mutants in yeast helped to identify two major control points in the eukaryotic cell cycle (reviewed in 1, 2). The first operates prior to S phase, the time during which DNA is replicated, and is called 'START'. The second controls the onset of M phase, the stage of nuclear division. The *cdc2* gene attracted considerable interest when it was found to be essential for passage through both major cell cycle transition points in *S. pombe* (reviewed in 3). When it was sequenced, it was found to be very similar to the *S. cerevisiae CDC28* gene. CDC28 was first found to be essential for transition through START, and subsequently for entry into mitosis as well (reviewed in 4). The predicted protein sequences of both Cdc2 and CDC28 indicated that they function as protein kinases. By expressing the CDC28 protein in *S. pombe* strains deficient for Cdc2, it was found that *CDC28* and *cdc2* are functional homologues (5). That is to say, the *CDC28* gene product could rescue temperature-sensitive mutations in *cdc2*. As *S. cerevisiae* and *S. pombe* are evolutionarily quite divergent, this result was the first indication that cell cycle control proteins might be similar in all eukaryotes. The pivotal role of Cdc2 in regulating the cell cycle of all eukaryotes was fully recognized when a functional homologue of CDC28/Cdc2 (hereafter referred to as Cdc2) was isolated from a human cDNA library through a functional complementation approach (6). Functional homologues have now been identified in many other eukaryotic species (reviewed in 7).

The consolidation of several lines of cell cycle research occurred when it was found that Cdc2 was a component of maturation promoting factor (MPF) (reviewed in 8). MPF was originally defined as an activity present in the cytosol of mature

Xenopus laevis eggs which, when injected into immature *Xenopus* oocytes, would promote the process of maturation (cell cycle transit from prophase of meiosis I through to metaphase of meiosis II) in the absence of protein synthesis (reviewed in 9). Hence, MPF activity is all that is required to induce resumption of meiosis. Highly purified preparations of MPF possess protein-serine/threonine kinase activity and are composed of two proteins. One is Cdc2 and the other is a member of the cyclin family of proteins (reviewed in 9, 10). Just as all eukaryotes possess a Cdc2 homologue, all eukaryotic cells in mitosis possess MPF activity. MPF is now more broadly defined as M phase promoting factor since it will induce mitosis as well as meiosis.

2.3 A family of Cdks

The protein kinase activity associated with Cdc2 is sharply periodic through the cell cycle, peaking as cells enter mitosis and falling away again as cells exit mitosis. This activity profile matches the timing of Cdc2 function in yeast, determined genetically, as well as the appearance and disappearance of MPF activity in higher eukaryotes. Thus, the role of Cdc2 as a general eukaryotic M-phase promoting factor is well established (reviewed in 11). However, in higher eukaryotic cells, Cdc2 does not appear to have a major role in G1 as it does in yeast (reviewed in 7, 12). Several lines of evidence led to this conclusion. For example, injection of neutralizing antibodies raised against Cdc2 into fibroblasts does not prevent DNA replication although it prevents entry into mitosis (13). Similarly, depleting Cdc2 from frog egg extracts inhibits M phase but not DNA replication (14). Also, a temperature-sensitive Cdc2 mutation identified in a hamster cell line arrests cells in G2, as expected, but does not prevent DNA replication (15).

The above results are now fully appreciated in light of the identification in higher eukaryotic cells of a number of Cdc2-related protein kinases (reviewed in 7, 12). Many of these new family members, if not all, require a cyclin subunit for activity, and thus have been designated Cdks (Table 1). The DNA clone for the first member of the growing family, Cdk2, was identified serendipitously in frogs because its mRNA was deadenylylated upon fertilization of eggs (16). It was subsequently identified from human cells based on its ability to complement temperature-sensitive Cdc2 mutations in yeast (17, 18) and because of its structural similarity with Cdc2 (19). Much is now known about the function of Cdk2. When it is depleted from frog cell-free extracts using antibodies specific for it, the extracts can no longer replicate DNA (14). However, mitosis in the extracts is not impaired, indicating that Cdk2 is not required for entry into mitosis. The requirement of Cdk2 activity for DNA replication has also been demonstrated in mammalian cells. Microinjection of anti-Cdk2 antibodies into fibroblasts inhibits DNA replication (20, 21). These experiments suggest that in higher eukaryotic cells, the functions performed by a single enzyme in yeast have been delegated amongst at least two and probably more enzymes.

Other Cdk family members have been discovered by homology-based screening

---

Table 1 Nomenclature of Cdks and cyclins

Cdks

Species	Nomenclature
Multicellular organisms	Cdc2
	Cdk2
	Cdk3
	Cdk4
	Cdk5
S. cerevisiae	CDC28
S. pombe	Cdc2

Cyclins

Species	Nomenclature	
	G1	Mitotic
Multicellular organisms	Cyclin C	Cyclin A
	Cyclins D1, D2, D3	Cyclin B1, B2
	Cyclin E	
S. cerevisiae	CLNs 1, 2, 3	CLBs 1, 2, 3..–6
	HCS26	
S. pombe	Puc1	Cdc13
		Cig 1
		Cig 2

---

techniques (22–27). At the time of writing, five *bona fide* Cdks are known (including Cdc2) with probably more to be uncovered. Little is yet known about the function or regulation of the more recently identified Cdks. Cdk3, like Cdk2, is able to rescue temperature-sensitive mutations in yeast Cdc2 whereas Cdks 4 and 5 cannot (26). Both Cdk4 and Cdk5 are proposed to function in the early stages of G1 (see below).

With these many additional family members, it can begin to look as if the complexity of the system might elude understanding for some time to come. However, an important point to bear in mind is that not all the Cdc2-related protein kinases are likely to function in cell cycle control. For example, the *S. cerevisiae PHO85* gene product is highly related to Cdc2 (51% identical) and yet is involved in regulating phosphate metabolism in yeast, not the cell cycle (28). The paradigm of a small catalytic subunit activated by binding to a transiently produced regulatory subunit could be widespread in intracellular signalling pathways.

3. Cyclins

Cyclins were first identified in sea urchin eggs as proteins which accumulated during interphase and were specifically degraded during mitosis (reviewed in 9,

29, 30). Two subclasses of cyclins (A and B) were initially recognized based on slightly different kinetics of accumulation and destruction. Cyclin B accumulates in somatic cells during G2 and is degraded during mitosis whereas cyclin A begins to accumulate much earlier (during S phase) and is degraded slightly before cyclin B (31). Cyclin A, in fact, is required for S phase in addition to its requirement for mitosis (32–34). Cyclin B, like Cdc2, has been conserved throughout eukaryotic evolution and, as mentioned above, is the second recognized component of MPF (35, 36). In general, however, there is considerably less conservation in sequence between cyclins of yeast and higher eukaryotes (and between the *S. cerevisiae* and *S. pombe* cyclins) as compared with Cdks.

3.1 A proliferation of cyclins

The cyclin family now extends well beyond cyclins A and B, both in terms of differential expression during the cell cycle and in sequence diversity (Table 1). In contrast to the 'mitotic cyclins' A and B, other cyclins accumulate in G1 and persist for different periods of the cell cycle (reviewed in 4, 37, 38). The first G1 cyclins were identified in *S. cerevisiae* and are called CLNs and HCS26 (39–41). In meta-zoans, G1 cyclins have been classified into C, D, and E categories because their sequences are significantly different from the CLNs (42–47). Only one G1 cyclin gene has been reported in *S. pombe* (*puc1*) and it appears to fall into a distinct category based on its sequence (48). Although there does not appear to be a cognate of cyclin A in the yeasts, there are B-type cyclins in *S. cerevisiae* (CLB 3 and 4) which have similar kinetics of expression to that of cyclin A, and might perform a similar function (reviewed in 4).

In addition to the numerous categories of cyclins, there are multiple members of most cyclin types. For example, there are several different cyclin Bs, Ds, and CLNs and they are distinguished from one another with numbers following their desig-nations, i.e. B1, B2. In *S. pombe*, there are three known cyclin B-related proteins encoded by the *cdc13* (49, 50), *cig1* (51, 52) and *cig2* (53) genes. In *S. cerevisiae*, there are at least six cyclins which are related to cyclin B and are termed CLBs (4, 54–56). There is evidence in every organism to indicate that individual members of the cyclin classes, both G1 and mitotic, are functionally distinct. For example, each cyclin D has its own pattern of cell cycle and tissue expression (reviewed in 38). Of the four mitotic CLBs, the loss of CLB1 function has the most profound effect on the process of meiosis (57).

All cyclins have a stretch of ~150 amino acids which is similar in sequence amongst all members of the family and is termed the 'cyclin box' (29). New mem-bers are now added to the cyclin family based on the presence of a cyclin box, and certain of the cyclins do not actually change in abundance during the cell cycle. Not surprisingly, the cyclin box appears to be the primary region through which cyclins bind Cdks (58, 59). Beyond the cyclin box, the primary sequences of the various cyclins diverge considerably. Despite this, there does appear to be a degree of functional conservation between all cyclins. *S. cerevisiae* cells are viable in the absence of any two but not three of CLNs 1–3 suggesting that some aspect of CLN function is redundant (60, 61). Moreover, cyclins C, D, and E can functionally substitute for the absence of CLNs 1–3 in yeast when overexpressed (42–45). That is, in fact, one method by which they were identified. Even the mitotic cyclins, when overexpressed, can substitute for the G1 cyclins in yeast (42, 45).

3.2 Specificities of interaction

With the various cyclins and the various CDKs, the possibilities for combinatorial control appear endless. However, there appear to be some limitations in the selection of binding partners. In vertebrate cells, cyclin B is complexed primarily to Cdc2 (62). Cdc2 also associates better with cyclin B than cyclin A *in vitro* (63). Cdk2 is more promiscuous and forms complexes with cyclin A (31, 63–65), cyclin Ds (66) and cyclin E (67, 68). However, these complexes appear to be temporally distinct. The Cdk2/cyclin D complexes are found earliest, followed by Cdk2/cyclin E in late G1 and, finally, Cdk2 complexes primarily with cyclin A during S phase (12, 38). While the binding partners of Cdk3 and cyclin C are not known, Cdk4 has been shown to bind cyclin Ds exclusively (69). An emerging theme is that certain complexes are the predominant activators of different transition points in the cell cycle. The cyclin B/Cdc2 complex acts as the primary promoter of mitosis while other Cdk complexes are responsible for triggering events earlier in the cell cycle.

4. Other subunits
4.1 Suc1 and CSK1

Cdks can be found in complexes containing proteins other than cyclins. A protein which has long been known to bind certain Cdks was first identified in *S. pombe* as the *suc1* gene product, p13 (70). This small protein complemented only certain temperature-sensitive mutations in Cdc2, suggesting that it formed a protein: protein complex with Cdc2 (71), and this notion was borne out in biochemical studies (72). The *S. cerevisiae* homologue of Suc1, CSK1, was identified in the same way (73). Despite many studies investigating its role in the yeast cell cycle, the function of Suc1 has remained an enigma. When it is produced in bacteria and bound to a solid matrix, it exhibits a strong and selective affinity for Cdc2 (72), Cdk2 (26, 74), and Cdk3 (26). Thus, it has been used extensively in cell cycle studies as an affinity reagent to isolate Cdk complexes. Cdk complexes bound to Suc1 beads exhibit protein kinase activity, and Suc1 binding does not interfere with cyclin binding.

The *suc1* and *CSK1* genes are essential (73, 75), and there is evidence from studies in both yeast systems that their function is required more than once in the cell cycle. In *S. pombe*, cells lacking Suc1 arrest in a mitotic state with high Cdc2 kinase activity, mitotic spindles and condensed chromosomes (76). In contrast, overexpression of Suc1 causes a G2 arrest (70). These results suggest that Suc1 functions before mitosis and also at the end of mitosis. Similar experiments performed in *S. cerevisiae* with the CSK1 gene have suggested that it functions at more than one point in the cell cycle (77). Presumably, Suc1/CSK1 act through association with Cdks but it has been difficult to demonstrate association between endogenous Suc1 and Cdks (70, 78). Homologues of Suc1/CSK1 have been identified in many eukaryotes (78, 79). Injection of either antibodies raised against Suc1 or Suc1 itself causes mitotic abnormalities, but the mechanism of action is not understood (13).

4.2 Retinoblastoma protein and p107

Some Cdks are known to associate with yet other proteins during the G1 and S phases. Although there is certainly more to be learned about the formation and function of these complexes, there are several intriguing results which provide clues to the regulation of the G1/S phase transition. For example, *in vitro*, cyclins D2 and D3 associate stably with the retinoblastoma protein, pRb, and the pRb-related protein p107, and can activate Cdk4 to phosphorylate pRb (69, 80, 81). The hypophosphorylated form of pRb inhibits the G1/S transition (reviewed in 82). Cdk2 and cyclins A and E have also been found in complexes which contain pRb or p107 (82). The mechanism whereby pRb and p107 influence the G1/S transition is not completely clear but probably involves their association with the transcription factor E2F and related proteins (reviewed in 83). E2F promotes transcription of genes required for S phase. Known targets include c-myc (84, 85) and the gene encoding dihydrofolate reductase (86). It has been proposed that phosphorylation of pRb by Cdks disrupts the E2F:pRb complex, freeing E2F to promote transcription (reviewed in 82).

4.3 Other associations

Cyclin Ds also interact with proliferating cell nuclear antigen (PCNA), a subunit of ∂ DNA polymerase, and Cdk4 associates with an unknown 21 kDa protein (66). The significance of these associations has not been elucidated. In yeast, an uncharacterized 43 kDa protein has been reported to bind Cdc2 and inhibit its protein kinase activity towards the exogenous substrate, histone H1 (87).

5. Regulation

The activities of the Cdk/cyclin complexes are periodic through the cell cycle. This periodicity is achieved by regulation at many levels. Strict regulation is required to coordinate the many events of the cell cycle, making sure that they occur in the proper order. In particular, the integrity of the DNA must be maintained by ensuring the coupling of S phase with mitosis.

5.1 Transcriptional and translational control

In yeast, although Cdc2 activity fluctuates, the amount of Cdc2 mRNA and protein are constant through the cell cycle and during exit from the cell cycle (88–90). In contrast, in animal cells, Cdc2 mRNA and protein levels can vary depending on the growth state. In general, Cdc2 levels are very low in non-proliferating tissues such as brain (91), and are undetectable in senescent human diploid fibroblasts (92). In fibroblasts which have stopped dividing due to serum deprivation, Cdc2 mRNA and protein are also undetectable (93–95). When these cells are stimulated to re-enter the cycle, both Cdc2 mRNA and protein are synthesized at the beginning of S phase (93–95). Cdk2 mRNA and protein levels also increase as quiescent cells re-enter the cycle (64, 65). These increases occur significantly earlier than for Cdc2 and, unlike Cdc2, Cdk2 is detectable at low levels in quiescent cells (64, 65). This profile is consistent with the proposed timing of Cdk2 function in the G1 and S phases. In populations of cycling animal cells, Cdc2 levels are constant but synthesis is periodic (96). The mechanism by which constant levels of Cdc2 are achieved given different synthetic rates has not been established.

5.2 Regulation by phosphorylation

5.2.1 Activatory phosphorylation

The predominant modes of Cdc2 regulation are the binding of a cyclin partner and accompanying phosphorylation and dephosphorylation events (reviewed in 97). As mentioned above, monomeric Cdc2 is inactive. Binding to a cyclin subunit is essential for Cdc2 activity (63, 98), but the complex must also be phosphorylated on a conserved threonine residue at position 161 or 167, depending on the species, to become active (99–101). Phosphorylation at this residue is mediated by a protein kinase termed CAK for 'Cdk activating kinase' (102). Little is currently known about CAK specificity or regulation, but it is certain to be studied in great detail in the coming years. Cdk2 is also activated by CAK (103). However, this phosphorylation, although stimulatory, is not critical for Cdk2/cyclin A activity *in vitro* as it is for Cdc2/cyclin B activity (103). CAK phosphorylation of Cdc2 and Cdk2 is cell-cycle regulated. Phosphorylation increases during S and G2 and falls off as cells complete mitosis (98, 100, 101, 104).

5.2.2 Inhibitory phosphorylation

The activity of Cdc2 complexed with cyclins A or B is restrained by phosphorylation at two residues, Thr14 and Tyr15, that lie within a domain involved in binding ATP known as the glycine loop (98, 101, 104, 105). These phosphorylation events inhibit the activity of the enzyme but do not block ATP binding *in vitro*. Efficient phosphorylation of these sites *in vitro* depends upon cyclin binding (98, 106, 107), and the timing of these phosphorylation events coincides with that of CAK-dependent phosphorylation. Thus, Cdc2 and Cdk2 achieve the necessary pre-requisites for activation (cyclin binding and CAK phosphorylation) but are held in

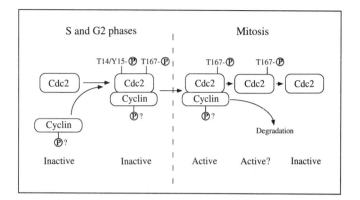

Fig. 1 Model of Cdc2 activation and inactivation cycle mediated by association with cyclin and phosphorylation events.

an inactive state by phosphorylation at Thr14 and Tyr15. Activation of the complexes at the appropriate time in the cell cycle is then accomplished by dephosphorylation of Thr14/Tyr15 (Fig. 1). This mechanism is apparently utilized for the abrupt activation of the Cdc2/cyclin B complex at the onset of mitosis. In contrast, inhibition of cyclin A complexes by tyrosine phosphorylation is not complete. These complexes are active to some extent throughout S phase and G2, and their activities rise and fall more gradually (31, 108–110).

The timing of Cdc2/cyclin B activation and entry into mitosis is controlled by a gene network identified first in *S. pombe* (Fig. 2). Genetic and biochemical analyses have shown that the rate limiting step for activation of Cdc2 in *S. pombe* and vertebrate cells is Thr14/Tyr15 dephosphorylation (97). In *S. pombe*, Thr 14 phosphorylation is not detectable under normal growth conditions and the level of Tyr15 phosphorylation alone controls Cdc2 activity (105). The regulatory gene network identified genetically modulates the phosphorylation state of these sites, and thus the timing of mitosis.

Dephosphorylation and activation of Cdc2 and Cdk2 is mediated by the Cdc25

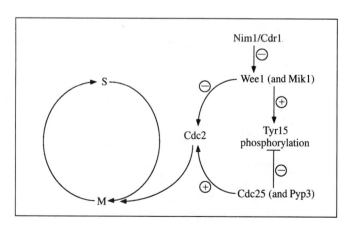

Fig. 2 Model of Cdc2 regulation at the G2/M transition by the Wee1, Mik1, and Nim1/Cdr1 protein kinases, and the Cdc25 protein phosphatase.

protein phosphatase (111–15), first shown genetically to be a dose-dependent activator of mitosis (116). In higher eukaryotic cells, three Cdc25-related proteins have been identified (there will probably be more). They appear to act at different points in the cell cycle and presumably regulate different cell cycle transition points (reviewed in 117). Unlike most known protein phosphatases which are specific for either serine/threonine or tyrosine phosphates, Cdc25 is able to dephosphorylate both Thr14 and Tyr15 (118, 119). Cdc25 activity also appears to be regulated by its phosphorylation state (120–122). The protein kinases and phosphatases which regulate Cdc25 activity are not known. However, Cdc25 can be phosphorylated and activated by Cdc2 *in vitro* (120). In the absence of Cdc25 in *S. pombe*, Tyr15 dephosphorylation can be mediated directly or indirectly by another tyrosine phosphatase termed Pyp3 when it is overexpressed (123). This phosphatase has considerable similarity to low molecular weight protein-tyrosine phosphatases of higher eukaryotes.

The protein-tyrosine kinase primarily responsible for Tyr15 phosphorylation is Wee1 (107, 124). It was identified genetically as a dose-dependent inhibitor of mitosis in *S. pombe* (125). A functional homologue of Wee1 has also been identified in human cells. It remains to be determined whether there are multiple Wee1-related protein kinases in higher eukaryotes (126–129). Wee1 does not phosphorylate Thr14, at least *in vitro* (107), and the identity of the protein kinase responsible for Thr14 phosphorylation is actively being pursued. Mik1 is a *S. pombe* protein kinase similar in structure to Wee1 (130). In the absence of Wee1, Mik1 mediates phosphorylation of Cdc2 at Tyr15 (130). In the absence of both Tyr15 protein kinases, Cdc2 is activated far before the cell is ready to enter mitosis (before DNA replication is complete). This premature mitotic entry is lethal (130). Besides its ability to cooperate with Wee1, little information is currently available regarding Mik1's regulation or substrate specificity. Nim1/Cdr1 was identified genetically as an upstream inhibitor of Wee1 (131, 132), and biochemical evidence supports a direct role for its phosphorylation-mediated inhibition of Wee1 (133–135). Still other protein kinases and phosphatases have been identified genetically as upstream regulators of Cdc2 activation, but their exact biochemical function has yet to be delineated (reviewed in 1, 2). The complexity of this network reflects the importance of the timely activation of Cdc2, which ensures an orderly mitosis, and also the probable role of Cdc2 as a focal point in signal transduction cascades which culminate in the decision to enter or delay mitosis.

One of the most important signals relayed to Cdc2 is the state of the DNA. Activation of Cdc2 and the onset of mitosis are normally prevented in the presence of damaged or unreplicated DNA (reviewed in 136). The prevention of Cdc2 activation in the presence of unreplicated DNA in *S. pombe* and frog extracts is accomplished by maintaining Tyr15 in a phosphorylated state (137, 138). This mechanism could work either by inhibiting Cdc25 activity or activating Wee1 activity. Understanding the precise biochemical signals upstream of Cdc2 inactivation in the presence of damaged and unreplicated DNA is presently an area of intense research.

5.3 Regulation of other Cdks

Cdks do not appear to be phosphorylated at Thr14/Tyr15 to an appreciable level during G1, nor apparently are they phosphorylated to a significant extent by CAK at this time (101, 104). As mentioned above, Cdk2 bound to cyclin A has significant activity in the absence of CAK phosphorylation (103). Although it has not yet been investigated, this is likely to be true of other Cdk/G1 cyclin complexes. Indeed, the available data suggest that Cdks during G1 are regulated primarily by the abundance of cyclins rather than by posttranslational modifications. One of the clearest examples of this is the effect of cyclin E overexpression in mammalian fibroblasts (139). Overexpression of cyclin E shortens the G1 phase of the cell cycle by driving the cells into S phase. Because G1 is abbreviated, these cells have less time to grow (G1 is the primary growth phase), and their average cell size is reduced. This result shows that cyclin E accumulation is rate limiting for the G1/S transition much as tyrosine dephosphorylation of Cdc2 is rate limiting for entry into mitosis. In many tumour cell lines the mRNA levels of cyclins A and/or D are significantly elevated (38). This suggests that deregulation of cyclin A and/or D expression can also alter the G1/S transition controlled by Cdks.

5.4 Switching off the Cdks—cyclin destruction

Inactivation of Cdks is accomplished by disrupting the complex between Cdks and cyclins. Cyclins A and B contain a nine amino acid 'destruction box' near their N-terminus which is the site of ubiquitin conjugation during mitosis (140). Proteolysis of cyclins A and B proceeds through a multi-ubiquitinylated intermediate (140). Truncation of the destruction box results in stable cyclin proteins which are capable of binding and activating Cdc2. These stabilized forms inhibit exit from mitosis (141–144). The N-terminal portion of cyclins A and B, containing the destruction box, will also prevent exit from mitosis when overexpressed in cell-free lysates, presumably by competing for the destruction machinery (145, 146). The nature of the cyclin–ubiquitin conjugating enzyme and the mechanism of its regulation during mitosis are unknown at this time. G1 cyclins, in contrast to cyclins A and B, are inherently short-lived proteins. The half-life of cyclin D1, for example, is only 20–40 min (147, 148). G1 cyclins contain so-called PEST sequences which target proteins for rapid turnover (149). Since the half life of G1 cyclins is unaffected by Cdk binding, their accumulation depends upon an increased rate of synthesis (reviewed in 38).

In addition to cyclin proteolysis, two lines of evidence suggest that dephosphorylation of Cdc2 at Thr167 is necessary for exit from mitosis. First, *S. pombe* cells expressing a Cdc2 mutant in which a glutamic acid replaces Thr167 arrest in a pseudomitotic state characterized by an abundance of mitotic spindles (100). Since Glu mimics, to some extent, a phosphorylated residue, this result suggests that the inability to dephosphorylate Thr167 causes a delay or block in mitosis. Second, inhibition of protein phosphatase 1 in frog extracts by okadaic acid does not prevent cyclin proteolysis but does prevent Cdc2 kinase inactivation (150). Under these circumstances, Cdc2 is phosphorylated on the Thr167 equivalent (150). Thus, monomeric Cdks phosphorylated at Thr167 might be active as protein kinases.

6. Structure

6.1 Interaction domains

As mentioned above, Cdks interact with at least two other proteins, a cyclin and Suc1. To understand which residues are important for these interactions, Cdc2 was subjected to charged to alanine scanning mutagenesis (99, 151). This strategy is based on the notion that clusters of charged residues are likely to be found on the surfaces of proteins rather than in their interior. By altering the charged residues to alanines, any protein–protein interactions involving the charged residues are likely to be disrupted. Charged residues important for cyclin A binding appear to be within the N-terminal half of Cdc2 (99). Within this region is a domain (residues 42–57) conserved in all Cdc2 proteins, the so-called PSTAIRE motif. This motif is conserved exactly in Cdc2, Cdk2, and Cdk3 but is slightly altered in other Cdks. This motif appears to be involved in cyclin A binding since mutations within the PSTAIRE domain prevent cyclin A association (99), and antibodies directed against a peptide corresponding to this region do not co-immunoprecipitate cyclin B (62). Residues important for Suc1 binding were also determined by this method. They mapped to several different regions of the protein including the glycine loop in the N-terminus and near the APE motif in the core of the protein (151).

6.2 Crystal structure of Cdk2

The charged to alanine scanning mutagenesis is most informative when the results can be compared with the three-dimensional structure of the protein. Recently, the crystal structure of Cdk2 was solved within a resolution of 2.4 Å (152). The residues which were implicated by charged to alanine scanning mutagenesis in cyclin A binding are adjacent to each other in the tertiary structure (152). However, residues implicated in Suc1 binding are scattered throughout the tertiary structure and, therefore, it is still unclear where the Suc1 protein binds to Cdc2 (152).

The sites of inhibitory phosphorylation in the tertiary structure are in positions likely to affect either ATP orientation or substrate binding. The CAK phosphorylation site is present on a loop which blocks substrate binding in the apoenzyme (152). Presumably, phosphorylation of this residue and/or cyclin binding would alter the structure to allow access to substrates. In support of this, there are several conserved basic residues located in a region which could bind the phosphorylated threonine, opening the substrate binding site. A similar loop is present in cyclic AMP-dependent protein kinase (see Chapter 1). Indeed, the tertiary structure of Cdk2 is very similar to that of the catalytic subunit of cAMP-dependent protein kinase (152). What differences there are might be attributable to the different catalytic states of the crystallized enzymes, since the structure of the active form of cAMP-dependent protein kinase was obtained (153) whereas the Cdk2 structure represents the inactive and unmodified monomer (152). A full understanding of the structural effects of cyclin binding and regulatory phosphorylation events awaits the crystallization and subsequent structural analysis of other Cdk forms.

7. Localization

The question of Cdc2 localization has been addressed in several systems by indirect immunofluorescent methods as well as cell fractionation experiments (13, 154–157). Unfortunately, many of the earlier studies utilized antibodies raised against the conserved PSTAIRE domain rather than antibodies specific for individual Cdks. Nevertheless, the available evidence suggests that Cdc2 is distributed throughout the cell with a portion in the nucleus and a portion associated with centrosomes (spindle pole bodies in yeast). These structures are likely to contain substrates relevant to an enzyme responsible for engineering entry into mitosis.

What about the whereabouts of proteins which regulate Cdc2? The intracellular localization of the Wee1 and CAK protein kinases is unknown but the Cdc25 protein phosphatase is detected in the nucleus (158, 159). In frog extracts in which nuclear transport has been inhibited, tyrosine dephosphorylation of Cdc2 is prevented, consistent with the observed nuclear distribution of Cdc25 (122).

Cyclins have distinct localization patterns in a variety of organisms (147, 148, 160–162). Cyclin A is detected in the nucleus throughout the cell cycle, and a co-localization with chromatin has been reported (160, 161). Cyclin B, on the other hand, accumulates around the nucleus until prophase of mitosis when it then enters and a portion co-localizes with the mitotic spindle (160–162). Nuclear entry appears to be dependent on phosphorylation of cyclin B which perhaps uncovers a nuclear localization signal. Cdc2 phosphorylates these sites *in vitro*, but the identity of the relevant kinase in cells is unknown. Cyclin D1 is found in the nucleus during G1 but moves out of the nucleus when S phase begins (147, 148). The specificity of cyclin localization compared with that of Cdc2 has led to the suggestion that cyclins target Cdks to particular intracellular sites and substrates. The only direct evidence for this has been obtained in *S. pombe*. Here, the loss of cdc 13 (a cyclin B) function prevents Cdc2 from accumulating in the nucleus (154). A matter of some interest is whether cyclins have intrinsic targeting sequences or whether another protein(s) is (are) responsible for targeting Cdk/cyclin complexes.

8. Substrates
8.1 Important caveats

The task of identifying physiologically relevant substrates for protein-serine/threonine kinases is always a difficult one. In the case of the Cdk family, the job is exceptionally complicated. Although careful kinetic analyses have been done in only a few situations, it is clear that, at least *in vitro*, the substrate specificities of Cdks are very similar (reviewed in 163). When examined carefully, it is clear that the cyclin binding partner can affect the substrate specificity of Cdks *in vitro*, and/or the rate of phosphorylation of a particular substrate (69, 81, 110, 164). Even with careful kinetic analyses, however, it is going to be very difficult to sort out which proteins are targets of which Cdk/cyclin complex. The substrate specificity of each Cdk/cyclin complex might not be unique. A certain amount of redundancy might help to ensure the fidelity of cell cycle transitions.

To date, most substrate studies have concentrated on Cdc2 complexed with cyclin B (as MPF) or with cyclin A. These complexes, like most protein kinases, can phosphorylate a wide assortment of proteins *in vitro* (Table 2). To determine

Table 2 Proposed Cdk substrates

Substrate	References
Structural proteins	
Nuclear lamins	163[a], 172, 174–176
Vimentin	163[a], 170
Caldesmon	163, 171
Desmin	194
Myosin regulatory LC	169
Neurofilament H	163[a]
Chromatin associated proteins	
HMGI/Y, P1	182
Nucleolin, No38	180, 181
Histone H1	163[a]; also reviewed in 177, 178
Kinases and phosphatases	
Casein kinase II (a/b)	163[a], 196
c-Abl	195
c-Src	198, 199
Cdc25	120
Regulatory subunit of cAMP-dependent protein kinase	193
DNA binding proteins	
SWI5	192
c-Myb	185
c-Fos/jun	163[a]
SV40 large T	197
p53	186
RPA	187, 188
Others	
pRb/p107	163[a]
Cyclin B	163[a]
RNA polymerase II	173
EF1-γ	163[a]
Lamin B receptor	163[a]
Rab1/Rab4	163[a]
FAR1	190, 191

[a] Original references are provided within this reference.

whether the substrates are functionally relevant, three minimal criteria must be met. First, the sites phosphorylated by Cdc2 *in vitro* should be identical to those phosphorylated in cells. Second, the level of phosphorylation through the cell cycle should parallel the activity of Cdc2. Third, phosphorylation should alter the protein's function in a manner consistent with its cell cycle function. However, even meeting all of these criteria does not prove that Cdc2 phosphorylates the protein *in vivo*.

One complication is that the substrate specificity of Cdc2 overlaps with that of at least one other protein kinase family, the MAP kinase family. Both classes of protein kinase are considered to be 'proline-directed', and can phosphorylate similar or identical sites in substrate proteins (165, see Chapter 4). This is an especially important consideration in instances in which these other protein kinases are activated at the same time as Cdc2. This occurs, for example, during frog and echinoderm oocyte maturation (166, 167). Thus, although Cdc2 may phosphorylate a particular site *in vitro*, another protein kinase might be responsible for the phosphorylation event in the cell.

A second complication is that although a common recognition sequence for Cdc2 in substrates is S/T–P–X–K/R (163, 168), not all known Cdc2 substrates contain this sequence (Table 3). For example, the regulatory myosin II light chain is phosphorylated by the mitotic form of Cdc2 on serines that lack an adjacent proline (169). Sites phosphorylated by Cdc2 in vimentin (170), caldesmon (171), lamin C (172), and RNA polymerase II (173) have an adjacent proline but not a nearby basic residue. Thus, potential substrates should not be ruled out on the basis that they do not contain this consensus sequence (see Chapter 2 for further discussion of substrate specificities).

8.2 Likely cellular targets of Cdks

8.2.1 Chromatin-associated proteins

Despite the above mentioned complications, several probable *in vivo* substrates for Cdc2 have been identified. Most of these substrates are involved in the structural reorganization of the cell during mitosis and include such proteins as nuclear lamins (172, 174–176), myosin light chain (169), and vimentin (170). For each of these proteins, studies have documented an alteration of function caused by Cdc2 phosphorylation. The activity of Cdks during all phases of the cell cycle is generally measured using histone H1 as an exogenous substrate (see for example 110). Whether histone H1 is a physiologically relevant substrate of Cdks, however, remains unclear (reviewed in 177, 178). It has long been proposed that histone H1 phosphorylation promotes chromosome condensation, but unequivocal results have been elusive, possibly due to the difficulties of working with chromatin. There is also more recent evidence that chromosome condensation can occur in the absence of histone H1 phosphorylation (179).

Unfortunately, a histone H1-like protein has not been identified in yeast, where this problem could be addressed genetically. In any case, histone H1 is a con-

Table 3 Cdc2 phosphorylation sites

Substrate protein	Target sequence	References
c-abl	RAAEQKDAPD**T**PELLHTKGLG	195
	SDALDSEPAV**S**PLLPRKERGP	
Caldesmon	SMWEKGNVFS**S**PGGTGTPNKE	171
	NVFSSPGGTG**T**PNKETAGLKV	
	SSRINEWLTN**T**PEGNKSPARK	
	WLTNTPEGNK**S**PAPKPSDLRP	
Casein kinase IIβ subunit	LQLQAASNFK**S**PVKTIR	196
HMG-I	GSQKEPSEVP**T**PKRPRGRPKG	182, 200
	PLASKQEKDG**T**EKRGRGRPRK	
Nuclear lamins	IRGTPGGTPL**S**PTRISRLQEK	172, 174, 175
	SGAQASSTPL**S**PTRITRLQEK	
	GEEERLRLSP**S**PTSQRSRGRA	
	PTSQRSRGRA**S**HSSQTQGGPT	
Myosin II light chain	.S**S**KKAKTKTTK	169
	.S**S**KKAKTKTTKK	
	.SSKKAKTK**T**TKKRPQRATS	
Nucleolin	AVTPGKKAAA**T**PAKKAVTPAK	180, 181
p53	RAALPNNTSS**S**PQPKKKPLDG	186
pp60[c-src]	THHGGFPASQ**T**PNKTAAPDTH	198, 199
	NKTAAPDTHR**T**PSRSFGTVAT	
	GGFNTSDTVT**S**PQRAGALAGG	
RNA polymerase II	SPTSPSYSPT**S**PSYSPTSPSY	173
SV40 large T	DEATADSQHS**T**PPKKKRKVED	197
Vimentin	PSTSRSLYSS**S**PGGAYVTRSS	170
	TTSTRTYSLG**S**ALRPSTSTRS	
	SPGGAYVTRS**S**AVRLRSSMPG	
Desmin	SQSYS**S**SQRVSSYRRT	194
	SYRRTFGGGT**S**PVFPRASFGS	
	TLSTFRTTRV**T**PLRTYQSAYQ	
CONSENSUS	+1+2+3 **(S/T)**(P) (X) (K/R)	

venient substrate with which to measure Cdk kinase activity, since Cdks do not autophosphorylate and histone H1 is available commercially. Other chromatin-associated and nucleolar proteins are substrates for Cdc2 *in vitro* and are hyper-phosphorylated during mitosis (180, 182–184), possibly contributing to the process of chromosome condensation and nucleolar breakdown, respectively. Several transcription factors are also hyperphosphorylated during mitosis, and can be phosphorylated by Cdc2 *in vitro* (163). It has been proposed that this might be part of a general mechanism of clearing them from DNA to prepare for chromosome condensation (168). This notion has been borne out in the case of c-Myb since phosphorylation of c-Myb by Cdc2 *in vitro* reduces its affinity for DNA (185).

Probable substrates important for cell cycle progression in G1 and S phase have also emerged. They include pRb and the pRb-related protein, p107, in vertebrate cells (reviewed in 163). As discussed above, phosphorylation of pRb is correlated with its release of the cell cycle specific transcription factor, E2F, and the concomitant activation of E2F (82). p53 is also phosphorylated by Cdc2 (186). The consequence of this phosphorylation is not understood but could inhibit p53's affinity for DNA. The replication factor, RPA, has also been reported to be phosphorylated and activated by Cdks (187, 188), and other proteins involved in the initiation of replication are likely to be targets of Cdk phosphorylation as well.

8.2.2 Cyclins

Most cyclins can also be phosphorylated by Cdc2 *in vitro* in immune complex kinase assays (29). Whether cyclins are physiologically relevant targets *in vitro* is not yet clear. In the one instance that this has been studied in detail, mutants of cyclin B lacking presumptive Cdc2 phosphorylation sites functioned normally (189). However, there remains the possibility that phosphorylation of cyclins regulates some aspect of their function such as their binding to Cdc2, subcellular localization, or proteolysis.

8.2.3 Genetically implicated targets in yeast

Few substrates of Cdc2 have been identified in yeast. The FAR1 protein is necessary for mating-factor induced cell cycle arrest. It is only present at functional levels prior to Cdc2 activation at the G1/S phase transition and it becomes phosphorylated in a Cdc2-dependent manner (190, 191). It has been proposed that Cdc2-mediated phosphorylation might target FAR1 for destruction via ubiquitinylation as part of a mechanism which renders the cell resistant to mating factors in other phases of the cell cycle (190). SWI5, a transcription factor involved in mating-type switching, is another substrate of Cdc2 in *S. cerevisiae*. Phosphorylation of SWI5 prevents its entry into the nucleus in all but the G1 phase of the cell cycle (192). In *S. pombe*, there are no clearly identified substrates for Cdc2. However, as discussed above, the Cdc25 phosphatase which activates Cdc2 at the G2/M transition is positively regulated by phosphorylation (120). Feedback activation of Cdc25 by Cdc2 could lead to the abrupt activation of Cdc2 observed at the onset of mitosis.

9. Conclusions

In this chapter, I have tried to briefly describe some of the salient features of the Cdk protein kinase family. The most well-known members of this family regulate important transitions in the eukaryotic cell cycle. They are small protein kinases, and are regulated by their interaction with short-lived regulatory subunits, cyclins. Protein kinases with a similar structure and mode of regulation are still being uncovered and, most likely, will function in other aspects of cellular metabolism as well. The activity of each Cdk oscillates during the cell cycle and this oscillation is controlled not only by its association with a member of the cyclin family of proteins, but also by a network of protein kinases and protein phosphatases. Cdks target a wide variety of proteins involved in coordinating the major events of the cell cycle. These substrates are found throughout the cell and are involved in structural reorganization during mitosis, the transition from G1 into S phase and the initiation and maintenance of S phase. The combined efforts to understand Cdk function in a variety of organisms should continue to yield rapid advances in our understanding of eukaryotic cell cycle regulation.

References

1. Forsburg, S. L. and Nurse, P. (1991) Cell cycle regulation in the yeasts *Saccharomyces cerevisiae* and *Schizosaccharomyces pombe*. *Annu. Rev. Cell Biol.*, **7**, 227.
2. Fantes, P. A. (1989) Cell cycle controls. In *Molecular biology of the fission yeast* (ed. A. Nasim, P. Young, and B. F. Johnson), p. 128. Academic, Press, San Diego.
3. MacNeill, S. A. and Nurse, P. (1989) Genetic interactions in the control of mitosis in fission yeast. *Curr. Genet.*, **19**, 1.
4. Nasmyth, K. (1993) Control of the yeast cell cycle by the Cdc28 protein kinase. *Curr. Opin. Cell Biol.*, **5**, 166.
5. Beach, D., Durkacz, B., and Nurse, P. (1982) Functionally homologous cell cycle control genes in budding and fission yeast. *Nature*, **300**, 706.
6. Lee, M. G. and Nurse, P. (1987) Complementation used to clone a human homologue of the fission yeast cell cycle control gene *cdc2*. *Nature*, **327**, 31.
7. Norbury, C. and Nurse, P. (1992) Animal cell cycles and their control. *Annu. Rev. Biochem.*, **61**, 441.
8. Kirschner, M. (1992) The cell cycle then and now. *Trends Biochem. Sci.*, **17**, 281.
9. Minshull, J. (1993) Cyclin synthesis: who needs it? *Bioessays*, **15**, 149.
10. Freeman, R. S. and Donoghue, D. J. (1991) Protein kinases and protooncogenes: biochemical regulators of the eukaryotic cell cycle. *Biochemistry*, **30**, 2293.
11. Nurse, P. (1990) Universal control mechanism regulating onset of M-phase. *Nature*, **344**, 503.
12. Pines, J. (1993) Cyclins and cyclin-dependent kinases: take your partners. *Trends Biochem. Sci.*, **18**, 195.
13. Riabowol, K., Draetta, G., Brizuela, L., Vandre, D., and Beach, D. (1989). The cdc2 kinase is a nuclear protein that is essential for mitosis in mammalian cells. *Cell*, **57**, 393.
14. Fang, F. and Newport, J. W. (1991) Evidence that the G1–S and G2–M transitions are controlled by different cdc2 proteins in higher eukaryotes. *Cell*, **66**, 731.
15. Th'ng, J. P. H., Wright, P. S., Hamaguchi, J., Lee, M. G., Norbury, C. J., Nurse, P., and Bradbury, E. M. (1990) The FT210 cell line is a mouse G2 phase mutant with a temperature-sensitive CDC2 gene product. *Cell*, **63**, 313.
16. Paris, J., Le Guellec, R., Couturier, A., Le Guellec, K., Omilli, F., Camonis, J. *et al.* (1991). Cloning by differential screening of a *Xenopus* cDNA coding for a protein highly homologous to cdc2. *Proc. Natl. Acad. Sci. USA*, **88**, 1039.
17. Elledge, S. J. and Spottswood, M. R. (1991) A new human p34 protein kinase, CDK2, identified by complementation of a *cdc28* mutation in *Saccharomyces cerevisiae*, is a homolog of *Xenopus* Eg1. *EMBO J.*, **10**, 2653.
18. Ninomiya-Tsuji, J., Nomoto, S., Yasuda, H., Reed, S. I., and Natsumoto, K. (1991)

Cloning of a human cDNA encoding a CDC2-related kinase by complementation of a budding yeast *cdc28* mutation. *Proc. Natl. Acad. Sci. USA*, **88**, 9006.

19. Tsai, L.-H., Harlow, E., and Meyerson, M. (1991) Isolation of the human *cdk2* gene that encodes the cyclin A- and adenovirus E1A-associated p33 kinase. *Nature*, **353**, 174.

20. Pagano, M., Pepperkok, R., Lukas, J., Baldin, V., Ansorge, W., Bartek, J., and Draetta, G. (1993) Regulation of the cell cycle by the cdk2 protein kinase in cultured human fibroblasts. *J. Cell Biol.*, **121**, 101.

21. Tsai, L.-H., Lees, E., Faha, B., Harlow, E., and Riabowol, K. (1993) The cdk2 kinase is required for the G1-to-S transition in mammalian cells. *Oncogene*, **8**, 1593.

22. Hanks, S. K. (1987) Homology probing: identification of cDNA clones encoding members of the protein-serine kinase family. *Proc. Natl. Acad. Sci. USA*, **84**, 388.

23. Hellmich, M. R., Pant, H. C., Wada, E., and Battey, J. F. (1992) Neuronal cdc2-like kinase: a cdc2-related protein kinase with predominantly neuronal expression. *Proc. Natl. Acad. Sci. USA*, **88**, 10867.

24. Lew, J., Winkfein, R. J., Paudel, H. K., and Wang, J. H. (1992) Brain proline-directed protein kinase is a neurofilament kinase which displays high sequence homology to p34^{cdc2}. *J. Biol. Chem.*, **267**, 25922.

25. Matsushime, H., Ewen, M. E., Strom, D. K., Kato, J.-Y., Hanks, S. K., Roussel, M. F., and Sherr, C. J. (1992) Identification and properties of an atypical catalytic subunit (p34^{PSK-J3}/cdk4) for mammalian D type G1 cyclins. *Cell*, **71**, 323.

26. Meyerson, M., Enders, G. H., Wu, C.-L., Su, L-K., Gorka, C., Nelson, C. *et al.* (1992) A family of human cdc2-related protein kinases. *EMBO J.*, **11**, 2909.

27. Okuda, T., Cleveland, J. L., and Downing, J. R. (1992) *PCTAIRE-1* and *PCTAIRE-3*, two members of a novel *cdc2/CDC28*-related protein kinase gene family. *Oncogene*, **7**, 2249.

28. Toh-e, A., Tanaka, K., Uesono, Y., and Wickner, R. B. (1988) *PHO85*, a negative regulator of the PHO system, is a homolog of the protein kinase gene, *CDC28*, of *Saccharomyces cerevisiae*. *Mol. Gen. Genet.*, **214**, 162.

29. Hunt, T. (1991) Cyclins and their partners: from a simple idea to complicated reality. *Seminars Cell Biol.*, **2**, 213.

30. Pines, J. (1991) Cyclins: wheels within wheels. *Cell Growth and Diff.*, **2**, 305.

31. Pines, J. and Hunter, T. (1990) Human cyclin A is adenovirus E1A-associated protein p60 and behaves differently from cyclin B. *Nature*, **346**, 760.

32. Girard, F., Strausfeld, U., Fernandez, A., and Lamb, N. J. C. (1991) Cyclin A is required for the onset of DNA replication in mammalian fibroblasts. *Cell*, **67**, 1169.

33. Pagano, M., Pepperkok, R., Verde, F., Ansorge, W., and Draetta, G. (1992) Cyclin A is required at two points in the human cell cycle. *EMBO J.*, **11**, 961.

34. Zindy, F., Lamas, E., Chenivess, X., Sobczak, J., Wang, J., Fesquet, D. *et al.* (1992) Cyclin A is required in S phase in normal epithelial cells. *Biochem. Biophys. Res. Comm.*, **182**, 1144.

35. Gautier, J., Minshull, J., Lohka, M., Glotzer, M., Hunt, T., and Maller, J. L. (1990) Cyclin is a component of maturation-promoting factor from *Xenopus*. *Cell*, **60**, 487.

36. Labbe, J. C., Capony, J. P., Caput, D., Cavadore, J. C., Derancourt, J., Kagahad, M. *et al*. (1989) MPF from starfish oocytes at first meiotic metaphase is a heterodimer containing one molecule of cdc2 and one molecule of cyclin B. *EMBO J.*, **8**, 3053.

37. Reed, S. (1991) G1-specific cyclins: in search of an S-phase-promoting factor. *Trends Genet.*, **7**, 95.

38. Sherr, C. J. (1993) Mammalian G1 cyclins. *Cell*, **73**, 1059.

39. Hadwiger, J. A., Wittenberg, C., Richardson, H. E., De Barros Lopes, M., and Reed, S. I. (1989) A family of cyclin homologs that control the G1 phase in yeast. *Proc. Natl. Acad. Sci. USA*, **86**, 6255.

40. Nash, R., Tokiwa, G., Anand, S., Erickson, K., and Futcher, A. B. (1988) The WHI1+ gene of *Saccharomyces cerevisiae* tethers cell division to cell size and is a cyclin homolog. *EMBO J.*, **7**, 4335.

41. Ogas, J., Andrews, B. J., and Herskowitz, I. (1991) Transcriptional activation of CLN1, CLN2, and a putative new G1 cyclin (HCS26) by SWI4, a positive regulator of G1-specific transcription. *Cell*, **6**, 1015.

42. Koff, A., Cross, F., Fisher, A., Schumacher, J., Leguellec, K., Philippe, M., and Roberts, J. M. (1991) Human cyclin E, a new cyclin that interacts with two members of the CDC2 gene family. *Cell*, **66**, 1217.

43. Lahue, E. E., Smith, A. V., and Orr-Weaver, T. L. (1991) A novel cyclin gene from *Drosophila* complements *CLN* function in yeast. *Genes and Dev.*, **5**, 2166.

44. Leopold, P. and O'Farrell, P. H. (1991) An evolutionarily conserved cyclin homolog from *Drosophila* rescues yeast deficient in G1 cyclins. *Cell*, **66**, 1207.

45. Lew, D. J., Dulic, V., and Reed, S. I. (1991) Isolation of three novel human cyclins by rescue of G1 cyclin (Cln) function in yeast. *Cell*, **66**, 1197.

46. Matsushime, H., Roussel, M. F., Ashmun, R. A., and Sherr, C. J. (1991) Colony-stimulating factor 1 regulates novel cyclins during the G1 phase of the cell cycle. *Cell*, **65**, 701.

47. Xiong, Y., Connolly, T., Futcher, B., and Beach, D. (1991) Human D-type cyclin. *Cell*, **65**, 691.

48. Forsburg, S. L. and Nurse, P. (1991) Identification of a G1-type cyclin *puc1*+ in the fission yeast *Schizosaccharomyces pombe*. *Nature*, **351**, 6323.

49. Booher, R. and Beach, D. (1988) Involvement of *cdc13*+ in mitotic control in *Schizosaccharomyces pombe*: possible interaction of the gene product with microtubules. *EMBO J.*, **7**, 2321.

50. Hagan, I., Hayles, J., and Nurse, P. (1988) Cloning and sequencing of the cyclin-related *cdc13*+ gene and a cytological study of its role in fission yeast mitosis. *J. Cell Sci.*, **91**, 587.

51. Bueno, A., Richardson, H., Reed, S. I., and Russell, P. (1991) A fission yeast B-type cyclin functioning early in the cell cycle. *Cell*, **66**, 149.

52. Bueno, A., Richardson, H., Reed, S. I., and Russell, P. (1993) Erratum. *Cell*, **73**, 1050.

53. Bueno, A. and Russell, P. (1993) Two fission yeast B-type cyclins, Cig2 and Cdc13, have different functions in mitosis. *Mol. Cell. Biol.*, **13**, 2286.

54. Richardson, H., Lew, D. J., Henze, M., Sugimoto, K., and Reed, S. I. (1992) Cyclin B homologs in *Saccharomyces cerevisiae* function in S phase and in G2. *Genes Dev.*, **6**, 2021.

55. Surana, U., Robitsch, H., Price, C., Schuster, T., Fitch, I., Futcher, A. B., and Nasmyth, K. (1991) The role of CDC28 and cyclins during mitosis in the budding yeast *S. cerevisiae*. *Cell*, **65**, 145.

56. Epstein, C. B. and Cross, F. R. (1992) CLB5: a novel B cyclin from budding yeast with a role in S phase. *Genes Dev.*, **6**, 1695.

57. Grandin, N. and Reed, S. I. (1993) Differential function and expression of *Saccharomyces cerevisiae* B-type cyclins in mitosis and meiosis. *Mol. Cell. Biol.*, **13**, 2113.

58. Kobayashi, H., Stewart, E., Poon, R., Adamczewski, J. P., Gannon, J., and Hunt, T. (1992) Identification of the domains in cyclin A required for binding to, and activation of, p34^{cdc2} and p32^{cdk2} protein kinase subunits. *Mol. Biol. Cell*, **3**, 1279.

59. Lees, E. M. and Harlow, E. (1993) Sequences within the conserved cyclin box of human cyclin A are sufficient for binding to and activation of cdc2 kinase. *Mol. Cell. Biol.*, **13**, 1194.

60. Cross, F. R. (1990) Cell cycle arrest caused by CLN gene deficiency in *Saccharomyces cerevisiae* resembles START-I arrest and is independent of the mating-pheromone signalling pathway. *Mol. Cell. Biol.*, **10**, 6482.

61. Richardson, H. E., Wittenberg, C., Cross, F., and Reed, S. I. (1989) An essential G1 function for cyclin-like proteins in yeast. *Cell*, **59**, 1127.

62. Pines, J. and Hunter, T. (1989) Isolation of a human cyclin cDNA: evidence for cyclin mRNA and protein regulation in the cell cycle and for interaction with p34^{cdc2}. *Cell*, **58**, 833.

63. Desai, D., Gu, Y., and Morgan, D. O. (1992) Activation of human cyclin-dependent kinases *in vitro*. *Mol. Biol. Cell*, **3**, 571.

64. Elledge, S. J., Richman, R., Hall, F. L., Williams, R. T., Lodgson, N., and Harper, J. W. (1992) *CDK2* encodes a 33-kDa cyclin A-associated protein kinase and is expressed before *CDC2* in the cell cycle. *Proc. Natl. Acad. Sci. USA*, **89**, 2907.

65. Rosenblatt, J., Gu, Y., and Morgan, D. O. (1992) Human cyclin-dependent kinase 2 is activated during the S and G2 phases of the cell cycle and associates with cyclin A. *Proc. Natl. Acad. Sci. USA*, **89**, 2824.

66. Xiong, Y., Zhang, H., and Beach, D. (1992) D type cyclins associate with multiple protein kinases and the DNA replication and repair factor PCNA. *Cell*, **71**, 505.

67. Dulic, V., Lees, E., and Reed, S. I. (1992) Association of human cyclin E with a periodic G1–S phase protein kinase. *Science*, **257**, 1958.

68. Koff, A., Giordano, A., Desai, D., Yamashita, K., Harper, J. W., Elledge, S. *et al.* (1992) Formation and activation of a cyclin E-cdk2 complex during the G1 phase of the human cell cycle. *Science*, **257**, 1989.

69. Kato, J., Matsushime, H., Hiebert, S. W., Ewen, M. E., and Sherr, C. J. (1993) Direct binding of cyclin D to the retinoblastoma gene product (pRb) and pRb phosphorylation by the cyclin D-dependent kinase CDK4. *Cell*, **75**, 331.

70. Hindley, J., Phear, G., Stein, M., and Beach, D. (1987) *Suc1*$^+$ encodes a predicted 13-kilodalton protein that is essential for cell viability and is directly involved in the division cycle of *Schizosaccharomyces pombe*. *Mol. Cell. Biol.*, **7**, 504.

71. Hayles, J., Beach, D., Durkacz, B., and Nurse, P. (1986) The fission yeast cell cycle control gene *cdc2*: isolation of a sequence *suc1* that suppresses *cdc2* mutant function. *Mol. Gen. Genet.*, **202**, 291.

72. Brizuela, L., Draetta, G., and Beach, D. (1987) p13^{suc1} acts in the fission yeast cell division cycle as a component of the p34^{cdc2} protein kinase. *EMBO J.*, **6**, 3507.

73. Hadwiger, J. A., Wittenber, C., Mendenhall, M. D., and Reed, S. I. (1989) The *Saccharomyces cerevisiae* CSK1 gene, a homolog of the *Schizosaccharomyces pombe suc1* + gene, encodes a subunit of the Cdc28 protein kinase complex. *Mol. Cell. Biol.*, **9**, 2034.

74. Gabrielli, B. G., Roy, L. M., Gautier, J., Philippe, M., and Maller, J. L. (1992) A *cdc2*-related kinase oscillates in the cell cycle independently of cyclins G2/M and *cdc2*. *J. Biol. Chem.*, **267**, 1969.

75. Hayles, J., Aves, S., and Nurse, P. (1986) *suc1* is an essential gene involved in both the cell cycle and growth in fission yeast. *EMBO J.*, **5**, 3373.

76. Moreno, S., Hayles, J., and Nurse, P. (1989) Regulation of p34^{cdc2} protein kinase during mitosis. *Cell*, **58**, 361.

77. Tang, Y. and Reed, S. I. (1993) The Cdk-associated protein Cks1 functions both in G1 and G2 in *Saccharomyces cerevisiae*. *Genes Dev.*, **7**, 822.

78. Draetta, G., Brizuela, L., Potashkin, J., and Beach, D. (1987) Identification of p34 and p13, human homologs of the cell cycle regulators of fission yeast encoded by *cdc2+* and *suc1+*. *Cell*, **50**, 319.

79. Richardson, H. E., Stueland, C. S., Thomas, J., Russell, P., and Reed, S. I. (1990) Human cDNAs encoding homologs of the small p34$^{Cdc28/Cdc2-}$ associated protein of *Saccharomyces cerevisiae* and *Schizosaccharomyces pombe*. *Genes Dev.*, **4**, 1332.

80. Dowdy, S. F., Hinds, P. W., Louie, K., Reed, S. I., Arnold, A., and Weinberg, R. A. (1993) Physical interaction of the retinoblastoma protein with human D cyclins. *Cell*, **73**, 499.

81. Ewen, M. E., Sluss, H., Sherr, C., Matsushime, H., Kato, J., and Livingston, D. M. (1993) Functional interactions of the retinoblastoma protein with mammalian D-type cyclins. *Cell*, **73**, 487.

82. Hollingsworth, R. E. Jr., Chen, P.-L., and Lee, W.-H. (1993) Integration of cell cycle control with transcriptional regulation by the retinoblastoma protein. *Curr. Opin. Cell Biol.*, **5**, 194.

83. Nevins, J. R. (1992) E2F: A link between the Rb tumor suppressor protein and viral oncoproteins. *Science*, **258**, 424.

84. Hiebert, S. W., Lipp, M., and Nevins, J. R. (1989) E1A-dependent transactivation of the human MYC promoter is mediated by the E2F factor. *Proc. Natl. Acad. Sci. USA*, **86**, 3594.

85. Thalmeier, K., Synovzik, H., Mertz, R., Winnacker, E.-L., and Lipp, M. (1989) Nuclear factor E2F mediates basic transcription and *trans*-activation by E1a of the human *MYC* promoter. *Genes Dev.*, **3**, 527.

86. Blake, M. C. and Azizkhan, J. C. (1989) Transcription factor E2F is required for efficient expression of the hamster dihydrofolate reductase gene *in vitro* and *in vivo*. *Mol. Cell. Biol.*, **9**, 4994.

87. Mendenhall, M. D. (1993) An inhibitor of p34^{CDC28} protein kinase activity from *Saccharomyces cerevisiae*. *Science*, **259**, 216.

88. Durkacz, B., Carr, A., and Nurse, P. (1986) Transcription of the *cdc2* cell cycle control gene of the fission yeast *Schizosaccharomyces pombe*. *EMBO J.*, **5**, 369.

89. Mendenhall, M. D., Jones, C. A., and Reed, S. I. (1987) Dual regulation of the yeast CDC28-p40 protein kinase complex: cell cycle, pheromone, and nutrient limitation effects. *Cell*, **50**, 927.

90. Simanis, V. and Nurse, P. (1986) The cell cycle control gene *cdc2+* of fission yeast encodes a protein kinase potentially regulated by phosphorylation. *Cell*, **45**, 261.

91. Draetta, G., Beach, D., and Morna, E. (1988) Synthesis of p34, the mammalian homolog of the yeast *cdc2+/CDC28* protein kinase, is stimulated during adenovirus-induced proliferation of primary baby rat kidney cells. *Oncogene*, **2**, 553.

92. Stein, G. H., Drullinger, L. F., Robetorye, R. S., Pereira-Smith, O. M., and Smith, J. R. (1991) Senescent cells fail to express *cdc2*, *cycA*, and *cycB* in response to mitogen stimulation. *Proc. Natl. Acad. Sci. USA*, **88**, 11012.

93. Dalton, S. (1992) Cell cycle regulation of the human *cdc2* gene. *EMBO J.*, **11**, 1797.

94. Lee, M. G., Norbury, C. J., Spurr, N. K., and Nurse, P. (1988) Regulated expression and phosphorylation of a possible mammalian cell-cycle control protein. *Nature*, **333**, 676.

95. Morla, A. O., Draetta, G., Beach, D., and Wang, J. Y. J. (1989) Reversible tyrosine

phosphorylation of cdc2: dephosphorylation accompanies activation during entry into mitosis. *Cell*, **58**, 193.

96. McGowan, C. H., Russell, P., and Reed, S. I. (1990). Periodic biosynthesis of the human M-phase promoting factor catalytic subunit p34 during the cell cycle. *Mol. Cell Biol.*, **10**, 3847.

97. Solomon, M. J. (1993) Activation of the various cyclin/cdc2 protein kinases. *Curr. Opin. Cell Biol.*, **5**, 180.

98. Solomon, M. J., Glotzer, M., Lee, T. H., and Kirschner M. W. (1990) Cyclin activation of p34^{cdc2}. *Cell*, **63**, 1013.

99. Ducommun, B., Brambilla, P., Felix, M.-A., Franza, B. R. Jr., Karsenti, E., and Draetta, G. (1991) cdc2 phosphorylation is required for its interaction with cyclin. *EMBO J.*, **10**, 3311.

100. Gould, K. L., Moreno, S., Owen, D. J., Sazer, S., and Nurse, P. (1991) Phosphorylation at Thr167 is required for *Schizosaccharomyces pombe* p34^{cdc2} function. *EMBO J.*, **10**, 3297.

101. Norbury, C., Blow, J., and Nurse, P. (1991) Regulatory phosphorylation of the p34^{cdc2} protein kinase in vertebrates. *EMBO J.*, **10**, 3321.

102. Solomon, M. J., Lee, T., and Kirschner, M. W. (1992) Role of phosphorylation in p34^{cdc2} activation: identification of an activating kinase. *Mol. Biol. Cell*, **3**, 13–27.

103. Connell-Crowley, L., Solomon, M. J., Wei, N., and Harper, J. W. (1993) Phosphorylation independent activation of human cyclin-dependent kinase 2 by cyclin A *in vitro*. *Mol. Biol. Cell*, **4**, 79.

104. Krek, W. and Nigg, E. A. (1991) Differential phosphorylation of vertebrate p34^{cdc2} kinase at the G1/S and G2.M transitions of the cell cycle: identification of major phosphorylation sites. *EMBO J.*, **10**, 305.

105. Gould, K. L. and Nurse, P. (1989) Tyrosine phosphorylation of the fission yeast cdc2 protein kinase regulates entry into mitosis. *Nature*, **342**, 39.

106. Meijer, L., Azzi, L., and Wang, J. Y. J. (1991) Cyclin targets p34^{cdc2} for tyrosine phosphorylation. *EMBO J.*, **10**, 1545.

107. Parker, L. L., Atherton-Fessler, S., Lee, M. S., Ogg, S., Falk, J. L., Swenson, K. I., and Piwnica-Worms, H. (1991) Cyclin promotes the tyrosine phosphorylation of p34^{cdc2} in a wee1$^+$ dependent manner. *EMBO J.*, **10**, 1255.

108. Clarke, P. R., Leiss, D., Pagano, M., and Karsenti, E. (1992) Cyclin A- and cyclin B-dependent protein kinases are regulated by different mechanisms in *Xenopus* egg extracts. *EMBO J.*, **11**, 1751.

109. Gu, Y., Rosenblatt, J., and Morgan, D. O. (1992) Cell cycle regulation of CDK2 activity by phosphorylation of Thr160 and Tyr15. *EMBO J.*, **11**, 3995.

110. Minshull, J., Golsteyn, R., Hill, C. S., and Hunt, T. (1990) The A- and B-type cyclin associated cdc2 kinases in *Xenopus* turn on and off at different times in the cell cycle. *EMBO J.*, **9**, 2865.

111. Gabrielli, B. G., Lee, M. S., Walker, D. H., Piwnica-Worms, H., and Maller, J. (1992) Cdc25 regulates the phosphorylation and activity of the *Xenopus* cdk2 protein kinase complex. *J. Biol. Chem.*, **267**, 18040.

112. Gautier, J., Solomon, M. J., Booher, R. N., Bazan, J. F., and Kirschner, M. W. (1991) cdc25 is a specific tyrosine phosphatase that directly activates p34^{cdc2}. *Cell*, **67**, 197.

113. Kumagai, A. and Dunphy, W. G. (1991) The cdc25 protein controls tyrosine dephosphorylation of the cdc2 protein in a cell-free system. *Cell*, **64**, 903.

114. Lee, M. S., Ogg, S., Xu, M., Parker, L. L., Donoghue, D. J., Maller, J. L., and Piwnica-

Worms, H. (1992) cdc25+ encodes a protein phosphatase that dephosphorylates p34^{cdc2}. *Mol. Biol. Cell*, **3**, 73.

115. Millar, J. B. A., McGowan, C. H., Lenaers, G., Jones, R., and Russell, P. (1991) p80^{cdc25} mitotic inducer is the tyrosine phosphatase that activates p34^{cdc2} kinase in fission yeast. *EMBO J.*, **10**, 4301.

116. Russell, P. and Nurse, P. (1986) cdc25+ functions as an inducer in the mitotic control of fission yeast. *Cell*, **45**, 145.

117. Millar, J. B. A. and Russell, P. (1991) The cdc25 M-phase inducer: an unconventional protein phosphatase. *Cell*, **66**, 407.

118. Honda, R., Ohba, Y., Nagata, A., Okayama, H., and Yasuda, H. (1993) Dephosphorylation of human p34^{cdc2} kinase on both Thr-14 and Tyr-15 by human cdc25B phosphatase. *FEBS Lett.*, **318**, 331.

119. Sebastian, B., Kakizuka, A., and Hunter, T. (1993) Cdc25M2 activation of cyclin-dependent kinases by dephosphorylation of threonine-14 and tyrosine-15. *Proc. Natl. Acad. Sci. USA*, **90**, 3521.

120. Hoffman, I., Clarke, P. R., Marcote, M. J., Karsenti, E., and Draetta, G. (1993) Phosphorylation and activation of human cdc25-C by cdc2-cyclin B and its involvement in the self-amplification of MPF at mitosis. *EMBO J.*, **12**, 53.

121. Izumi, T., Walker, D. H., and Maller, J. L. (1992) Periodic changes in phosphorylation of the *Xenopus* cdc25 phosphatase regulate its activity. *Mol. Biol. Cell*, **3**, 927.

122. Kumagai, A. and Dunphy, W. G. (1992) Regulation of the cdc25 protein during the cell cycle in *Xenopus* extracts. *Cell*, **70**, 139.

123. Millar, J. B. A., Lenaers, G., and Russell, P. (1992) *Pyp3* PTPase acts as a mitotic inducer in fission yeast. *EMBO J.*, **11**, 4933.

124. Featherstone, C. and Russell, P. (1991) Fission yeast p107^{wee1} mitotic inhibitor is a tyrosine/serine kinase. *Nature*, **349**, 808.

125. Russell, P. and Nurse, P. (1987) Negative regulation of mitosis by *wee1+*, a gene encoding a protein kinase homolog. *Cell*, **49**, 559.

126. Honda, R., Ohba, Y., and Yasuda, H. (1992) The cell cycle regulator, human p50^{wee1}, is a tyrosine kinase and not a serine/tyrosine kinase. *Biochem. Biophys. Res. Commun.*, **186**, 1333.

127. Igarashi, M., Nagata, A., Jinno, S., Suto, K., and Okayama, H. (1991) *Wee1*+-like gene in human cells. *Nature*, **353**, 80.

128. McGowan, C. H. and Russell, P. (1993) Human Wee1 kinase inhibits cell division by phosphorylating p34^{cdc2} exclusively on Tyr15. *EMBO J.*, **12**, 75.

129. Parker, L. L. and Piwnica-Worms, H. (1992) Inactivation of the p34^{cdc2}-cyclin B complex by the human WEE1 tyrosine kinase. *Science*, **257**, 1955.

130. Lundgren, K., Walworth, N., Booher, R., Dembski, M., Kirschner, M., and Beach, D. (1991) mik1 and wee1 cooperate in the inhibitory tyrosine phosphorylation of cdc2. *Cell*, **64**, 1111.

131. Russell, P. and Nurse, P. (1987) The mitotic inducer *nim1* + functions in a regulatory network of protein kinase homologs controlling the initiation of mitosis. *Cell*, **49**, 569.

132. Feilotter, H., Nurse, P., and Young, P. (1991) Genetic and molecular analysis of *cdr1/nim1* in Schizosaccharomyces pombe. *Genetics*, **127**, 309.

133. Coleman, T. R., Tang, Z., and Dunphy, W. G. (1993) Negative regulation of the Wee1 protein kinase by direct action of the Nim1/Cdr1 mitotic inducer. *Cell*, **72**, 919.

134. Parker, L. L., Walter, S. A., Young, P. G., and Piwnica-Worms, H. (1993) Phosphoryl-

ation and inactivation of the mitotic inhibitor Wee1 by the *nim1/cdr1* kinase. *Nature*, **363**, 736.

135. Wu, L. and Russell, P. (1993) Nim1 kinase promotes mitosis by inactivating Wee1 tyrosine kinase. *Nature*, **363**, 738.

136. Enoch, T. and Nurse, P. (1991) Coupling M phase and S phase: controls maintaining the dependence of mitosis on chromosome replication. *Cell*, **65**, 921.

137. Enoch, T., Gould, K. L., and Nurse, P. (1991) Mitotic checkpoint control in fission yeast. *Cold Spring Harb. Sym. Quant. Biol.*, **56**, 409.

138. Smythe, C. and Newport, J. W. (1992) Coupling of mitosis to the completion of S phase in *Xenopus* occurs via modulation of the tyrosine kinase that phosphorylates p34^{cdc2}. *Cell*, **68**, 787.

139. Ohtsubo, M. and Roberts, J. M. (1993) Cyclin-dependent regulation of G1 in mammalian fibroblasts. *Science*, **259**, 1908.

140. Glotzer, M., Murray, A. W., and Kirschner, M. W. (1991) Cyclin is degraded by the ubiquitin pathway. *Nature*, **349**, 132.

141. Gallant, P. and Nigg, E. (1992) Cyclin B2 undergoes cell cycle-dependent nuclear translocation and, when expressed as a non-destructible mutant, causes mitotic arrest in HeLa cells. *J. Cell Biol.*, **117**, 213.

142. Ghiara, J. B., Richardson, H. E., Sugimoto, K., Henze, M., Lew, D. J., and Reed, S. I. (1991) A cyclin B homologue in *S. cerevisiae*: chronic activation of the CDC28 protein kinase by cyclin prevents exit from mitosis. *Cell*, **65**, 163.

143. Lorca, T., Labbe, J.-C., Devault, A., Fesquet, D., Strausfeld, U., Nilsson, J. *et al.* (1992) Cyclin A-cdc2 kinase does not trigger but delays cyclin degradation in interphase extracts of amphibian eggs. *J. Cell Sci.*, **102**, 55.

144. Luca, F. C., Shibuya, E. K., Dohrmann, C. E., and Ruderman, J. V. (1991) Both cyclin AΔ60 and BΔ97 are stable and arrest cells in M-phase, but only cyclin BΔ97 turns on cyclin destruction. *EMBO J.*, **10**, 4311.

145. Holloway, S. L., Glotzer, M., King, R. W., and Murray, A. W. (1993) Anaphase is initiated by proteolysis rather than by the inactivation of maturation-promoting factor. *Cell*, **73**, 1393.

146. Van Der Velden, H. M. W. and Lohka, M. J. (1993) Mitotic arrest caused by the amino terminus of *Xenopus* cyclin B2. *Mol. Cell. Biol.*, **13**, 1480.

147. Baldin, V., Lukas, J., Marcote, M. J., Pagano, M., and Draetta, G. (1993) Cyclin D1 is a nuclear protein required for cell cycle progression in G1. *Genes Dev.*, **7**, 812.

148. Sewing, A., Burger, C., Brusselbach, S., Schalk, C., Lucibello, F. W., and Muller, R. (1993) *J. Cell Sci.*, **104**, 545.

149. Rogers, W., Wells, R., and Rechsteiner, M. (1986) Amino acid sequences common to rapidly degraded proteins: the PEST hypothesis. *Science*, **234**, 364.

150. Lorca, T., Labbe, J.-C., Devault, A., Fesquet, D., Capony, J.-P., Cavadore, J.-C. *et al.* (1992) Dephosphorylation of cdc2 on threonine 161 is required for cdc2 kinase inactivation and normal anaphase. *EMBO J.*, **11**, 2381.

151. Ducommun, B., Brambilla, P., and Draetta, G. (1991) Mutations at sites involved in Suc1 binding inactivate Cdc2. *Mol. Cell Biol.*, **11**, 6177.

152. De Bondt, H. L., Rosenblatt, J., Jancarik, J., Jones, H. D., Morgan, D. O., and Kim, S.-H. (1993) Crystal structure of cyclin-dependent kinase 2. *Nature*, **363**, 595.

153. Zheng, J., Knighton, D. R., Ten Eyck, L. F., Karlsson, R., Xuong, N.-H., Taylor, S. S., and Sowadski, J. M. (1993) Crystal structure of the catalytic subunit of cAMP-dependent protein kinase complexed with MgATP and peptide inhibitor. *Biochemistry*, **32**, 2154.

154. Alfa, C. E., Ducommun, B., Beach, D., and Hyams, J. S. (1990) Distinct nuclear and spindle pole body populations of cyclin-cdc2 in fission yeast. *Nature*, **347**, 680.

155. Bailly, E., Doree, M., Nurse, P., and Bornens, M. (1989) p34^{cdc2} is located in both nucleus and cytoplasm; part is centrosomally associated at G2/M and enters vesicles at anaphase. *EMBO J.*, **8**, 3985.

156. Leiss, D., Felix, M.-A., and Karsenti, E. (1992) Association of cyclin-bound, p34^{cdc2} with subcellular structures in *Xenopus* eggs. *J. Cell Sci.*, **102**, 285.

157. Wittenberg, C., Richardson, S. L., and Reed, S. I. (1987) Subcellular localization of a protein kinase required for cell cycle initiation in *Saccharomyces cerevisiae*: evidence for an association between the CDC28 gene product and the insoluble cytoplasmic matrix. *J. Cell Biol.*, **105**, 1527.

158. Girard, F., Strausfel, U., Cavadore, J.-C., Russell, P., Fernandez, A., and Lamb, N. J. C. (1992) cdc25 is a nuclear protein expressed constitutively throughout the cell cycle in nontransformed mammalian cells. *J. Cell Biol.*, **118**, 785.

159. Millar, J. B. A., Blevitt, J., Gerace, L., Sadhu, K., Featherstone, C., and Russell, P. (1991) p55^{CDC25} is a nuclear protein required for the initiation of mitosis in human cells. *Proc. Natl. Acad. Sci. USA*, **88**, 10500.

160. Pines, J. and Hunter, T. (1991) Human cyclins A and B1 are differentially located in the cell and undergo cell cycle-dependent nuclear transport. *J. Cell Biol.*, **115**, 1.

161. Maldonado-Codina, G. and Glover, D. M. (1992) Cyclins A and B associate with chromatin and the polar regions of spindles, respectively, and do not undergo complete degradation at anaphase in syncytial *Drosophila* embryos. *J. Cell Biol.*, **116**, 967.

162. Ookata, K., Hisanaga, S.-I., Okano, T., Tachibana, K., and Kishimoto, T. (1992) Relocation and distinct subcellular localization of p34^{cdc2}-cyclin B complex at meiosis reinitiation in starfish oocyte. *EMBO J.*, **11**, 1763.

163. Nigg, E. (1993) Targets of cyclin-dependent protein kinases. *Curr. Opin. Cell Biol.*, **5**, 187.

164. Peeper, D. S., Parker, L. L., Ewen, M. E., Toebes, M., Hall, F. L., Xu, M. *et al.* (1993) A- and B-type cyclins differentially modulate substrate specificity of cyclin-cdk complexes. *EMBO J.*, **12**, 1947.

165. Peter, M., Sanghera, J. S., Pelech, S. L., and Nigg, E. A. (1992) Mitogen-activated protein kinases phosphorylate nuclear lamins and display sequence specificity overlapping that of mitotic protein kinase p34^{cdc2}. *Eur. J. Biochem.*, **205**, 287.

166. Cicirelli, M. F., Pelech, S. L., and Krebs, E. G. (1988) Kinase activation during the burst in protein phosphorylation that precedes meiotic cell division in *Xenopus* oocytes. *J. Biol. Chem.*, **263**, 2009.

167. Pelech, S. L., Tombes, R. M., Meijer, L., and Krebs, E. G. (1988) Activation of myelin basic protein kinases during echinoderm oocyte maturation and egg fertilization. *Dev. Biol.*, **130**, 28.

168. Moreno, S. and Nurse, P. (1990) Substrates for p34^{cdc2}: *in vivo veritas*? *Cell*, **61**, 549.

169. Satterwhite, L. L., Lohka, M. J., Wilson, K. L., Scherson, T. Y., Cisek, L. J., Corden, J. L., and Pollard, T. D. (1992) Phosphorylation of myosin-II regulatory light chain by cyclin-p34^{cdc2}: a mechanism for the timing of cytokinesis. *J. Cell Biol.*, **118**, 595.

170. Chou, Y.-H., Ngai, K.-L., and Goldman, R. (1991) The regulation of intermediate filament reorganization in mitosis. *J. Biol. Chem.*, **266**, 7325.

171. Mak, A. S., Carpenter, M., Smillie, L. B., and Wang, J. H. (1991) Phosphorylation of caldesmon by p34^{cdc2} kinase. *J. Biol. Chem.*, **266**, 19971.

172. Ward, G. E. and Kirschner, M. W. (1990) Identification of cell cycle-regulated phosphorylation sites on nuclear lamin C. *Cell*, **61**, 561.

173. Cisek, L. J. and Corden, J. L. (1989) Phosphorylation of RNA polymerase by the murine homologue of the cell-cycle control protein cdc2. *Nature*, **339**, 679.

174. Heald, R. and McKeon, F. (1990) Mutations of phosphorylation sites in lamin A that prevent nuclear lamina disassembly in mitosis. *Cell*, **61**, 579.

175. Peter, M., Nakagawa, J., Doree, M., Labbe, J. C. and Nigg, E. A. (1990) *In vitro* disassembly of the nuclear lamina and M phase—specific phosphorylation of lamins by cdc2 kinase. *Cell*, **61**, 591.

176. Lüscher, B., Brizuela, L., Beach, D., and Eisenman, R. N. (1991) A role for the p34^{cdc2} kinase and phosphatases in the regulation of phosphorylation and disassembly of lamin B2 during the cell cycle. *EMBO J.*, **10**, 865.

177. Bradbury, E. M. (1992) Reversible histone modifications and the chromosome cell cycle. *Bioessays*, **14**, 9.

178. Reeves, R. (1992) Chromatin changes during the cell cycle. *Curr. Opin. Cell Biol.*, **4**, 413.

179. Lin, R., Cook, R. G., and Allis, C. D. (1991) Proteolytic removal of core histone amino termini and dephosphorylation of histone H1 correlate with the formation of condensed chromatin and transcriptional silencing during *Tetrahymena* macronuclear development. *Genes Dev.*, **5**, 1601.

180. Belenguer, P., Caizergues-Ferrer, M., Labbe, J.-C., Doree, M., and Amalric, F. (1990) Mitosis-specific phosphorylation of nucleolin by p34^{cdc2} protein kinase. *Mol. Cell. Biol.*, **10**, 3607.

181. Peter, M., Nakagawa, J., Doree, M., Labbe, J. C., and Nigg, E. A. (1990) Identification of major nucleolar proteins as candidate mitotic substrates of cdc2 kinase. *Cell*, **60**, 791.

182. Meijer, L., Ostvold, A.-C., Walaas, S. I., Lund, T., and Laland, S. G. (1991) High-mobility-group proteins P1, I and Y as substrates of the M-phase-specific p34^{cdc2}/cyclin^{cdc13} kinase. *Eur. J. Biochem.*, **196**, 557.

183. Gottesfeld, J. M., Wolf, V. J., Dang, T., Forbes, D. J., and Hartle, P. (1994) Mitotic repression of RNA polymerase III transcription in vitro mediated by phosphorylation of a TFIIIB component. *Science*, **263**, 81.

184. Reeves, R., Langan, T. A., and Nissen, M. S. (1991) Phosphorylation of the DNA-binding domain of nonhistone high-mobility group I protein by cdc2 kinase: reduction of binding affinity. *Proc. Natl. Acad. Sci. USA*, **88**, 1671.

185. Lüscher, B. and Eisenman, R. N. (1992) Mitosis-specific phosphorylation of the nuclear oncoproteins Myc and Myb. *J. Cell Biol.*, **118**, 775.

186. Bischoff, J. R., Friedman, P. N., Marshak, D. R., Prives, C., and Beach, D. (1990) Human p53 is phosphorylated by p60-cdc2 and cyclin B-cdc2. *Proc. Natl. Acad. Sci. USA*, **87**, 4766.

187. Dutta, A. and Stillman, B. (1992) cdc family kinases phosphorylate a human cell DNA replication factor, RPA, and activate DNA replication. *EMBO J.*, **11**, 2189.

188. Fotedar, R. and Roberts, J. M. (1992) Cell cycle regulated phosphorylation of RPA-3 occurs within the replication initiation complex. *EMBO J.*, **11**, 2177.

189. Izumi, T. and Maller, J. L. (1991) Phosphorylation of *Xenopus* cyclins B1 and B2 is not required for cell cycle transitions. *Mol. Cell. Biol.*, **11**, 3860.

190. McKinney, J. D., Chang, F., Heintz, N., and Cross, F. R. (1993) Negative regulation of *FAR1* at the start of the yeast cell cycle. *Genes Dev.*, **7**, 833.

191. Peter, M., Gartner, A., Horecka, J., Ammerer, G., and Herskowitz, I. (1993) FAR1

links the signal transduction pathway to the cell cycle machinery in yeast. *Cell*, **73**, 747.

192. Moll, T., Tebb, G., Surana, U., Robitsch, H., and Nasmyth, K. (1991) The role of phosphorylation and the CDC28 protein kinase in cell cycle-regulated nuclear import of the *S. cerevisiae* transcription factor SWI5. *Cell*, **66**, 743.

193. Keryer, G., Luo, A., Cavadore, J.-C., Erlichman, J., and Bornens, M. (1993) Phosphorylation of the regulatory subunit of type IIβ cAMP-dependent protein kinase by cycline B/p34^{cdc2} kinase impairs its binding to microtubule-associated protein 2. *Proc. Natl. Acad. Sci. USA*, **90**, 5418.

194. Kusubata, M., Matsuoka, Y., Tsujimura, K., Ito, H., Ando, S., Kamijo, M. *et al.* (1993) cdc2 kinase phosphorylation of desmin at three serine/threonine residues in the amino-terminal head domain. *Biochem. Biophys. Res. Commun.*, **190**, 927.

195. Kipreos, E. T. and Wang, J. Y. J. (1990) Differential phosphorylation of c-abl in cell cycle determined by *cdc2* kinase and phosphatase activity. *Science*, **248**, 217.

196. Litchfield, D. W., Lozeman, F. J., Cicirelli, M. F., Harrylock, M., Ericsson, L. H., Piening, C. J., and Krebs, E. G. (1991) Phosphorylation of the β subunit of casein kinase II in human A431 cells. *J. Biol. Chem.*, **266**, 20380.

197. McVey, D., Brizuela, L., Mohr, I., Marshak, D. R., Gluzman, Y., and Beach, D. (1989) Phosphorylation of large tumour antigen by cdc2 stimulates SV40 DNA replication. *Nature*, **341**, 503.

198. Morgan, D. O., Kaplan, J. M., Bishop, J. M., and Varmus, H. E. (1989) Mitosis-specific phosphorylation of p60^{c-src} by p34^{cdc2} associated protein kinase. *Cell*, **57**, 775.

199. Shenoy, S., Choi, J.-K., Bagrodia, S., Copeland, T. D., Maller, J. L., and Shalloway, D. (1989) Purified maturation promoting factor phosphorylates pp60^{c-src} at the sites phosphorylated during fibroblast mitosis. *Cell*, **57**, 763.

200. Nissen, M. S., Langan, T. A., and Reeves, R. (1991) Phosphorylation by cdc2 kinase modulates DNA binding activity of high mobility group I nonhistone chromatin protein. *J. Biol. Chem.*, **266**, 19945.

6 | Receptor protein-tyrosine kinases

TONY HUNTER and RICHARD A. LINDBERG

1. Introduction

Cells change their activities in response to signals from their surrounding environment. Single cell organisms (including those from which all multicellular organisms evolved) respond mainly to nutrient cues, many of which are able to cross the cell membrane. As multicellular life evolved, the much more constant milieu that bathed the individual cells reduced the need for cells to respond to nutritional signals, but, instead, the need to modulate cellular activities in an integrated fashion with other cells resulted in the evolution of complex intercellular signalling pathways.

Extracellular signalling molecules used in cell–cell communication essentially control all the activities of the vertebrate cell. The signals carried by these molecules are often transduced across the cellular membrane by transmembrane receptors. One common type of surface receptor has intrinsic protein-tyrosine kinase (PTK) activity and this belongs to the receptor protein-tyrosine kinase (RPTK) family. These receptors are probably involved in the regulation of many cellular activities that have yet to be discovered, but it is already clear that the RPTKs are involved in the control of cellular differentiation programmes and cell growth. Because of these activities, this class of molecules is also important in the genesis of many neoplasias. RPTKs have been found in all multicellular eukaryotic organisms, from nematodes to man.

RPTKs are activated by polypeptide ligands commonly known as growth factors, but more properly called cytokines. Most known ligands for RPTKs are secreted, soluble proteins that were discovered because they were growth or differentiation modulating factors, but the list also includes the classical hormone, insulin, and it is speculated that membrane-bound and membrane-spanning proteins as well as extracellular matrix proteins will also prove to be ligands for RPTKs. In addition to many well-characterized growth factor/RPTK pairs, such as EGF and the EGF RPTK, the recent matching of a number of orphan receptors with cognate ligands has been invaluable in answering questions about the function of RPTKs. The ability to study biochemical pathways that are modulated by ligand activation of RPTKs has contributed greatly to the understanding of how a cell can be directed by extracellular signals.

How the ligand-binding signal is actually transduced across the plasma membrane by RPTKs can be divided into discrete steps that we are just beginning to understand (see below). Signalling involves ligand-mediated receptor dimerization, which results in transphosphorylation of the receptor subunits within a dimer and activation of the RPTK catalytic domains for the phosphorylation of cytoplasmic substrates. The identity of a number of RPTK substrates has been uncovered in the past few years. In several cases tyrosine phosphorylation has been shown to modulate the activity of these substrates, and some understanding of how the ligand-mediated RPTK signal is propagated into known cellular responses is being gained.

The use of new cloning techniques has rapidly increased the number of known RPTKs in recent years. This has resulted in a rapid expansion of knowledge of the structure and function of this class of receptors. In this chapter we discuss the deduced structures, classification of the members of this expanding receptor family, and some of the advances that have been reported subsequent to other recent reviews (1–6).

2. Structure

The classification of RPTKs is based on structural characteristics rather than on function. Therefore, before discussing the composition and classification of this family, we will review what is known about the structure of RPTKs.

2.1 Overall basic structure

RPTKs are type I transmembrane proteins, with their N-termini outside the cell and single membrane-spanning regions. The structural features that are common to all RPTKs are the transmembrane domain that divides the molecule into a ligand-binding domain and a cytoplasmic domain. The cytoplasmic portion has a conserved protein kinase catalytic domain. Starting at the N-terminal end, the generic RPTK has a signal peptide that ensures that the protein will be targeted to the secretory pathway. This is followed by an extracellular domain of several hundred amino acids that contain N-linked glycosylation sites, a distinctive pattern of cysteine residues, and often a characteristic array of structural motifs. The transmembrane domain consists of ~24 hydrophobic residues that are usually succeeded by several basic residues that function as a stop-transfer signal. On the cytoplasmic side of the membrane there is a juxtamembrane region, which is usually ~50 residues in length, and which in some cases is known to have important regulatory functions. Next follows the catalytic domain, which is related to the catalytic domains of the cytoplasmic protein-tyrosine kinases and the protein-serine/threonine kinases (see Section 2.4), and is about 250 residues in length, excluding inserts. The phosphotransfer function lies entirely within this region. The region C-terminal to the catalytic domain is of variable length, and can be up to 200 residues. The functions of this C-terminal tail vary among members of the RPTKs.

2.2 Molecular topology of receptor protein-tyrosine kinases

The shared motifs among all the RPTKs are the signal peptide, the transmembrane domain, and the catalytic domain. The ligand-binding domains are variable and consist of combinations of various structural motifs. Figure 1 shows the topology of 13 vertebrate RPTKs. These are shown because they are unique with respect to each other, and each is a representative of a subfamily whose members have a similar molecular topology. The catalytic domains of the members of each subfamily are more closely related to each other than to other RPTK catalytic domains.

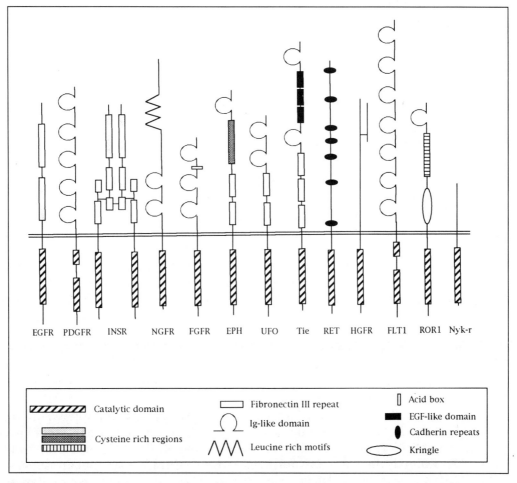

Fig. 1 Structural topology of receptor protein-tyrosine kinases. Compiled from refs 2, 7, 80–83. EGFR, epidermal growth factor receptor; PDGFR, platelet-derived growth factor receptor; INSR, insulin receptor; NGFR, nerve growth factor receptor; FGFR, fibroblast growth factor receptor; HGFR, hepatocyte growth factor receptor.

2.3 The ligand-binding domain

The most distinctive feature in the RPTKs depicted in Fig. 1 is the ligand-binding domain. The combinations of various repeated motifs give a distinctive 'finger-print' for each subfamily, and this allows one to predict which subfamily a new RPTK may fall into and perhaps what its ligand might be. The EGF receptor has two cysteine-rich regions that are related to each other and to a similar region in the insulin receptor. The exact function of these domains is not known, but based on the mapping of the EGF-binding site, these domains are probably involved in maintenance of structural integrity as opposed to being directly involved in ligand contact, at least in the EGF receptor. The PDGF receptor has five immunoglobulin-like (Ig) domains. The insulin receptor, in addition to the EGF receptor-related cysteine-rich domain, has three fibronectin type III (FNIII) repeats (FNIII and Ig repeats have an overall similar structure). The insulin receptor is derived from a precursor that is proteolytically cleaved to yield a mature receptor in which the α and β subunits are covalently linked by disulphide bonds, and functions as a heterotetrameric molecule. The NGF receptor contains two Ig-like domains, and a recently defined leucine-rich motif, which has been speculated to function in cell adhesion (7). The FGF receptor has three Ig-like domains, but there are also forms of the receptor that can be generated by alternative splicing that contain two of these domains (8–10). EPH is the prototype of a family for which no ligands have been found. The ligand-binding domain of EPH has an Ig-like domain, a cysteine-rich region that is unrelated to those in the EGF and insulin receptors, and two FNIII repeats. UFO has two FNIII repeats and two Ig-like domains. The Tie ligand binding domain has two Ig-like domains that flank three EGF-like repeats and three FNIII repeats. The only feature noted thus far in RET is the presence of seven repeats found in cadherin domains. Since the cadherins are cell surface adhesion receptors that interact homotypically, this suggests that RET could also be involved in cell adhesion responses. The HGF receptor is a heterodimer that, like the insulin receptor, is formed by proteolytic cleavage of a precursor, resulting in a receptor with two subunits held together by a disulphide linkage. FLT1 is a member of the PDGFR subfamily, and has seven Ig-like domains instead of five. The ROR1 RPTK has a unique cysteine-rich region, an Ig-like domain, and a kringle motif in its ligand-binding domain.

The functions of these motifs, individually or in combination, are not yet clear. It would be expected that some of them function in the actual protein–protein inter-action that occurs when a ligand molecule binds. However, RPTK extracellular domains are significantly larger than is necessary for binding polypeptide ligands (c.f. the TGFβ receptor that has an extracellular domain of only ∼ 140 residues in length). Moreover, from the limited mapping of RPTK ligand-binding sites, these appear to lie in relatively short regions of the extracellular domain, suggesting that RPTK extracellular domains have other functions. Ligand-dependent dimerization of two receptor molecules is critical for RPTK signalling, and some motifs in the extracellular domain may serve this function. However, functions such as adhesion between cells or interactions with extracellular matrix should still be considered as possible functions for RPTK extracellular domains.

2.4 The conserved catalytic domain

All PTKs share a conserved catalytic domain, which can act as an independent entity. Sequences on either side of this domain and some large inserts within the domain itself are dispensable for activity, and are likely to be regulatory or to have functions that are not involved in phosphotransfer. The minimal catalytic domain is ~ 250 amino acids and has been delineated by sequence similarities between different protein kinases (11). Experiments to determine the boundaries of the catalytic domain by deleting in from the ends and testing for activity have shown that the region required for activity is concordant with the boundaries determined by homology.

The similarity within the catalytic domain for the RPTKs ranges from 32% up to 95% identity for what are probably distinct gene products. Sequence alignments that would include the entire RPTK family are only useful when done with the catalytic domain sequences, since this is the only region that has homology among all the members. Such alignments that include all PTKs have been published several times. The most recent alignment of PTKs revealed that there are 13 residues conserved in all of the 54 known sequences (12). Several other residues would be invariant if not for the ERBB3 RPTK, which may not be active as a protein kinase, and the divergence of some of the PTKs from lower organisms, such as *Dictyostelium* and *Drosophila*.

The recent elucidation of the crystal structure of the cAMP-dependent protein kinase (PKA) catalytic subunit has given new insights or confirmed already suspected notions on the architecture of a protein kinase catalytic domain (13, 14; see Chapter 1). Although the structure of a PTK has not yet been accomplished, many of the features that are dictated by residues conserved in both protein-serine/threonine kinases and PTKs will probably be very similar. Using the information from alignments and the three-dimensional structure solved for PKA (see Chapter 1), it is possible to predict the functions of some of the conserved residues in PTKs. Eleven RPTKs are aligned in Fig. 2. The eleven sequences represent the major subfamilies and two unique RPTKs, RET and Klg; the numbering system for the amino acids corresponds to that of human PDGF receptor β. In the following discussion the corresponding number for PKA is given in parentheses. The structure is made up of two lobes separated by a cleft. Mg^{2+} ATP and the protein substrate are brought together through interaction with specific binding sites in the cleft, which allows phosphotransfer to be catalysed. The smaller lobe is the N-terminal region and is responsible for binding Mg^{2+} ATP. The motif responsible is the GXGXXG(21 amino acids)K635 (K72). The glycine fold cradles the phosphate moieties of the nucleotide and the lysine residue, which had previously been shown to become covalently modified by the ATP analogue p-fluorosulphonyl-benzoyl adenosine, and later, to be required for ATP binding, interacts with the

```
          600                                                                    671
PDGFR-B   LVLGRTLGSGAFGQVVEATAHGLSHSQ..ATMKVAVKMLKSTARSS.EKQALMSELKIMSHLGPHLNVVNLLGACTKG
    RET   LVLGKTLGEGEFGKVVKATAFHLKGRA..GYTTVAVKMLKENASPS.ELRDLLSPFNVLKQV.NHPHVIKLYGACSQD
    Tie   DTFEDLIGEGNFGQVIRAMIKKDG.....LKMNAAIKMLKEYASEN.DHRDFAGELEVLCKLGHHPNIINLLGACKNR
   INSR   ITLLRELGQGSFGMVYEGNARDIIKGE..AETRVAVKTVNESASLR.ERIEFLNEASVMKGF.TCHHVVRLLGVVSKG
   NGFR   IVLKWELGEGAFGKVFLAECHNLLPEQ..DKMLVAVKALKEASES..ARQDFQREVELLTML.QHQHIVRFFGVCTEG
   HGFR   VHFNEVIGRGHFGCVYHGTLLDNDG....KKIHCAVKSLNRITDIG.EVSQFLTEGIIMKDF.SHPNVLSLLGICLRS
    UFO   VALGKTLGEGEFGAVMEGQLNQDD.....SILKVAVKTMKIAICTRSELEDFLSEAVCMKEF.DHPNVMRLIGVCFQG
   EGFR   FKKIKVLGSGAFGTVYKGLWIPEGEK...VKIPVAIKELREATSPK.ANKEILDEAYVMASV.DNPHVCRLLGICLT.
    EPH   LMVDTVIGEGEFGEVYRGTLRLPSQ....DCKTVAIKTLKDTSPGG.QWWNFLREATIMGQF.SHPHILHLEGVVTKR
    Klg   LQTITTIGRGEFGEVFLAKAKGAEDAE..GEALVLVKSLQTRDEQ..LQLDFRREAEMFGKL.NHVNVVRLLGLCREA
  FGFR1   LVLGKPLGEGCFGQVVLAEAIGLDKDKPNRVTKVAVKMLKSDATEK.DLSDLISEMEMMKMIGKHKNIINLLGACTQD
               I                          II              III              IV
```

```
          672                                                                    805
PDGFR-B   ......GPIYIITEYCRYGDLVDYLHRNKHTFLQHHSDKRRPPSAELYSNALPVGLPL{72}LSYMDLVGLSYMDLVG
    RET   ......GPLLLIVEYAKYGSLRGFLRESRKVGPGYLGSGGSRNSSSLDHPDERA........LTMGDLISLTMGDLIS
    Tie   ......GYLYIAIEYAPYGNLLDFLRKSRVLETDPAFAREHGTAST...............LSSRQLLRLSSRQLLR
   INSR   ......QPTLVVMELMAHGDLKSYLRSLRPEAENNPGRPP....................PTLQEMIQPTLQEMIQ
   NGFR   ......RPLLMVFEYMRHGDLNRFLRSHGPDAKLLAGG.EDVAPGP...............LGLGLLALGLGQLLA
   HGFR   E.....GSPLVVLPYMKHGDLRNFIRNETHN........................PTVKDLIGPTVKDLIG
    UFO   SERESFPAPVVILPFMKHGDLHSFLLYSRLGGQPVY.....................LPTQMLVKLPTQMLVK
   EGFR   ......STVQLITQLMPFGCLLDYVREHKDN......................IGSQYLLNIGSQYLLN
    EPH   ......KPIMIITEFMENGALDAFLREREDQ......................LVPGQLVALVPGQLVA
    Klg   ......EPHYMVLEYVDLGDLKQFLRISKSKDESLKPQP...............LSTKHKVSLSTKHKVS
  FGFR1   ......GPLYVIVEYASKGNLREYLQARRPPGLEYCYNPSHNPEEQ................LSSSKDLVSLSSKDLVS
              V                                             VIa
```

```
          806                                                                    877
PDGFR-B   FSYQVANGMEFLASKNCVHRDLAARNVLICEGK.LVKICDFGLARDIMRD..SNYISKGSTFLPLKWMAPESI.FNSL
    RET   FAWQISQGMQYLAEMKLVHRDLAARNILVAEGR.KMKISDFGLSRDVYEE..DSYVKRSQGRIPVKWMAIESL.FDHI
    Tie   FASDAANGMQYLSEKQFIHRDLAARNVLVGENL.ASKIADFGLSRGE.....EVYVKKTMGRLPVRWMAIESLNYSVY
   INSR   MAAEIADGMAYLNAKKFVHRDLATRNCMVAHDF.TVKIGDFGMTRDIYET..DYYRKGGKGLLPVRWMPPESI.KDGV
   NGFR   VASQVAAGMVYLAGLHFVHRDLATRNCLVGQGL.VVKIGDFGMSRDIYST..DYYRVGGRTMLPIRWMPPESI.LYRK
   HGFR   FGLQVAKAMKYLASKKFVHRDLAARNCMLDEKF.TVKVADFGLARDMYDKEYYSVHNKTGAKLPVKWMALESL.QTQK
    UFO   FMADIASGMEYLSTKRFIHRDLAARNCMLNENM.SVCVADFGLSKKIYNG..DYYRQGRIAKMPVKWIAIESL.ADRV
   EGFR   WCVQIAKGMNYLEDRRLVHRDLAARNVLVKTPQ.HVKITDFGLAKLLGAE..EKEYHAEGGKVPIKWMALESI.LHRI
    EPH   MLQGIASGMKYLSNHNYVHRDLAARNILVNQNL.CCKVSDFGLTRLLDDF..DGTYETQGGKIPIRWTAPEAI.AHRI
    Klg   LCTQVALGMEHLSNGRFVHRDLAARNCLVSAQR.QVKVSALSLSKDVYN...SEYYHFRQAWIPLRWMPPESAV.LEDE
  FGFR1   CAYQVARGMEYLASKKCIHRDLAARNVLVTEDN.VMKIADFGLARDIHHI..DYYKKTTNGRLPVKWMAPEAL.FDRI
              VIb                           VII                 VIII
```

```
          878                                                                    958
PDGFR-B   YTTLSDVWSFGILLWEIFTLGGTPYPELPMNEQF.YNAIKRGYRMAQPAHASDEIYEIMQKCWEEKFEIRPPFSQLVLLL
    RET   YTTQSDVWSFGVLLWEIVTLGGNPYPGIPPERLF.NLLKT.GHRMERPDNCSEEMYRLMLQCWKQEPDKRPVFADISKDL
    Tie   TT.KSDVWSFGVLLWEIVSLGGTPYCGMT.CAELYEKLPQ.ADRMEQPRNCDDEVYELMRQCWRDRPYERPPFAQIALQL
   INSR   FTTSSDMWSFGVVLWEITSLAEQPYQGLSNEQVL.KFVMD.GGYLDQPDNCPERVTDLMRMCWQFNPNMRPTFLEIVNLL
   NGFR   FTTESDVWSFGVVLWEIFTYGKQPWYQLSNTEAI.DCITQ.GRELERPRACPPEVYAIMRGCWQREPQQRHSIKDVHARL
   HGFR   FTTKSDVWSFGVVLWELMTRGAPPYPDVNTFDIT.VYLLQ.GRRLLQPEYCPDPLYEVMLKCWHPKAEMRPSFSELVSRI
    UFO   YTSKSDVWSFGVTMWEIATRGQTPYPGVENSEIY.DYLRQ.GNRLKQPADCLDGLYALMSRCWELNPQDRPSFTELREDL
   EGFR   YTHQSDVWSYGVTVWELMTFGSKPYDGIPASEIS.SILEK.GERLPQPPICTIDVYMIMVKCWMIDADSRPKFRELIIEF
    EPH   FTTASDVWSFGIVMWEVLSFGDKPYGEMS.NQEV.MKSIEDGYRLPPPVDCPAPLYELMKNCWAYDRARRPHFQKLQAHL
    Klg   FSTKSDVWSFGVLMWEVFTQGEMPYAPLADDEVLAGLKSG.KTKLPQPEGCPSRLTKLMQRCWAPSPKDRPSFSELAAAL
  FGFR1   YTHQSDVWSFGVLLWEIFTLGGSPYPGVPVEELF.KLLKE.GHRMDKPSNCTNELYMMMRDCWHAVPSQRPTFKQLVEDL
              IX                          X                   XI
```

Fig. 2 Amino acid sequence alignment of the catalytic domains of 11 receptor protein-tyrosine kinases. Highly conserved residues among all protein kinases are indicated by black boxes with white lettering. The numbering is for human PDGF receptor β. The sequences were obtained from the protein kinase data base (15) and aligned by eye.

α and β phosphates. Moving towards the C-terminus, the next residue conserved in all protein kinases is E651 (E91), which is also believed to be involved in coordinating Mg^{2+} ATP. In the peptide-binding domain is the region that forms the catalytic loop, HRDLAARN (residues 824–831 for the PDGF receptor; 164–171 for PKA). The aspartate is thought to be the catalytic base, and LAAR functions in peptide accessibility and recognition. This region is highly conserved in both protein kinase families, yet discriminates between PTKs and protein-serine/threonine kinases (e.g. this sequence is LKPE in PKA). The aspartate of DFG (residues 844–846 for PDGFR; 184–186 for PKA) functions in the chelation of Mg^{2+}. E873 (E208) and R948 (R280) are thought to form ion bridges that stabilize the two lobes and D885 (D220) stabilizes the catalytic loop. When the structure of a PTK catalytic domain is resolved it will probably confirm these similarities, but more interestingly, it will highlight the differences between the two families, and provide insight into how the different phosphoamino acid acceptor specificities are achieved.

The sequences of the catalytic domains of 48 RPTKs have been analysed as described by Hanks and Quinn (15) and the results are represented as a molecular phylogenetic tree (Fig. 3). The tree is an unrooted relatedness tree that clusters similar sequences and indicates relative relatedness by branch lengths. For reference, the cytoplasmic PTKs would branch off the EPH subfamily branch. The 48 vertebrate RPTKs chosen for this analysis represent what are putatively distinct gene products (see below) and represent the distinct vertebrate RPTKs whose sequences are in the protein kinase data bank (15).

3. Receptor protein-tyrosine kinase classification

We have classified the RPTKs based on structure. The classification proposed here makes use of the types of structural analyses shown in Figs 1 and 3. Fortunately, there is a good concordance between the two sets of information. The phylogenetic tree shown in Fig. 3 clusters the RPTKs that have similar catalytic domain sequences. The results of the computer program that compares the sequences are reflected graphically in the tree. Breaking the long branches of the tree to separate a group of RPTKs (or single RPTKs) from more distantly related ones is one way to visualize subfamily groups. These groupings would be very similar if instead one used overall molecular topology, as shown in Fig. 1, to group the subfamilies.

The nomenclature used for this review considers all protein kinases (including protein-serine/threonine kinases) as a superfamily and the PTKs as a family. Groupings of PTKs that have similar overall structure and cluster in phylogenetic trees are referred to as subfamilies. We have grouped the RPTKs into subfamilies, which are shown in Table 1. Many of the receptors listed in Table 1 have been isolated as cDNAs multiple times. When this occurs in the same species the identity is obvious, but when closely-related genes are isolated from different species it is difficult to determine whether they are orthologues if little is known about the gene or the gene product. The ideal way to determine if two genes are orthologous is to isolate all the family members from both species; however, this

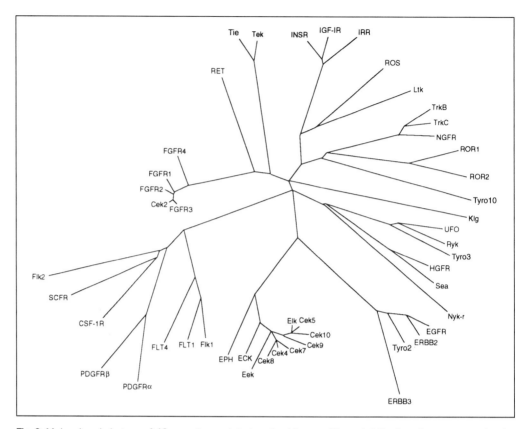

Fig. 3 Molecular phylogeny of 48 receptor protein-tyrosine kinases. The catalytic domain sequences for the receptors were obtained from the protein kinase data base or by personal communication and analysed as described previously (15). The receptor protein-tyrosine kinases shown are distinct vertebrate gene products (i.e. none are orthologues). The receptors are named as in the data base, or as the receptor for a known ligand, following conventions set forth previously (84). The sequences are human, except for those with lower case letters. IGF-IR, insulin-like growth factor-I receptor; CSF-1R, colony stimulating factor-1 receptor; SCFR, stem cell factor receptor.

is not always accomplished. Chromosomal mapping or cross-species genomic Southern hybridizations with probes from both species are ways to determine whether the genes are orthologous. However, in most cases only sequence data are available. In some cases the alternative names obviously represent the same RPTK, e.g. EGF receptor, HER, and c-*erb*B. In other cases genes from different species have been grouped together under the assumption that they are from orthologous genes, e.g. the FGFR2 group. cDNAs for this gene have been isolated from three species: human, TK14 and K-SAM; mouse, Bek; and chicken, Cek3. The catalytic domains for these are about 99% identical and are considered to represent products of orthologous genes. With other genes it is not so clear. Cek2 is 95% identical to human FGFR3 in the catalytic domain and could be the chicken homologue, but with only these data prediction is difficult. Sequences outside the catalytic domain

Table 1 Vertebrate receptor protein-tyrosine kinases divided into subfamilies[a]

PDGFR subfamily
PDGF receptor α (5)
PDGF receptor β (5)
SCF receptor, c-*kit* (5)
CSF-1 receptor, c-*fms* (5)
Flk2, Flt3 (5)

FLT1 (7)
Flk1, KDR, Nyk (7)
FLT4 (7)

FGFR subfamily
FGF receptor 1, FLG, Cek1
FGF receptor 2, Bek, K-SAM, Cek3, TK14
FGF receptor 3, FLG2
FGF receptor 4, TKF
Cek2

INSR subfamily
Insulin receptor
IGF-1 receptor
IRR
ROS
Ltk

EGFR subfamily
EGF receptor, HER, c-*erb*B, *X-mrk*
Neu, ERBB2, HER2
ERBB3, HER3
Tyro2, ERBB4

NGFR subfamily
NGF receptor, TrkA
TrkB
TrkC

HGFR subfamily
HGF receptor, MET
Sea

EPH subfamily
EPH
ECK
Elk, Cek6
Eek
Cek4, Mek4, HEK
Cek5, Nuk
Cek7
Cek8, Sek
Cek9
Cek10

UFO subfamily
UFO, AXL, Ark
c-*ryk* c-*eyk*
Tyro3

TIE
Tek

RET

Klg

Nyk-r, Ryk, Vik, Nbtk-1

ROR1
ROR2

Tyro10

[a] The receptors are grouped according to structural characters and sequence homology in their catalytic domains. The number of Ig-like repeats in members of the PDGFR subfamily is indicated in parentheses. References not in Hunter (84) are: Flk2 (85); Flt3 (86); Flk1 (87); KDR (88); Nyk (89); FLT4 (90); FGFR3 (91); FGFR4 (92); TK14 (93); FLG2 (94); TrkC (95); ECK (96); Eek (21); Cek4/Mek4 (22); HEK (97); Cek5 (23); Nuk (Tony Pawson, personal communication); Sek (98); UFO (99); AXL (80); Ark (100); c-*ryk* (Ran Jia and Hidesaburo Hanafusa, personal communication); TIE (81); Tek (101); Klg (102); Nyk-r (103); Ryk (104); Vik (105); Nbtk-1 (106); ROR 1 and 2 (83); Tyro2, Tyro3, and Tyro10 (Cary Lai, personal communication); Cek6, Cek7, Cek8, Cek9, and Cek10 (107); ERBB4 (108).

should also be considered. The same caveats apply here too, but since these are commonly more divergent between subfamily members than the catalytic domains, sequences outside the catalytic domain may prove useful in determining whether two RPTKs from different species with closely related catalytic domains are orthologues or not. The predictions in Table 1 as to which RPTKs are homologues from different species should be considered tentative.

4. Receptor protein-tyrosine kinase function

RPTKs are known to act as receptors for growth factors, for differentiation factors, and for factors that stimulate metabolic responses. The functions of the RPTKs are governed by three principles. First, expression of almost all RPTKs is restricted to specific cell types in the organism, and this depends upon the nature of the RPTK gene promoter and enhancer. Second, in the cells where a particular RPTK is expressed, its function is dictated by the activating ligands that bind to its extracellular domain, and this in turn will depend on the availability and distribution of the cognate ligands in the organism. Third, the response to ligand-dependent RPTK activation in a given cell depends upon the intracellular proteins that are phosphorylated. The second and third properties are intrinsic to the structure of the protein, and thus these functions would be expected to parallel the structural relationships within RPTK subfamilies, whose members may be anticipated to have similar functions. This is evidently true in most cases where ligands are known, but is not universal. For example, the closely-related insulin and IGF-1 receptors have different functions; activation of the insulin receptor induces a metabolic response, whereas activation of the IGF-1 receptor elicits a growth response.

The functions of RPTKs are most readily apparent where a ligand is known. However, because of the rapid pace of identification of new RPTKs, the cognate ligands are lacking for almost half the known RPTKs. With regard to ligand binding, even though the conserved structural motifs are readily apparent, extracellular domain sequences are more divergent within a subfamily than the catalytic domain sequences, with sequence identities ranging from as low as ~30% to ~80%. Nevertheless, for most of the RPTK subfamilies there is evidence that they bind ligands that are themselves members of families (see below). For some RPTK subfamilies, no ligands have been identified, and therefore we cannot be certain that these orphan RPTKs will interact with ligand families, even though this is a strong prediction.

The identities of substrates for different RPTKs are still being uncovered. It is becoming apparent that many of the physiologically relevant substrates are proteins that bind with high affinity to RPTK autophosphorylation sites via a specific domain (see below). This binding is sequence specific with the residues immediately to the C-terminal side of the phosphotyrosine being most important. Therefore, the extent to which autophosphorylation sites and the precise sequences around them are conserved between members of a subfamily will dictate whether or not individual subfamily members phosphorylate the same spectrum of substrates. Since autophosphorylation sites are commonly located outside the catalytic domain proper in regions that are less highly conserved, even closely related RPTKs will bind and phosphorylate distinct but usually overlapping sets of substrates. For instance, this has been found to be the case for the PDGF and CSF-1 receptors, which are members of the PDGF receptor subfamily.

It is clear that the cell type in which a RPTK is expressed affects the cellular response to receptor activation. For instance, in neuronal cells NGF elicits a differentiation response upon binding to the NGF receptor (Trk), whereas when the NGF receptor is expressed ectopically in fibroblasts, NGF induces a growth response. This may in part be due to the different repertoire of substrates available in different cell types, as well as to differences in the programmed responses of cells to activation of the same signal pathway. In other cases, activation of an ectopically expressed RPTK may result in the same response as in its normal host cell. For instance, CSF-1 treatment of fibroblasts expressing the CSF-1 receptor elicits a growth response, as it does in myeloid precursor cells. This implies that many of the RPTK-activated signalling pathways are conserved between different cell types (see below). Conversely, different RPTKs expressed in the same cell can elicit different responses through phosphorylation of different substrates, which in turn trigger different signal pathways.

We will start by discussing the vertebrate RPTKs. As indicated above, for many RPTKs homologues have been cloned from several species, and we will refer to the human RPTK where possible. A few RPTKs have been only identified in non-human mammalian or avian species, but the authors believe that each of the genes listed in Table 1 is distinct, and will be found in the genome of a single vertebrate species. Representatives of the major RPTK subfamilies have been found in a wide variety of vertebrates, including fish, amphibians, birds, and mammals.

4.1 Platelet-derived growth factor receptor subfamily

The receptors in the PDGF receptor subfamily are not as closely related within their catalytic domains as some of the RPTK subfamilies, but they do cluster on one branch of the RPTK catalytic domain relatedness tree (Fig. 3). Moreover, they all contain a kinase insert in their catalytic domain, and they all possess various numbers of Ig-like domains in their ligand-binding domains. The prototypical member, PDGF receptor, has five Ig repeats, as do the CSF-1 and SCF receptors and Flk2 (also known as Flt3). The ligands of the two PDGF receptors and the CSF-1 and SCF receptors are related (see Section 5), and their binding sites map to Ig repeats 1 and 2 of their receptors. Flk2 is expressed most highly in the brain and haematopoetic cells, but its ligand is unknown. However, it seems likely that the Flk2 ligand will be related to the PDGF cytokine family. The genes corresponding to the CSF-1 and SCF receptors were originally identified as retroviral transforming genes. Flk1, FLT1, and FLT4, the other three RPTKs in this subfamily, have seven Ig repeats, and, since they cluster on a secondary branch and all bind VEGF or VEGF-related ligands, one might consider them as a separate subfamily. A unique feature of the PDGF receptor subfamily is the presence of a split kinase domain, in which there is an insert of 65–97 amino acids (compared to Src) between the ATP-binding portion of the catalytic domain and the substrate-binding portion (between subdomains V and VIA). This insert is quite widely diverged between subfamily members, and, because it contains several autophosphorylation sites, the kinase insert is important for substrate recognition (see Section 7).

4.2 Fibroblast growth factor receptor subfamily

The FGF receptor subfamily is a highly related group, as seen in the tree in Fig. 3. This subfamily is characterized by three Ig-like repeats in the ligand-binding domain. It has been claimed that these receptors have a kinase insert like the PDGF receptor subfamily, but, although the group has 14 more residues in this region of the catalytic domain than in SRC, this length is not significantly different than in RET, the NGF receptor subfamily, or the insulin receptor subfamily. In addition, no significant function has been attributed to this region in the FGF receptor subfamily. Therefore, the molecular topology shown in Fig. 1 does not show a kinase insert. Since four human FGF receptors have been reported, we know there are at least that many in this subfamily and probably more. These receptors transmit signals mediated by at least eight related growth factors (see below) that give rise to a myriad of biological responses that include angiogenesis and meso-derm induction in embryos. The use of alternate promoters gives rise to forms of FGFR1 and FGFR2 that lack the first Ig-like repeat, but which bind ligand normally. Alternate splicing of the second half of the third repeat in FGFR1 and FGFR2 gives rise to receptors with altered ligand-binding specificity. Together, these results suggest that the ligand-binding site of this subfamily lies in the third repeat. Each RPTK in this subfamily binds αFGF and at least one other member of the FGF family of cytokines, with the exact specificity depending on which alternate exon sequence is present in the third Ig repeat. Although not all of the FGFs have been tested against every FGFR, the picture that emerges for the FGFR and FGF families is that each receptor binds a subset of the growth factors and vice versa.

4.3 Insulin receptor subfamily

The insulin receptor subfamily includes the namesake, the closely-related IGF-1 receptor, and the next three closest RPTK relatives, IRR, ROS, and Ltk. The mature insulin and IGF-1 receptors are disulphide-bonded heterotetramers containing two copies each of the α and β subunits, which are derived by proteolytic cleavage from a precursor molecule. The insulin and IGF-1 receptors both bind insulin and IGF-1, which are related factors. However, the IGF-1 receptor has a much higher affinity for IGF-1 than insulin, and in consequence the majority of the metabolic effects of insulin are mediated through the insulin receptor. Conversely, the growth-stimulatory effects of IGF-1 are elicited via the IGF-1 receptor. A recently dis-covered substrate for the insulin receptor, IRS-1, has been found to function as an adapter protein that serves to bind and activate phosphatidylinositol 3' kinase (PI3-kinase). IRS-1 contains many tyrosine phosphorylation sites, several of which are in the proper context to bind PI3-kinase. The use of an intermediate protein for binding PI3-kinase instead of a PI3-kinase binding site in the receptor itself is unique amongst the receptors studied thus far, but could be connected to the fact that the insulin receptor is distinct from other RPTKs in its ability to induce a metabolic rather than a growth response.

The ROS RPTK is the closest vertebrate RPTK to the *Drosophila* Sevenless RPTK, which is involved in the differentiation of the R7 photoreceptor cell in each ommatidion in the eye. Although the ligand for ROS is unknown, the ligand for Sevenless is the *boss* protein, which is present as a surface protein containing seven transmembrane domains on the R8 photoreceptor cell that lies adjacent to the R7 cell expressing the Sevenless RPTK. So far this is the only known example of a ligand for an RPTK that is obligatorily a cell surface protein, although several soluble growth factors that activate RPTKs are derived from precursors with transmembrane domains, and are active when anchored in this fashion (e.g. an anchored form of TGFα can activate the EGF receptor). Multiple forms of the Ltk RPTK exist that differ in their extracellular domains. One form of Ltk with a short extracellular domain of ~100 residues is located predominantly in the endoplasmic reticulum, and is activated by disulphide-bond crosslinking under conditions of oxidative stress, apparently in the absence of a ligand.

4.4 Epidermal growth factor receptor subfamily

The EGF receptor subfamily currently comprises four members. These receptors have similar overall topology and significant relatedness in amino acid sequence. This subfamily is of special interest because both the EGF receptor and ERBB2 have been implicated in a high percentage of some types of human cancers by virtue of the presence of amplified copies of their genes (reviewed in 1, 2). Moreover, the retroviral v-*erb*B oncogene was derived from the EGF receptor gene. An interesting feature of ERBB3 is the fact that several residues conserved in all PTKs are absent, including the aspartate that has been proposed to be the catalytic base in the PKA catalytic subunit, which in ERBB3 is an asparagine. There is some evidence that ERBB3 can autophosphorylate on tyrosine (16), but until it has been convincingly demonstrated that ERBB3 can phosphorylate substrates on tyrosine, one should consider the possibility that it has a function that does not require PTK activity. ERBB4 (also known as Tyro2) is a recently described member of this subfamily, which based on catalytic domain similarity, is most similar to the ERBB2 and EGF receptors. The ligands for ERBB3 and ERBB4 are not yet known, but several related factors that bind and activate the EGF receptor have been identified, including EGF and TGFα (see below). Recently, a family of proteins generated by alternative splicing has been reported to act as ligands for ERBB2. The EGF-binding site in the extracellular domain of the EGF receptor has been mapped by expressing chicken/human EGF receptor chimeras (17) and by chemical crosslinking studies to the region between the two cysteine-rich domains, with some contribution from the N-terminal 100 residues of the receptor.

4.5 Nerve growth factor receptor subfamily

The NGF receptor subfamily has three members that share structure and sequence similarities and also bind ligands of similar nature (see below). The original receptor in this family was called proto-Trk after the human oncogene product, Trk, which is a cytosolic non-muscle tropomyosin-receptor kinase fusion protein. Subsequently, the Trk-related RPTKs, TrkB and TrkC, were identified and proto-Trk was re-named TrkA. TrkA is now known to be the high affinity NGF receptor. There is significant similarity in the external domains of this family, and the discovery that Trk binds NGF quickly led to the demonstration that TrkB and TrkC bind NGF-related neurotrophins; TrkB binds BDNF, NT-3, and NT-4/5, and TrkC binds NT-3. TrkB appears to be promiscuous in its ligand-binding specificity, but there is some evidence that BDNF is the most important physiological ligand for TrkB. All three Trk family RPTKs are expressed in neurones, predominantly in the central nervous system. In neuronally-derived cells activation of Trk RPTKs by their cognate ligands elicits a differentiation response, which includes neurite outgrowth. The neurotrophins, as their name implies, may play a role in directional neurite extension, and they also promote the survival of neurones in culture. There is some indirect evidence that the binding site for NGF on Trk lies close to the membrane.

4.6   Hepatocyte growth factor receptor subfamily

The HGF receptor has been shown to be identical to the *MET* gene product. This receptor was discovered as a transforming gene in the NIH3T3 transfection assay, and has subsequently been implicated in naturally occurring tumours as a result of amplification. MET is synthesized as a precursor that is proteolytically cleaved into two subunits that are held together by a disulphide bond in a fashion analogous to the insulin receptor (18) (see Fig. 1). The receptor mediates the effects of a cytokine known as HGF or scatter factor. HGF was originally characterized as a growth factor for hepatocytes and other epithelially-derived cells, whereas scatter factor was identified as a motogenic factor able to stimulate the movement or scattering of epithelial cells (reviewed in 19). The same epithelial cells can respond mitogenically or motogenically depending on the conditions, but it is not clear exactly what governs these two different responses. Presumably, at least some of the substrates phosphorylated by the activated HGFR RPTK are different in the two situations, but what these substrates are remains to be determined (see Table 3 for substrates known to associate with HGFR). Little is known about Sea, which was discovered as a transduced retroviral oncogene, but the similarity between these two receptors indicates that they may have similar functions and bind related ligands.

4.7   EPH subfamily

The EPH subfamily currently consists of ten members and is the largest of the RPTK subfamilies. There has not been any report of a ligand found for any member of this RPTK subfamily, and their functions remain unknown. Several of the receptors in this subfamily are expressed most highly in adult brain, i.e. Elk (20), Eek (21), Cek4 (22), and Cek5 (23), and Cek5 is known to be expressed in neurones. This implies a function other than in mitogenesis or differentiation. However, these receptors might be involved in such processes earlier in development, and they could play a role in differentiated function in the adult. The EPH and ECK RPTKs are not expressed in neural tissues, and ECK is most highly expressed in epithelial cells. None of these RPTKs has been implicated in human cancer, but overexpression of EPH in NIH3T3 cells makes these cells tumourigenic in nude mice. The function of this group of RPTKs will be difficult to decipher until ligands are found. However, based on the conserved nature of their external domains, it is likely that they bind related ligands.

4.8 UFO subfamily

UFO was isolated as a transforming gene in the NIH3T3 cell transfection assay from the leukaemic cells of a patient with a chronic myeloproliferative disorder. AXL, identical to UFO but isolated independently, was isolated by the same NIH3T3 cell assay from two chronic myelogenous leukaemia patients. This RPTK is expressed in many cell types, and wild type UFO can transform when overexpressed in rodent fibroblasts. Another member of this subfamily, Ryk, was identified as the cellular counterpart of a retroviral oncogene (24). Tyro3 was discovered in a PCR-based screen for novel PTKs, and is expressed predominantly in the adult brain and in primitive haematopoietic cells. Tyro3 can apparently transform rodent fibroblasts when overexpressed (Cary Lai, personal communication). These data indicate that the RPTKs in this subfamily have an oncogenic propensity. Ligands for this subfamily have not been identified, but given the fact that unmutated RPTKs of this subfamily can transform rodent fibroblasts, it is possible that the ligands are expressed by these cells resulting in autocrine transformation.

4.9 Other receptors

The remaining RPTKs are not closely related to any of the subfamilies described above. Ligands have not been identified for any of these RPTKs. Tie was identified in a PCR-based screen for novel PTKs. The unique feature of this RPTK is the presence of three EGF-like repeats in the ligand binding domain. Tie is expressed most highly in endothelial cells and some haematopoetic cell lines of early lineage. A mouse RPTK, Tek, which is closely related to Tie, has also been isolated recently but Tek is distinct from Tie, even though Tek is also expressed in cells of the endothelial lineage. RET was originally identified by transfection of NIH3T3 cells as an oncogene that had been activated by a transfection-induced rearrangement. A cadherin repeat containing domain has recently been noted in the RET ligand-binding domain. Cadherins form a major class of calcium-dependent cell–cell adhesion molecules. The presence of cadherin repeats in RET could mean that calcium-dependent interaction of RET with other cadherin-like molecules is a mechanism of receptor activation. Klg was isolated from a chicken embryonic cDNA library with a v-*sea* probe. Klg stands for kinase like gene, so-named because it was noted that the highly conserved DFG motif was absent, leading to the speculation that Klg lacks intrinsic PTK activity. Although the catalytic domain is not closely related to any other receptors, Klg does have seven Ig loops in its ligand-binding domain, like some of the members of the PDGF receptor subfamily. Nyk-r, Ryk, Vik, and Nbtk-1 were all found in PCR-based screens and appear to be products of the same gene. This RPTK has a rather short extracellular domain (~200 amino acids) and some unusual substitutions at highly conserved residues in the catalytic domain. However, none of the substitutions is in residues as critical as those altered in Klg or ERBB3, and Nyk-r would be expected to have protein kinase activity. Two RPTKs, ROR1 and ROR2, were discovered in a screen designed to find receptors in the Trk subfamily. Their closest relatives are the Trks, but it seems that the RORs represent a new subfamily. In addition to a unique extracellular region, they too have some unusual aspects in their catalytic domains. Most notably, they have a lysine instead of the conserved arginine adjacent to the catalytic base. All other PTKs (except ERBB3) have HRD in this region of the catalytic loop; the RORs have HKD. Despite this the RORs do have intrinsic PTK activity. The RORs are expressed much more highly in the embryo than the adult. The last receptor, Tyro10, is represented by a long singular branch and, other than partial sequence data, little is known about this RPTK.

4.10 Non-vertebrate RPTKs

In addition to the vertebrate RPTKs, several RPTKs have been identified in invertebrate species. In most cases, it is hard to be certain whether these are the true homologues of vertebrate RPTKs, although some clearly fall into particular RPTK subfamilies. In *Drosophila melanogaster* these are DER (25), DILR (26), *sevenless* (27), *torso* (28), D*trk* (29), and DFGFR (30). In *Caenorhabditis elegans*, a single RPTK gene is known, *let-23* (31). The functional insights that have been gained from the genetic analysis of these RPTKs are discussed below.

5. Receptor protein-tyrosine kinase ligands

The key to understanding the biological function of an RPTK is to identify the cognate ligand(s) and determine how a cell responds when that ligand is present. Candidate ligands are presented in Table 2 and are grouped into families that correspond to the receptor subfamilies in Table 1. Most of these ligands are called growth factors, since growth has historically been a convenient assay for finding factors that modulate a cell's activity. However, cytokine is a better collective name for the RPTK ligands, since, while tyrosine phosphorylation pathways commonly stimulate cell division, many RPTK ligands elicit responses other than growth.

The families shown in Table 2 have intrafamily similarities, but in some cases it is very weak. For example, there is only 18% identity between VEGF and PDGF B. However, these two ligands do share both a homodimeric structure and the

Table 2 Ligands for receptor protein-tyrosine kinases[a]

PDGF family	**EGF family**
PDGF A	EGF
PDGF B	TGFα
SCF, steel factor, Kit ligand	NDF, heregulin, GGF
CSF-1	Amphiregulin
VEGF	HB-EGF
PLGF	SDGF
	Betacellulin
FGF family	
aFGF	**NGF family**
bFGF	NGF
int-2	BDNF
K-FGF	NT-3
FGF-5	NT-4/5
FGF-6	
KGF	**HGF family**
FGF-8	HGF
Insulin family	
Insulin	
IGF-1	
IGF-2	

[a] The ligands are grouped by primary structure relatedness. General references: 5, 19, 109–111.

cysteines involved in disulphide bridges. The molecules in this family exist as homodimers, with the exception of PDGF, which can be found as an active AB heterodimer. VEGF appears to be a ligand for at least two of the three clustered PDGFR subfamily RPTKs (Fig. 3) that have seven Ig-like domains, Flt1 (32) and Flk1 (33, 34). PLGF is related to VEGF and can be expected to bind one or more of these RPTKs as well. Of the known classes of RPTK ligand, the members of the PDGF family are the only ones that are covalently dimeric. The significance of this is not yet clear, although as indicated below this may play a role in receptor dimerization.

The members of the EGF family are ligands for the receptors in the EGF receptor subfamily. Most of those listed are capable of binding to the EGF receptor except for NDF (35) (also called heregulin (36), GGF (37), and ARIA (38)) which is reported to activate the ERBB2 protein. Other members of this subfamily may all be physiological ligands for the EGF receptor, and there may be additional members of this receptor family that have yet to be discovered. The NDF gene gives rise to multiple differentially spliced forms, and these may have different biological activities (37). All the members of the EGF family are derived from precursors that have a single transmembrane domain. The sequences for the mature factor, which contains an EGF repeat, lie close to the transmembrane domain in the precursor, and the soluble factor is derived by processing of the precursor at specific sites. In some cases these precursors have very long N-terminal extensions (e.g. the EGF precursor is ~ 100 kDa) and conserved cytoplasmic domains, and it is possible that these membrane-anchored precursors have a physiological signalling function.

The neurotrophin family members have recently been found to be the ligands for the Trk subfamily of RPTKs. NGF has been shown to regulate the PTK activity of the *trk* protooncogene product, which is now referred to as the NGF receptor. Subsequently, it has been shown that overlapping sets of these factors bind the three receptors (reviewed in 5 and references therein). NGF functions as a non-covalently associated dimer. Exactly which neurotrophins act as physiological regulators of the different Trk family RPTKs remains to be determined.

HGF, also known as scatter factor, is a ligand for the *met* protooncogene product. The structure of HGF is unique among the growth factors (39). It has homology with plasminogen, the precursor of the serine protease plasmin. The structure has four kringle domains and a domain that has homology with the serine protease catalytic subunit. However, because residues known to be part of the catalytic triad are altered in HGF, it is believed that HGF does not have protease activity. HGF has a signal peptide and the mature form is found as two disulphide-linked peptides that are derived by proteolysis.

Some of the ligands listed in Table 2 bind more than one RPTK, and vice versa. The FGF family members and the FGF receptor subfamily are good examples. In contrast, other ligands have been found to bind only one RPTK and, as in the case of CSF-1, the cognate receptor binds only one known ligand. In general, crosstalk occurs only within one family of cytokines and one subfamily of RPTKs. The existence of families of related cytokines that interact with subfamilies of RPTKs raises the critical questions of which ligands regulate which receptors *in vivo*, and whether there is redundancy. Although many ligands within a family may have the ability to bind and activate a specific RPTK *in vitro*, the temporal and spatial aspects of expression, processing, and activation of the ligands will clearly dictate this specificity *in vivo*. The use of homologous recombination to inactivate genes in the mouse, including RPTK and ligand genes, coupled with the use of transgenes expressing dominant negative forms of certain growth factors and their receptors, is beginning to answer some of these questions.

6. Receptor protein-tyrosine kinase activation

Signal transduction begins when the ligand binds the receptor. The generalized series of events is as follows:

(1) After the ligand binds, the receptor dimerizes. This is presumed to occur by a ligand-induced conformational change in the external domain that results in dimerization of receptor. Emerging evidence suggests that only a single ligand molecule is bound per receptor dimer. In the case of ligands that are dimers, such as PDGF, dimerization may in part be driven by the ligand binding to two

receptor molecules. The binding of dimeric ligands generates a truly symmetrical receptor dimer, but in the case of monomeric ligands such as EGF, there must be two distinct binding sites on the ligand and the interaction sites on the two receptor molecules must also be at least partly different.

(2) Receptor dimerization leads to intermolecular autophosphorylation. The ligand-induced dimerization of the extracellular domains necessarily results in juxtaposition of the cytoplasmic domains. It is thought that contact between the two cytoplasmic domains induces a conformational change that stimulates catalytic activity, which leads to mutual transphosphorylation between the two receptor molecules in a dimer. For each RPTK, autophosphorylation occurs at a distinct set of sites, most of which lie outside the catalytic domain.

(3) Transphosphorylation results in activation of the dimeric RPTK for phosphorylation of cytoplasmic substrates. Based largely on the use of 'dominant-negative' kinase-inactive mutant RPTKs, it has been concluded that transphosphorylation is essential for substrate phosphorylation. When 'dominant-negative' kinase-inactive mutant RPTKs are coexpressed with a wild type RPTK, signalling by the wild type RPTK is suppressed in proportion to the relative levels of the wild type and mutant RPTKs. The suppression is thought to occur because the kinase-inactive RPTK molecules form dimers with the wild type molecules and become phosphorylated, but cannot in turn phosphorylate their wild type partner. This result also implies that a secondary conformational change occurs upon transphosphorylation, which allows the catalytic domain to phosphorylate substrate molecules.

(4) The autophosphorylated dimer actively recruits substrates that have an increased affinity for the receptor due to its autophosphorylation (see below). The use of kinase-inactive RPTK mutants demonstrates that RPTK signalling is absolutely dependent on the phosphorylation of specific cytoplasmic substrates.

There is a large body of evidence that supports the dimerization model of RPTK activation (discussed in recent reviews (2–4, 6)). For most RPTKs, it is possible to detect dimer formation following ligand binding by chemical crosslinking experiments. In the PDGF receptor system, the fact that PDGF heterodimers bind two different PDGF receptors has made it possible to observe the ligand-driven dimerization of receptors into heterodimers (40). In this system the PDGF heterodimer, PDGF-AB, and the PDGF homodimers, PDGF-AA and PDGF-BB, can be assayed for activities on PDGF α and β receptors independently, and on mixed populations. Because the A chain of PDGF binds only the α receptor, but the B chain binds both types of receptors, it was shown that dimerization occurred during receptor activation and that receptor homodimers and heterodimers were utilized by PDGF.

The mechanism whereby monomeric ligands induce receptor dimerization is intuitively less obvious than for dimeric ligands, but the evidence that apparently monomeric ligands like EGF cause receptor dimerization is very strong. In the growth hormone receptor system, where a single growth hormone molecule is present in a receptor dimer, the three-dimensional structure of the receptor/growth hormone dimer shows that growth hormone has two distinct receptor binding sites, which can bind receptor molecules independently (41). Although there is no experimental evidence that any monomeric RPTK ligand has two binding sites, this seems to be a reasonable possibility. If this is true, then it should be possible to create 'dominant-negative' mutant growth factor molecules in which one of the two receptor-binding sites has been mutated. Although there are three-dimensional structures of several RPTK ligands, to date there is no three-dimensional structure of an RPTK ligand binding domain nor of an RPTK/ligand complex, which will be needed to elucidate the precise mechanism of ligand-mediated receptor dimerization.

In support of the dimerization model it has been shown that the constitutively activated *neu* oncogene, which has a single point mutation in its transmembrane domain converting an uncharged valine to a charged glutamate, forms multimers in the absence of ligand (42). This could result in constitutive PTK activity in the absence of the ligand. Even more convincing are recent reports that PDGF, EGF, and SCF receptor mutants lacking catalytic activity act in a dominant negative fashion to attenuate signal transduction by wild-type receptors (43–46). The interpretation of this phenotype is that inactive heterodimers form upon ligand binding in which the wild-type receptor cannot be transphosphorylated by the inactive subunit, and therefore is not activated for substrate phosphorylation.

All evidence to date would indicate that signal transduction by RPTKs is dependent entirely on PTK activity. Mutations of many varieties, from truncations to single point mutations that block ATP binding and eliminate PTK activity, also eliminate most known cellular responses. However, there are scattered reports where a kinase-inactive mutant receptor has been found to elicit a specific cellular response. For example, a kinase-inactive EGF receptor has been shown to activate MAP kinase following EGF treatment even though it is incapable of stimulating a mitogenic response (47). The mechanism whereby this occurs is unclear, but it may occur by EGF-dependent activation of an EGF receptor-associated PTK (48).

7. Formation of an activated receptor protein-tyrosine kinase signalling complex

In recent years a new model describing the initial cytoplasmic events in signal transduction has emerged, which involves the formation of complexes between the autophosphorylated receptor and several intracellular signal transducing molecules, many of which are actually substrates for the receptor (1, 2, 49, 50). The underlying principle of this model is a novel mechanism whereby the affinity between an RPTK and a substrate is increased as a result of ligand binding. The majority of RPTK substrates that associate with the receptor (i.e. that can be co-immunoprecipitated) have a region of ~100 amino acids in common that is known as Src homology region 2 (SH2). The SH2 domain was originally identified as a region of similarity in v-Fps and v-Src that lies outside the catalytic domain (51). SH2 domains have subsequently been found in many proteins (reviewed in 49, 50, 52). Accumulating evidence (see below) favours a model in which an SH2 domain has affinity for a phosphotyrosine-containing motif on the receptor. In a typical RPTK, ligand binding leads to autophosphorylation of several tyrosines, and most if not all of these phosphotyrosine residues associate with SH2 domains of specific substrate molecules. The recruitment of SH2-containing signal-transducing molecules may transduce a signal by changing their cellular location, changing their activity as a result of binding to or of phosphorylation by the receptor, or by some other unknown mechanism.

The SH2-containing molecules that have been found to associate with RPTKs are shown in Table 3. There are a number of additional SH2 proteins known, and many of these have been found to be tyrosine phosphorylated in either growth factor-treated cells or cells transformed by an oncogenic PTK. Presumably, when tested these proteins will prove to associate with an activated RPTK or a secondarily-activated PTK. Many of the SH2-containing proteins also contain Src-homology 3 (SH3) domains. SH3 domains are found in a wide variety of proteins, some of which are associated with the cytoskeleton. There is accumulating evidence that the SH3 domains in SH2-containing proteins are involved in signal propagation (53). It is important to note that not all RPTK substrates contain SH2 domains. For instance, a number of cytoskeletal or submembranous proteins lacking SH2 domains are phosphorylated on tyrosine in response to growth factor treatment (e.g. ezrin, annexins I and II), and phosphorylation-dependent changes in the activities of these proteins may be involved in growth factor-mediated cell shape changes. There are numerous additional proteins that become phosphorylated on tyrosine, but they are not sufficiently characterized at the molecular level to be certain whether they contain SH2 domains.

PI3-kinase has been found to associate with many, but not all, RPTKs. The activity of this enzyme is easily detectable by assay in RPTK immunoprecipitates, and therefore it has been tested as a possible substrate more often than other substrates. Similarly, the fact that the PDGF receptor binds most of the SH2-containing proteins listed in Table 3 may in part reflect the amount of effort put into experiments with this receptor, rather than that it has more substrates than other RPTKs. However, the underlying principles for the interactions between all the substrates and RPTKs in Table 3 are the same.

The current model has gathered most of its momentum from the study of PI3-kinase association with RPTKs. PI3-kinase consists of a 110 kDa catalytic subunit and an 85 kDa regulatory subunit. The 85 kDa subunit of PI3-kinase was cloned by three groups (54–56) and was shown to contain two SH2 domains. Ligand-stimulated, autophosphorylated receptors associate with PI3-kinase. This association occurs through the binding of one or both of the SH2 domains in p85 and a specific phosphotyrosine on the receptor. The phosphotyrosines have been

Table 3 SH2-containing molecules that associate with receptor protein-tyrosine kinases[a]

SH2-containing substrate	Associated RPTK	References
PI-3 kinase (p85)	PDGFR	Reviewed in 1, 6
	CSF-1R	60, 112
	EGFR	113
	SCFR	59, 114, 115
	HGFR	116
	TrkA	117
PLCγ1	PDGFR	Reviewed in 1, 6
	EGFR	63
	SCFR	59, 114, 115
	FGFR	67, 68, 118
	HGFR	119
	TrkA	120
	TrkB	121
Ras GAP	PDGFR	Reviewed in 1
	HGFR	119
GRB2	PDGFR	53
(Sem-5, Drk)	EGFR	53
	CSF-1R	122
VAV	PDGFR	123
	EGFR	123, 124
c-Src	PDGFR	123
PTK subfamily	HGFR	119
	CSF-1R	125
SHC	EGFR	126
Nck	PDGFR	127, 128
	EGFR	127, 128
PTP1D	PDGFR	129, 130
(SYP, SH-PTP2)	EGFR	130
	ERBB2	130

[a] Not all RPTKs associate with all the SH2-containing proteins, and in several cases individual RPTKs have been tested for association with negative results. For example, the CSF-1 receptor does not associate with PLCγ1 or Ras GAP. PLC, phospholipase C; GAP, GTPase activating protein.

mapped to the kinase insert domains of the PDGFβ, CSF-1, and SCF receptors (57–60). The specific tyrosines involved, Y-740 and Y-751 of the human PDGFRβ and Y-721 of the mouse CSF-1R, have been identified by several methods. Site-specific mutagenesis of these tyrosine(s) to phenylalanine blocks PI3-kinase association. In addition, synthetic phosphopeptides that contain these phosphotyrosines block association in a manner dependent upon phosphorylation (57). In cells where ligand-dependent RPTK activation leads to association of PI3-kinase, there is an increase in 3' phosphoinositides. Mutation of the PI3-kinase binding sites in the PDGFR blocks the PDGF-dependent increase in 3' phosphoinositides. How PI3 kinase is activated by RPTKs is not fully understood, but there is a small increase in activity upon binding to an appropriate phosphotyrosine-containing peptide. Thus, PI3-kinase will be activated following binding to an activated RPTK. There also appears to be further activation upon tyrosine phosphorylation of one or both subunits of PI3-kinase. The function of 3' phosphoinositides is not yet understood, but at least in some cells a mutant PDGF receptor lacking both PI3-kinase binding sites fails to elicit a mitogenic response. It is presumed that 3'phosphoinositides act as second messengers, but their target(s) have not yet been identified.

Based on reported PI3-kinase binding sites, (Y)(V/M)(P/D/E)(M) was deduced as a consensus for PI3-kinase binding to RPTKs. Using recombinant p85 SH2 domains to select preferred sequences from a phosphotyrosine-containing peptide library, (Y)(M/I/L/V/E)(X)(M) has recently been defined as a broader consensus for PI3-kinase binding (61). The three-dimensional structures of several SH2 domains have been solved, in one case together with a tightly bound phosphotyrosine-containing peptide (62). Based on the experimental determination of phosphopeptide sequences that are bound preferentially, coupled with predictions from the three-dimensional structures of SH2 domains, it seems likely that we will soon be able to predict the proteins with which an SH2 protein will associate. The general rule is that the three residues to the C-terminal side of the phosphotyrosine are critical for defining the specificity of SH2 binding, and these residues can be seen to form direct contacts with the SH2 domain.

Perhaps the best understood example of an RPTK substrate is PLCγ1, which associates with several but not all activated RPTKs via its SH2 domains (Table 2), and is subsequently phosphorylated (63, 64). The activation of many RPTKs leads to increased PLC activity, and this in turn results in the generation of the second messengers, diacylglycerol and inositol-trisphosphate, through the hydrolysis of PIP_2. PLC activity is consistently increased in cells where PLCγ1 becomes phosphorylated on tyrosine in response to ligand treatment, and *in vitro* studies show that under some conditions tyrosine phosphorylated PLCγ1 has greater activity than unphosphorylated PLCγ1. The three sites of tyrosine phosphorylation on PLCγ1 have been mapped (65), and when a mutant PLCγ1 with phenylalanines at these sites is expressed, its activity is no longer stimulated by mitogens (66). The binding sites for PLCγ1 have been determined for several RPTKs, and a binding consensus sequence is beginning to emerge (61). Mutation of the RPTK PLCγ1 binding sites abolishes PLC activation when cells expressing the mutant receptors are treated with the cognate ligand (67, 68). Thus, association with autophosphorylated RPTKs results in binding, phosphorylation and activation of PLCγ1. An important aspect of PLCγ1 activation may be the fact that binding to the activated RPTK results in translocation to the plasma membrane, where the phospholipid substrates for PLCγ1 are located.

It is beyond the scope of this chapter to review what is known about the other

RPTK substrates and the multiple signalling pathways that are activated by RPTKs. However, the developing picture is that every RPTK has multiple autophosphorylation sites, each of which can bind one or perhaps a few SH2-containing proteins (see Table 3). The precise nature of the phosphorylation sites in an individual receptor will govern which SH2-containing proteins are bound. In RPTKs from the same subfamily many of the autophosphorylation sites lie in the least conserved regions of the cytoplasmic domain, and even when their position is conserved the exact sequence around these sites often differs. Thus even closely-related RPTKs may bind a different spectrum of SH2-containing proteins, resulting in a different cellular response. Not all RPTK-bound SH2 proteins are phosphorylated (e.g. GRB2 is not phosphorylated on tyrosine when associated with the PDGF or EGF receptors). However, either through tyrosine phosphorylation or via allosteric activation, it is believed that SH2-containing proteins are activated and transduce signals to other proteins. Based on mutagenesis studies of individual phosphorylation sites in the PDGFR and CSF-1R, it appears that SH2-containing proteins bind to individual phosphorylation sites and activate distinct signalling pathways, which then cooperate to provide an integrated cellular response. Such cooperation would be particularly important under physiological conditions where only a few receptor molecules per cell may be activated.

The site in the CSF-1R that is required for PI3-kinase binding was not mapped as a major *in vivo* phosphorylation site, raising the possibility that many RPTK tyrosines that have not been detected as autophosphorylation sites may actually be phosphorylated and bind to yet to be discovered SH2 proteins. Many proteins contain multiple SH2 domains, which could recruit other signalling molecules that are phosphorylated on tyrosine. In addition, RPTKs may activate non-receptor PTKs either through direct association, as is the case for Src family PTKs, or indirectly. The structure and function of non-receptor PTKs are reviewed in Chapter 7 of this volume. Tyrosine phosphorylation by activated non-receptor PTKs may play an important role in RPTK signal propagation, and could involve some of the same SH2-containing proteins as well as additional members of the family.

8. Genetics of receptor protein-tyrosine kinases

The consequences of loss of function mutations in RPTK genes have been observed in mouse, *Drosophila melanogaster*, and *Caenorhabditis elegans*. In mouse the *w* (white spotting) locus was originally associated with white spotting of the coat in hetero zygotes. Homozygotes have a more severe phenotype that includes no pigmentation, anaemia, sterility, and death. The implications are that the *w* locus gene product is required for the normal development of melanocytes and haematopoetic and germ cells. The *w* locus has been found to be the gene that codes for c-*kit*, which is now called the SCF receptor. Several *w* alleles have been sequenced, and mutations that diminish PTK activity cause the severest phenotype (69). Mutants in the *steel* locus have a similar phenotype to SCF receptor mutations, and it has been demonstrated that this locus encodes the ligand for SCF receptor (reviewed in 70). The originally unexpected dominant nature of the *w* mutant SCF receptors can be explained as discussed above by the ability of the kinase-inactive SCF receptor to form non-functional heterodimers with the normal SCF receptor in the same cell, thus attenuating the SCF signal. The mouse *patch* phenotype, which like *w* is also manifest by a localized coat colour defect, occurs as a result of the deletion of one copy of the PDGFα receptor. In this case, a 50% decrease in the level of the PDGFRα must be sufficient to cause this phenotype. The mouse *op* mutant, which develops osteopetrosis in the homozygous state, shows a severe deficiency in macrophages and osteoclasts. *op* encodes CSF-1, and it was expected that the development of macrophages would be affected by the absence of CSF-1, since CSF-1 is required for growth and differentiation of myeloid precursors. The failure to develop osteoclasts is more surprising, but osteoclast precursors are derived from the haematopoietic lineage and must require CSF-1.

In *Caenorhabditis elegans* the *let-23* gene encodes a RPTK that resembles the EGF receptor subfamily (31). The *let-23* gene product was shown to be required for vulval induction (see Chapter 8). This system has genetically linked the receptor to a gene encoding an SH2-containing protein, *sem-5* (reviewed in 53 and references therein), the *ras* gene, called *let-60*, and a *raf-1* homologue, called *lin-45*. The putative ligand for Let-23 is an EGF-related protein encoded by the *lin-3* gene.

Four *Drosophila* RPTK mutants have been found to affect development: (i) *torso* results in the lack of proper development of terminal structures in the embryo (28); (ii) *sevenless* mutations have only one observed effect on the developing fly, namely that photoreceptor cell R7 does not develop (27); (iii) *faint little ball* (an allele of the DER, an EGF receptor homologue) exhibits severe developmental effects when homozygous, and the embryo dies very early with many cell types and structures affected (71). Other mutant alleles of the DER locus are *torpedo* and *ellipse* (72, 73) — *ellipse* is of interest because it is a gain of function mutation that constitutively activates the DER; (iv) *breathless* (an allele of DFGFR, an FGFR homologue) exhibits a deficiency in tracheal development (30).

The conclusions that can be drawn from these genetic studies are that some RPTKs may have very specific functions, while others act more globally. Although the interpretation of genetic data may be obscured because RPTKs exist as large families and probably have overlapping functions, and the phenotypes seen in null mutations may only reveal a small subset of the functions of the particular gene product, it is clear that genetic approaches offer major benefits in studying RPTK function. As discussed above we can now study the function of mammalian gene products *in vivo* by making gene knockouts in the mouse. It is also possible to use dominant negative mutants of RPTKs and their ligands in a reverse genetic approach. New data from these methodologies and results from lower organisms should answer many questions about the function of RPTKs that cannot be addressed in other ways.

Genetic approaches are also providing fundamental insights into the signal pathways that are regulated by RPTKs. Cellular programmes for differentiation that involve the induction of perhaps hundreds of genes arc controlled by these receptors. The signalling pathways that traverse the cytoplasm and end by altering transcription in the nucleus are still elusive, but the realization that RPTK signal-ling pathways are highly conserved has led to rapid progress in the elucidation of a pathway that involves Sem-5, Sos (a Ras GTP exchange factor), Ras, Raf-1 (a protein-serine kinase), MAP kinase kinase (MEK), and MAP kinase. Activated MAP kinase can translocate into the nucleus, where it is able to phosphorylate and activate transcription factors, such as Elk1 (see Chapter 4). An RPTK-activated pathway of this sort has been found in *C. elegans*, *Drosophila* and vertebrates.

9. Perspectives

Several new approaches and methodologies have instigated the rapid increase in our knowledge about the functions of RPTKs in the past few years. The isolation of cDNAs that encode new RPTKs has become relatively straightforward. The most recent burst of new RPTKs is a result of the use of PCR (74–77). Based on the rate of discovery of new RPTKs, which is unabating, we may have identified less than half the distinct RPTKs in a single vertebrate species. Since there are currently ~50 known vertebrate RPTK genes, we would estimate the total number to be about 100 RPTK genes in humans. After taking into account recent reports showing that alternative splicing is common for RPTK genes (8–10, 78) it can be assumed that these 100 genes would give rise to 200–300 different RPTKs. The true number of vertebrate RPTK genes will ultimately be learned from the human genome project.

The continuing emergence of new RPTK subfamilies suggests that the functions of RPTKs are much broader than previously thought. The RPTKs and their sub-families that were discovered early on may be biased toward those that are most readily converted into oncogenes and/or function in pathways that stimulate cell division. It has been assumed for several years that there are other unknown, non-mitogenic functions, which would explain the expression of so many of these RPTKs in terminally-differentiated cells that have undergone their last division. In this regard, there may be some important pieces to the puzzle of RPTK function that we are lacking. In a recent review it was pointed out that the known biochemical responses of PC12 cells to NGF and EGF are the same, but the biological response is entirely different (79). NGF treatment initiates terminal differentiation, whereas EGF is mitogenic, and yet they elicit phosphorylation of many of the same pro-teins. This raises the question of whether the substrates that are being studied are the critical ones. Moreover, some of the substrate data being generated using cell lines that express high levels of receptor coupled with high concentrations of ligand may not reflect the normal physiological situation. Despite these concerns, the recent advances in the identification of substrates are exciting and at a mini-mum are leading to the elucidation of the RPTK-overstimulated pathways that are important to the development of many cancers.

The members of the RPTK family have many things in common; the critical question is what are the differences? All RPTKs will probably be mitogenic and phosphorylate some common substrates when overexpressed in cultured cells. However, it is doubtful that 200–300 RPTKs are required for this function, if only because this would seem to be more receptors than is necessary to account for the spatial and temporal differences in cell growth required in the development of a vertebrate organism. Undoubtedly, the discovery and characterization of new RPTK subfamilies will shed light on the functional differences between receptors. For instance, the EPH subfamily has at least 12 members, if partial sequence data are included (77). There are no known ligands for any of this group, so at present it is difficult to guess the function of these receptors. However, new information gleaned from the study of any receptor or receptor subfamily should continue to give insights into the general functions of RPTKs, as well as into the functions that are specific for that particular receptor or closely-related receptors. To accomplish this it will be necessary to continue to discover and match new ligand/receptor pairs.

Several new discoveries have shed light on the mechanics of RPTK activation and signal transduction in recent years. The recent identification of substrates that are likely to be links to pathways important to cellular responses have opened new avenues of study. The fact that these substrates bind with high affinity to the activated RPTKs implies that a change in the localization of the substrate could play a role in signalling. The idea that phosphotyrosine moieties on the receptor result in the increased affinity for signal-transducing molecules has potentially explained why RPTKs undergo ligand-dependent autophosphorylation. The continued dissection of the events that occur after ligand binding should give us a better understanding of how single cells are directed to regulate their activities, and how cells interact and cooperate with each other to form and maintain multicellular organisms.

Acknowledgements

The authors would like to thank Anne Marie Quinn for help with Figs 2 and 3, and Elena Pasquale, Cary Lai, Piotr Masiakowski, Ran Jia, Mohammed Shoyab, Kari Alitalo, and Tony Pawson for sharing unpublished data. This review was completed in May, 1993.

References

1. Cantley, L. C., Auger, K. R., Carpenter, C., Duckworth, B., Graziani, A., Kapeller, R., and Soltoff, S. (1991) Oncogenes and signal transduction. *Cell*, **64**, 281.
2. Ullrich, A. and Schlessinger, J. (1990) Signal transduction by receptors with tyrosine kinase activity. *Cell*, **61**, 203.
3. Yarden, Y. and Ullrich, A. (1988) Growth factor receptor tyrosine kinases. *Ann. Rev. Biochem.*, **57**, 443.
4. Williams, L. T. (1989) Signal transduction by the platelet-derived growth factor receptor. *Science*, **243**, 1564.
5. Chao, M. V. (1992) Neurotrophin receptors: a window into neuronal differentiation. *Neuron*, **9**, 583.

6. Schlessinger, J. and Ullrich, A. (1992) Growth factor signaling by receptor tyrosine kinases. *Neuron*, **9**, 383.

7. Schneider, R. and Schweiger, M. (1991) A novel modular mosaic of cell adhesion motifs in the extracellular domains of the neurogenic trk and trkB tyrosine kinase receptors. *Oncogene*, **6**, 1807.

8. Hou, J., Kan, M., McKeehan, K., McBride, G., Adams, P., and McKeehan, W. L. (1991) Fibroblast growth factor receptors from liver vary in three structural domains. *Science*, **251**, 665.

9. Reid, H. H., Wilks, A. F., and Bernard, O. (1990) Two forms of the basic fibroblast growth factor receptor-like mRNA are expressed in the developing mouse brain. *Proc. Natl. Acad. Sci. USA*, **87**, 1596.

10. Johnson, D. E., Lee, P. L., Lu, J., and Williams, L. T. (1990) Diverse forms of a receptor for acidic and basic fibroblast growth factors. *Mol. Cell. Biol.*, **10**, 4728.

11. Hanks, S. K., Quinn, A. M., and Hunter, T. (1988) The protein kinase family: conserved features and deduced phylogeny of the catalytic domains. *Science*, **241**, 42.

12. Hanks, S. K. (1991) Eukaryotic protein kinases. *Curr. Opin. Struct. Biol.*, **1**, 369.

13. Knighton, D. R., Zheng, J., Ten Eyck, L. F., Ashford, V. A., Xuong, N.-H., Taylor, S. S., and Sowadski, J. M. (1991) Crystal structure of the catalytic subunit of cyclic adenosine monophosphate-dependent protein kinase. *Science*, **253**, 407.

14. Knighton, D. R., Zheng, J., Ten Eyck, L. F., Xuong, N.-H., Taylor, S. S., and Sowadski, J. M. (1991) Structure of a peptide inhibitor bound to the catalytic subunit of cyclic adenosine monophosphate-dependent protein kinase. *Science*, **253**, 414.

15. Hanks, S. K. and Quinn, A. M. (1991) Protein kinase catalytic domain sequence database: identification of conserved features of primary structure and classification of family members. *Meth. Enzymol.*, **200**, 38.

16. Kraus, M. H., Fedi, P., Starks, V., Muraro, R., and Aaronson, S. A. (1993) Demonstration of ligand-dependent signaling by the *erbB-3* tyrosine kinase and its constitutive activation in human breast tumor cells. *Proc. Natl. Acad. Sci. USA*, **90**, 2900.

17. Lax, I., Bellot, F., Howk, R., Ullrich, A., Givol, D., and Schlessinger, J. (1989) Functional analysis of the ligand binding site of EGF-receptor utilizing chimeric chicken/human receptor molecules. *EMBO J.*, **8**, 421.

18. Giordano, S., Di Renzo, M. F., Ferracini, R., Chiado Piat, L., and Comoglio, P. M. (1988) p145, a protein with associated tyrosine kinase activity in a human gastric carcinoma cell line. *Mol. Cell. Biol.*, **8**, 3510.

19. Gherardi, E., and Stoker, M. (1991) Hepatocyte growth factor-scatter factor: mitogen, motogen and *Met. Cancer Cells*, **3**, 227.

20. Letwin, K., Yee, S. P. and Pawson, T. (1988) Novel protein-tyrosine kinase cDNAs related to *fps-fes* and *eph* cloned using anti-phosphotyrosine antibody. *Oncogene*, **3**, 621.

21. Chan, J. and Watt, V. (1991) *eek* and *erk*, new members of the *eph* subclass of receptor protein-tyrosine kinases. *Oncogene*, **6**, 1057.

22. Sajjadi, F. G., Pasquale, E. B., and Subramani, S. (1991) Identification of a new eph-related receptor tyrosine kinase gene from mouse and chicken that is developmentally regulated and encodes at least two forms of the receptor. *New Biol.*, **3**, 769.

23. Pasquale, E. B. (1991) Identification of chicken embryo kinase 5, a developmentally regulated receptor-type tyrosine kinase of the Eph familly. *Cell Reg.*, **2**, 523.

24. Jia, R., Mayer, B. J., Hanafusa, T., and Hanafusa, H. (1992) A novel oncogene, v-*ryk*, encoding a truncated receptor tyrosine kinase is transduced into the RPL30 virus without loss of viral sequences. *J. Virol.*, **66**, 5975.

25. Livneh, E., Glazer, L., Segal, D., Schlessinger, J., and Shilo, B.-Z. (1985) The Drosophila EGF receptor gene homolog: conservation of both hormone binding and kinase domains. *Cell*, **40**, 599.

26. Nishida, Y., Hata, M., Nishizuka, Y., Rutter, W. J., and Ebina, Y. (1986) Cloning of a *Drosophila* cDNA encoding a polypeptide similar to the human insulin receptor precursor. *Biochem. Biophys. Res. Commun.*, **141**, 474.

27. Hafen, E., Basler, K., Edstroem, J.-E., and Rubin, G. M. (1987) *Sevenless*, a cell-specific homeotic gene of *Drosophila*, encodes a putative transmembrane receptor with a tyrosine kinase domain. *Science*, **236**, 55.

28. Sprenger, F., Stevens, L. M., and Nüsslein-Volhard, C. (1989) The *Drosophila* gene *torso* encodes a putative receptor tyrosine kinase. *Nature*, **338**, 478.

29. Pulido, D., Campuzano, S., Koda, T., Modolell, J., and Barbacid, M. (1992) D*trk*, a *Drosophila* gene related to the *trk* family of neurotrophin receptors, encodes a novel class of neural cell adhesion molecule. *EMBO J.*, **11**, 391.

30. Glazer, L. and Shilo, B. Z. (1991) The *Drosophila* FGF-R homolog is expressed in the embryonic tracheal system and appears to be required for directed tracheal cell extension. *Genes Dev.*, **5**, 697.

31. Aroian, R. V., Koga, M., Mendel, J. E., Ohshima, Y., and Sternberg, P. W. (1990) The *let-23* gene necessary for *Caenorhabditis elegans* vulval induction encodes a tyrosine kinase of the EGF receptor subfamily. *Nature*, **348**, 693.

32. de Vries, C., Escobedo, J. A., Ueno, H., Houck, K., Ferrara, N., and Williams, L. T. (1992) The fms-like tyrosine kinase, a receptor for vascular endothelial growth factor. *Science*, **255**, 989.

33. Terman, B. I., Dougher-Vermazen, M., Carrion, M. E., Dimitrov, D., Armellino, D. C., Gospodarowicz, D., and Böhlen, P. (1992) Identification of the KDR tyrosine kinase as a receptor for vascular endothelial cell growth factor. *Biochem. Biophys. Res. Comm.*, **187**, 1579.

34. Millauer, G., Wizigmann-Voos, S., Schnürch, H., Martinez, R., Møller, N. P. H., Risau, W., and Ullrich, A. (1993) High affinity VEGF binding and developmental expression suggest Flk-1 as a major regulator of vasculogenesis and angiogenesis. *Cell*, **72**, 835.

35. Wen, D., Peles, E., Cupples, R., Suggs, S. V., Bacus, S. S., Luo, Y. *et al.* (1992) Neu differentiation factor: a transmembrane glycoprotein containing an EGF domain and an immunoglobulin homology unit. *Cell*, **69**, 559.

36. Holmes, W. E., Sliwkowski, M. X., Akita, R. W., Henzel, W. J., Lee, J., Park, J. W. *et al.* (1992) Identification of heregulin, a specific activator of p185^{erbB2}. *Science*, **256**, 1205.

37. Marchionni, M. A., Goodearl, A. D. J., Chen, M. S., Bermingham-McDonogh, O., Kirk, C., Hendricks, M. *et al.* (1993) Glial growth factors are alternatively spliced erbB2 ligands expressed in the nervous system. *Nature*, **362**, 312.

38. Falls, D. L., Rosen, K. M., Corfas, G., Lane, W. S., and Fischbach, G. D. (1993) ARIA, a protein that stimulates acetylcholine receptor synthesis, is a member of the Neu ligand family. *Cell*, **72**, 801.

39. Nakamura, T., Nishizawa, T., Hagiya, M., Seki, T., Shimonishi, M., Sugimura, A. *et al.* (1989) Molecular cloning and expression of human hepatocyte growth factor. *Nature*, **342**, 440.

40. Hammacher, A., Mellström, K., Heldin, C.-H., and Westermark, B. (1989) Isoform-specific induction of actin reorganization by platelet-derived growth factor suggests that the functionally active receptor is a dimer. *EMBO J.*, **8**, 2489.

41. Cunningham, B. C., Ultsch, M., De, V. A., Mulkerrin, M. G., Clauser, K. R., and Wells, J. A. (1991) Dimerization of the extracellular domain of the human growth hormone receptor by a single hormone molecule. *Science*, **254**, 821.

42. Weiner, D. B., Liu, J., Cohen, J. A., Williams, W. V., and Greene, M. I. (1989) A point mutation in the *neu* oncogene mimics ligand induction of receptor aggregation. *Nature*, **339**, 230.

43. Kashles, O., Yarden, Y., Fischer, R., Ullrich, A., and Schlessinger, J. (1991) A dominant negative mutation suppresses the function of normal epidermal growth factor receptors by heterodimerization. *Mol. Cell. Biol.*, **11**, 1454.

44. Ueno, H., Colbert, H., Escobedo, J. A., and Williams, L. T. (1991) Inhibition of PDGF β receptor signal transduction by coexpression of a truncated receptor. *Science*, **252**, 844.

45. Nocka, K., Tan, J. C., Chiu, E., Chu, T. Y., Ray, P., Traktman, P., and Besmer, P. (1990) Molecular bases of dominant negative and loss of function mutations at the murine c-*kit*/white spotting locus: W37, Wv, W41 and W. *EMBO J.*, **9**, 1805.

46. Reith, A. D., Ellis, C., Maroc, N., Pawson, T., Bernstein, A., and Dubreuil, P. (1993) 'W' mutant forms of the Fms receptor tyrosine kinase act in a dominant manner to suppress CSF-1 dependent cellular transformation. *Oncogene*, **8**, 45.

47. Campos-Gonzalez, R. and Glenney, J. R. (1992) Tyrosine phosphorylation of mitogen-activated protein kinase in cells with tyrosine kinase-negative epidermal growth factor receptors. *J. Biol. Chem.*, **267**, 14535.

48. Selva, E., Raden, D. L., and Davis, R. J. (1993) Mitogen-activated protein kinase stimulation by a tyrosine kinase-negative epidermal growth factor receptor. *J. Biol. Chem.*, **268**, 2250.

49. Pawson, T. and Gish, G. D. (1992) SH2 and SH3 domains: from structure to function. *Cell*, **71**, 359.

50. Mayer, B. J. and Baltimore, D. (1993) Signalling through SH2 and SH3 domains. *Trends Cell Biol.*, **3**, 8.

51. Sadowski, I., Stone, J. C., and Pawson, T. (1986) A noncatalytic domain conserved among cytoplasmic protein-tyrosine kinase modifies the kinase function and transforming activity of Fujinami sarcoma virus P130$^{gag\text{-}fps}$. *Mol. Cell. Biol.*, **6**, 4396.

52. Koch, C. A., Anderson, D., Moran, M. F., Ellis, C., and Pawson, T. (1991) SH2 and SH3 domains: elements that control interactions of cytoplasmic signaling proteins. *Science*, **252**, 668.

53. Lowenstein, E. J., Daly, R. J., Batzer, A. G., Li, W., Margolis, B., Lammers, R. *et al.* (1992) The SH2 and SH3 domain-containing protein GRB2 links receptor tyrosine kinases to ras signaling. *Cell*, **70**, 431.

54. Escobedo, J. A., Navankasattusas, S., Kavanaugh, W. M., Milfay, D., Fried, V. A., and Williams, L. T. (1991) cDNA cloning of a novel 85 kd protein that has SH2 domains and regulates binding of PI3-kinase to the PDGF beta-receptor. *Cell*, **65**, 75.

55. Otsu, M., Hiles, I., Gout, I., Fry, M. J., Ruiz-Larrea, F., Panayotou, G. *et al.* (1991) Characterization of two 85 kD proteins that associate with receptor tyrosine kinases, middle-T/pp60$^{c\text{-}src}$ complexes and PI3-kinase. *Cell*, **65**, 91.

56. Skolnik, E. Y., Margolis, B., Mohammadi, M., Lowenstein, E., Fischer, R., Drepps, A. *et al.* (1991) Cloning of PI3 kinase-associated p85 utilizing a novel method for expression/cloning of target proteins for receptor tyrosine kinases. *Cell*, **65**, 83.

57. Fantl, W. J., Escobedo, J. A., Martin, G. A., Turck, C. W., del Rosario, M., McCormick, F., and Williams, L. T. (1992) Distinct phosphotyrosines on a growth factor receptor bind to specific molecules that mediate different signalling pathways. *Cell*, **69**, 413.

58. Kashishian, A., Kazlauskas, A., and Cooper, J. A. (1992) Phosphorylation sites in the PDGF receptor with different specificities for binding GAP and PI3 kinase *in vivo*. *EMBO J.*, **11**, 1373.

59. Lev, S., Givol, D., and Yarden, Y. (1992) Interkinase domain of kit contains the binding site for phosphatidylinositol 3' kinase. *Proc. Natl. Acad. Sci. USA*, **89**, 678.

60. Reedijk, M., Liu, X., van der Geer, P., Letwin, K., Waterfield, M. D., Hunter, T., and Pawson, T. (1992) Tyr721 regulates specific binding of the CSF-1 receptor kinase insert to PI 3'-kinase SH2 domains: a model for SH2-mediated receptor-target interactions. *EMBO J.*, **11**, 1365.

61. Songyang, Z., Shoelson, S. E., Chaudhuri, M., Gish, G., Pawson, T., Haser, W. G. *et al.* (1993) SH2 domains recognize specific phosphopeptide sequences. *Cell*, **72**, 767.

62. Waksman, G., Shoelson, S. E., Pant, N., Cowburn, D., and Kuriyan, J. (1993) Binding of a high affinity phosphotyrosyl peptide to the Src SH2 domain: crystal structures of the complexed and peptide-free forms. *Cell*, **72**, 779.

63. Margolis, B., Rhee, S. G., Felder, S., Mervic, M., Lyall, R., Levitzki, A. *et al.* (1989) EGF induces tyrosine phosphorylation of phospholipase C-II: a potential mechanism for EGF receptor signaling. *Cell*, **57**, 1101.

64. Meisenhelder, J., Suh, P.-G., Rhee, S. G., and Hunter, T. (1989) Phospholipase C-γ is a substrate for the PDGF and EGF receptor protein-tyrosine kinases *in vivo* and *in vitro*. *Cell*, **57**, 1109.

65. Kim, J. W., Sim, S. S., Kim, U. H., Nishibe, S., Wahl, M. I., Carpenter, G., and Rhee, S. G. (1990) Tyrosine residues in bovine phospholipase C-γ phosphorylated by the epidermal growth factor receptor *in vitro*. *J. Biol. Chem.*, **265**, 3940.

66. Kim, H. K., Kim, J. W., Zilberstein, A., Margolis, B., Kim, J. G., Schlessinger, J., and Rhee, S. G. (1991) PDGF stimulation of inositol phospsholipid hydrolysis requires PLC-γl phosphorylation on tyrosine residues 783 and 1254. *Cell*, **65**, 435.

67. Mohammadi, M., Dionne, C. A., Li, W., Li, N., Spivak, T., Honegger, A. M. *et al.* (1992) Point mutation in FGF receptor eliminates phosphatidylinositol hydrolysis without affecting mitogenesis. *Nature*, **358**, 681.

68. Peters, K. G., Marie, J., Wilson, E., Ives, H. E., Escobedo, J., Rosario, M. D. *et al.* (1992) Point mutation of an FGF receptor abolishes phosphatidylinositol turnover and Ca^{2+} flux but not mitogenesis. *Nature*, **358**, 678.

69. Pawson, T. and Bernstein, A. (1990) Receptor tyrosine kinases: genetic evidence for their role in *Drosophila* and mouse development. *Trends Genet.*, **6**, 350.

70. Witte, O. N. (1990) Steel locus defines new multipotent growth factor. *Cell*, **63**, 5.

71. Schejter, E. D. and Shilo, B.-Z. (1989) The Drosophila EGF receptor homolog (DER) gene is allelic to *faint little ball*, a locus essential for embryonic development. *Cell*, **56**, 1093.

72. Price, J. V., Clifford, R. J., and Schüpbach, T. (1989) The maternal ventralizing locus *torpedo* is allelic to *faint little ball*, an embryonic lethal and encodes the *Drosophila* EGF receptor homolog. *Cell*, **56**, 1085.

73. Baker, N. E. and Rubin, G. M. (1989) Effect on eye development of dominant mutations in *Drosophila* homologue of the EGF receptor. *Nature*, **340**, 150.

74. Wilks, A. F. (1989) Two putative protein-tyrosine kinases identified by application of the polymerase chain reaction. *Proc. Natl. Acad. Sci. USA*, **86**, 1603.

75. Kamb, A., Weir, M., Rudy, B., Varmus, H., and Kenyon, C. (1989) Identification of genes from pattern formation, tyrosine kinase and potassium channel families by DNA amplification. *Proc. Natl. Acad. Sci. USA*, **86**, 4372.

76. Partanen, J., Mäkelä, T. P., Alitalo, R., Lehväslaiho, H., and Alitalo, K. (1990) Putative tyrosine kinases expressed in K-562 human leukemia cells. *Proc. Natl. Acad. Sci. USA*, **87**, 8913.

77. Lai, C. and Lemke, G. (1991) An extended family of protein-tyrosine kinase genes differentially expressed in the vertebrate nervous system. *Neuron*, **6**, 691.

78. Middlemas, D. S., Lindberg, R. A., and Hunter, T. (1991) *trk*B, a neural receptor protein-tyrosine kinase: evidence for full-length and two truncated receptors. *Mol. Cell. Biol.*, **11**, 143.

79. Chao, M. V. (1992) Growth factor signaling: where is the specificity? *Cell*, **68**, 995.

80. O'Bryan, J. P., Frye, R. A., Cogswell, P. C., Neubauer, A., Kitch, B., Prokop, C. *et al.* (1991) *axl*, a transforming gene isolated from primary human myeloid leukemia cells, encodes a novel receptor tyrosine kinase. *Mol. Cell. Biol.*, **11**, 5016.

81. Partanen, J., Armstrong, E., Makela, T. P., Korhonen, J., Sandberg, M., Renkonen, R. *et al.* (1992) A novel endothelial cell surface receptor tyrosine kinase with extracellular epidermal growth factor homology domains. *Mol. Cell. Biol.*, **12**, 1698.

82. Schneider, R. (1992) The human protooncogene *ret*: a communicative cadherin? *Trends in Biochem. Sci.*, **18**, 468.

83. Masiakowski, P. and Carroll, R. D. (1992) A novel family of cell surface receptors with tyrosine kinase-like domain. *J. Biol. Chem.*, **267**, 26181.

84. Hunter, T. (1991) Protein kinase classification. *Methods Enzymol.*, **200**, 3.

85. Matthews, W., Jordan, C. T., Wiegand, G. W., Pardoll, D., and Lemischka, I. R. (1991) A receptor tyrosine kinase specific to hematopoietic stem and progenitor cell-enriched populations. *Cell*, **65**, 1143.

86. Rosnet, O., Marchetto, S., deLapeyriere, O., and Birnbaum, D. (1991) Murine Flt3, a gene encoding a novel tyrosine kinase receptor of the PDGFR/CSF1R family. *Oncogene*, **6**, 1641.

87. Matthews, W., Jordan, C. T., Gavin, M., Jenkins, N. A., Copeland, N. G., and Lemischka, I. R. (1991) A receptor tyrosine kinase cDNA isolated from a population of enriched primitive hematopoietic cells and exhibiting close genetic linkage to c-*kit*. *Proc. Natl. Acad. Sci. USA*, **88**, 9026.

88. Terman, B. I., Carrion, M. E., Kovacs, E., Rasmussen, B. A., Eddy, R. L., and Shows, T. B. (1991) Identification of a new endothelial cell growth factor receptor tyrosine kinase. *Oncogene*, **6**, 1677.

89. Oelrichs, R. B., Reid, H. H., Bernard, O., Ziemiecki, A., and Wilks, A. (1993) NYF/FLK-1: a putative receptor protein tyrosine kinase isolated from E10 embryonic neuroepithelium is expressed in endothelial cells of the developing embryo. *Oncogene*, **8**, 11.

90. Aprelikova, O., Pajusola, K., Partanen, J., Armstrong, E., Alitalo, R., Bailey, S. *et al.* (1992) *FLT4*, a novel class III receptor tyrosine kinase in chromosome Sq33-qter. *Cancer Res.*, **52**, 746.

91. Keegan, K., Johnson, D. E., Williams, L. T., and Hayman, M. J. (1991) Isolation of an additional member of the fibroblast growth factor receptor family, FGFR-3. *Proc. Natl. Acad. Sci. USA*, **88**, 1095.

92. Partanen, J., Makela, T. P., Eerola, E., Korhonen, J., Hirvonen, H., Claesson, W. L., and Alitalo, K. (1991) FGFR-4, a novel acidic fibroblast growth factor receptor with a distinct expression pattern. *EMBO J.*, **10**, 1347.

93. Houssaint, E., Blanquet, P. R., Champion-Arnaud, P., Gesnel, M. C., Torriglia, A., and Courtois, Y. (1990) Related fibroblast growth factor receptor genes exist in the human genome. *Proc. Natl. Acad. Sci. USA*, **87**, 8180.

94. Avivi, A., Zimmer, Y., Yayon, A., Yarden, Y., and Givol, D. (1991) Flg-2, a new member of the family of fibroblast growth factor receptors. *Oncogene*, **6**, 1089.

95. Lamballe, F., Klein, R., and Barbacid, M. (1991) trkC, a new member of the trk family of tyrosine protein kinases, is a receptor for neurotrophin-3. *Cell*, **66**, 967.

96. Lindberg, R. and Hunter, T. (1990) cDNA cloning and characterization of *eck*, an epithelial cell receptor protein-tyrosine kinase in the *eph/elk* family of protein kinases. *Mol. Cell. Biol.*, **10**, 6316.

97. Wicks, I. P., Wilkinson, D., Salvaris, E., and Boyd, A. W. (1992) Molecular cloning of HEK, the gene encoding a receptor tyrosine kinase expressed by human lymphoid tumor cell lines. *Proc. Natl. Acad. Sci. USA*, **89**, 1611.

98. Gilardi-Hebenstreit, P., Nieto, M. A., Frain, M., Mattei, M.-G., Chestier, A., Wilkinson, D. G., and Charnay, P. (1992) An Eph-related receptor protein tyrosine kinase gene segmentally expressed in the developing mouse hindbrain. *Oncogene*, **7**, 2499.

99. Janssen, J. W. G., Schulz, A. S., Steenvoorden, A. C. M., Schmidberger, M., Strehl, S., Ambros, P. F., and Bartram, C. R. (1991) A novel putative tyrosine kinase receptor with oncogenic potential. *Oncogene*, **6**, 2113.

100. Rescigno, J., Mansukhani, A., and Basilico, C. (1991) A putative receptor tyrosine kinase with unique structural topology. *Oncogene*, **6**, 1909.

101. Dumont, D. J., Yamaguchi, T. P., Conlon, R. A., Rossant, J., and Breitman, M. L. (1992) *tek*, a novel tyrosine kinase gene located on mouse chromosome 4, is expressed in endothelial cells and their presumptive precursors. *Oncogene*, **7**, 1471.

102. Chou, Y.-H. and Hayman, M. J. (1991) Characterization of a member of the immunoglobulin gene superfamily that possibly represents an additional class of growth factor receptor. *Proc. Natl. Acad. Sci. USA*, **88**, 4897.

103. Paul, S. R., Mergerg, D., Finnerty, H., Morris, G. E., Morris, J. C., Jones, S. S. *et al.* (1992) Molecular cloning of the cDNA encoding a receptor tyrosine kinase-related molecule with a catalytic region homologous to c-met. *Int. J. Cell Cloning*, **10**, 309.

104. Hovens, C. M., Stacker, S. A., Andres, A.-C., Harpur, A. G., Ziemiecki, A., and Wilks, A. F. (1992) RYK, a receptor tyrosine kinase-related molecule with unusual kinase domain motifs. *Proc. Natl. Acad. Sci. USA*, **89**, 11818.

105. Kelman, Z., Simon-Chazottes, D., Guenet, J.-L., and Yarden, Y. (1993) The murine *vik* gene (chromosome 9) encodes a putative receptor with unique protein kinase motifs. *Oncogene*, **8**, 37.

106. Maminto, M. L. D., Williams, K. L., Nakagawara, A., Enger, K. T., Guo, C., Brodeur, G. M., and Deuel, T. F. (1992) Identification of a novel tyrosine kinase receptor-like molecule in neuroblastomas. *Biochem. Biophys. Res. Comm.*, **189**, 1077.

107. Sajjadi, F. G. and Pasquale, E. B. (1993) Five novel avian Eph-related tyrosine kinases are differentially expressed. *Oncogene*, **8**, 1807.

108. Plowman, G. D., Culouscou, J.-M., Whitney, G. S., Green, J. M., Carlton, G. W., Foy, L. *et al.* (1993) Ligand-specific activation of HER4/p180^{erbB4}, a fourth member of the epidermal growth factor receptor family. *Proc. Natl. Acad. Sci. USA*, **90**, 1746.

109. Baird, A. and Klagsbrun, M. (1991) The fibroblast growth factor family. *Cancer Cells*, **3**, 239.

110. Prigent, S. A. and Lemoine, N. R. (1992) The type-1 (EGFR-related) family of growth factor receptors and their ligands. *Progress in Growth Factor Res.*, **4**, 1.

111. Cross, M. and Dexter, T. M. (1991) Growth factors in development, transformation and tumorigenesis. *Cell*, **64**, 271.

112. Varticovski, L., Druker, B., Morrison, D., Cantley, L., and Roberts, T. (1989) The

colony stimulating factor-1 receptor associates with and activates phosphatidylinositol-3 kinase. *Nature*, **342**, 699.

113. Bjorge, J. D., Chan, T.-O., Antczak, M., Kung, H.-J. and Fujita, D. J. (1990) Activated type I phosphatidylinositol kinase is associated with the epidermal growth factor (EGF) receptor following EGF stimulation. *Proc. Natl. Acad. Sci. USA*, **87**, 3816.

114. Rottapel, R., Reedijk, M., Williams, D. E., Lyman, S. D., Anderson, D. M., Pawson, T., and Bernstein, A. (1991) The Steel/W transduction pathway: kit autophosphorylation and its association with a unique subset of cytoplasmic signaling proteins is induced by the Steel factor. *Mol. Cell. Biol.*, **11**, 3043.

115. Lev, S., Givol, D., and Yarden, Y. (1991) A specific combination of substrates is involved in signal transduction by the kit-encoded receptor. *EMBO J.*, **10**, 647.

116. Graziani, A., Gramaglia, D., Cantley, L. C., and Comoglio, P. M. (1991) The tyrosine-phosphorylated hepatocyte growth factor/scatter factor receptor associates with phosphatidylinositol 3-kinase. *J. Biol. Chem.*, **266**, 22087.

117. Soltoff, S. P., Rabin, S. L., Cantley, L., and Kaplan, D. R. (1992) Nerve growth factor promotes the activation of phosphatidylinositol 3-kinase and its association with the *trk* tyrosine kinase. *J. Biol. Chem.*, **267**, 17472.

118. Mohammadi, M., Honegger, A. M., Rotin, D., Fischer, R., Bellot, F., Li, W. *et al.* (1991) A tyrosine-phosphorylated carboxy-terminal peptide of the fibroblast growth factor receptor (Flg) is a binding site for the SH2 domain of phospholipase C-gamma 1. *Mol. Cell. Biol.*, **11**, 5068.

119. Bardelli, A., Maina, F., Gout, I., Fry, M. J., Waterfield, M. D., Comoglio, P. M., and Ponzetto, C. (1992) Autophosphorylation promotes complex formation of recombinant hepatocyte growth factor receptor with cytoplasmic effectors containing SH2 domains. *Oncogene*, **7**, 1773.

120. Obermeier, A., Halfter, H., Wiesmüller, K.-H., Jung, J., Schlessinger, J., and Ullrich, A. (1993) Tyrosine 785 is a major determinant of Trk-substrate interaction. *EMBO J.*, **12**, 933.

121. Middlemas, D. D., Meisenhelder, J., and Hunter, T. (1994) Identification of TrkB autophosphorylation sites and evidence that phospholipase C-γ1 is a substrate of the TrkB receptor. *J. Biol. Chem.*, **269**, 5458.

122. van der Geer, P. and Hunter, T. (1993) Mutation of tyrosine 697, a Grb2 binding site and tyrosine 721, a PI 3-kinase binding site, abrogates signal transduction by the murine CSF-1 receptor expressed in Rat-2 fibroblasts. *EMBO J.*, **12**, 5161.

123. Kypta, R. M., Goldberg, Y., Ulug, E. T., and Courtneidge, S. A. (1990) Association between the PDGF receptor and members of the src family of tyrosine kinases. *Cell*, **62**, 481.

124. Margolis, B., Hu, P., Katzav, S., Li, W., Oliver, J. M., Ullrich, A., Weiss, A., and Schlessinger, J. (1992) Tyrosine phosphorylation of *vav* proto-oncogene product containing SH2 domain and transcription factor motifs. *Nature*, **356**, 71.

125. Courtneidge, S. A., Dhand, R., Pilat, D., Twamley, G. M., Waterfield, M. D., and Roussel, M. F. (1993) Activation of Src family kinases by colony stimulating factor-1 and their association with its receptor. *EMBO J.*, **12**, 943.

126. Pelicci, G., Lanfrancone, L., Grignani, F., McGlade, J., Cavallo, F., Forni, G. *et al.* (1992) A novel transforming protein (SHC) with an SH2 domain is implicated in mitogenic signal transduction. *Cell*, **70**, 93.

127. Park, D. and Rhee, S. G. (1992) Phosphorylation of Nck in response to a variety of receptors, phorbol myristate acetate and cyclic AMP. *Mol. Cell. Biol.*, **12**, 5816.

128. Li, W., Hu, P., Skolnik, E. Y., Ullrich, A., and Schlessinger, J. (1992) The SH2 and SH3 domain-containing Nck protein is oncogenic and a common target for phosphorylation by different surface receptors. *Mol. Cell. Biol.*, **12**, 5824.

129. Feng, G.-S., Hui, C.-C., and Pawson, T. (1993) SH2-containing phosphotyrosine phosphatase as a target of protein-tyrosine kinases. *Science*, **259**, 1607.

130. Vogel, W., Lammers, R., Huang, J., and Ullrich, A. (1993) Activation of a phosphotyrosine phosphatase by tyrosine phosphorylation. *Science*, **259**, 1611.

7 | Non-receptor protein-tyrosine kinases

SARA A. COURTNEIDGE

1. Introduction

Protein tyrosine kinases can be broadly divided into two groups: receptor and non-receptor tyrosine kinases. The members of the first class have an extracellular, ligand binding domain, a transmembrane sequence, and a cytoplasmic domain within which is contained the catalytic sequences. The function of this class of kinases is to transduce a signal, originated by ligand binding, into the cytoplasm. Such a signal may result eventually in growth, movement and/or differentiation of the cell, depending on ligand, cell type, and context. This class of protein tyrosine kinases is the subject of Chapter 6. The non-receptor tyrosine kinases are diverse, and include membrane-associated, cytoplasmic and nuclear proteins. While some of these kinases are able, when appropriately mutated or deregulated, to transform cells, suggesting normal functions in signal transduction pathways that stimulate cell growth, others may be involved in other responses such as changes in cell shape and/or cell–substratum contacts, or in negative growth control.

Non-receptor protein-tyrosine kinases are listed in Table 1, which shows the nomenclature that will be used in this chapter, as well as other names by which the kinases are frequently referred to in the literature, and Fig. 1 shows the basic topography of these enzymes. Of the non-receptor kinases, by far the best studied is the Src family of tyrosine kinases, which will thus form the focus of the chapter. However, I will briefly describe the other non-receptor tyrosine kinases, particularly in order to emphasize their diversity. This chapter does not intend to be a comprehensive list of facts; rather, examples will be given that illustrate aspects of expression, regulation and function. Due to limitations of space, it is not possible to list all original references. Readers are referred to detailed reviews where these are available.

2. The Src family

One of the first viral transforming proteins to be described was a protein called $pp60^{v\text{-}src}$, the product of the oncogene (v-src) of the chicken retrovirus, Rous sarcoma virus. The v-src gene derives from a cellular gene, c-src, whose product

Table 1 Non-receptor protein-tyrosine kinases

Src family	Syk family
Src (pp60$^{c\text{-}src}$)	Syk (p72^{syk})
Yes (pp62$^{c\text{-}yes}$)	ZAP-70
Fyn (p59^{fyn})	
Yrk (p60^{yrk})	**Jak family**
Fgr (p55$^{c\text{-}fgr}$)	JAK1
Lck (p56^{lck})	JAK2
Hck (p59^{hck})	Tyk2
Lyn (p56^{lyn})	
Blk (p57^{blk})	**Fps family**
	Fps (p92$^{c\text{-}fes}$, p98$^{c\text{-}fps}$)
Csk (CYL, OB19)	Fer (NCP94, FER, flk)
(Tec)	
(Itk)	**Abl family**
(Atk)	Abl (p150$^{c\text{-}abl}$)
	Arg
Fak (pp125^{FAK})	

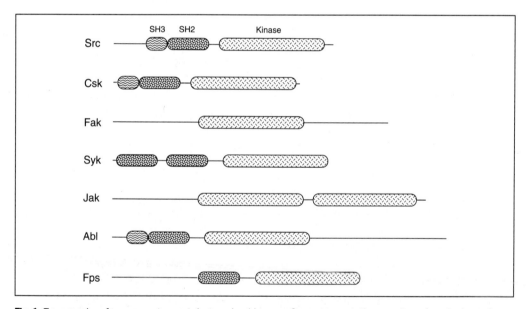

Fig. 1 Topography of non-receptor protein-tyrosine kinases. One representative member of each class of non-receptor protein-tyrosine kinase is shown, approximately to scale. The presence of kinase (catalytic), SH2, and SH3 domains are indicated by the shading, and as labelled (above the representation of Src).

will be referred to as Src here (in the older literature, it is frequently referred to as pp60$^{c\text{-}src}$), and is reviewed in (1). Both viral and cellular forms of Src have tyrosine kinase activity. Since the discovery of *src*, several related genes ('the Src family') have been identified, some first as viral oncogenes, and others by screening cDNA libraries from normal cells and tissues with appropriate probes (2). At the time of writing (January, 1993), the Src family of tyrosine kinases comprises nine members, grouped together because of the similarity of their sequences and their regulation. The current members of the family are Src, Yes (3), Fyn (4, 5), Yrk (6), Fgr (7, 8), Hck (9, 10), Lck (11, 12), Lyn (13), and Blk (14). Some of their main characteristics are shown in Table 2. The names are trivial, and usually reflect the ways in which they were discovered: some, for example Src and Yes, are named for the viruses from which they derive (Rous sarcoma virus, and Y73 and Esh sarcoma viruses respectively); others describe how they were identified, for example Fyn and Blk (**F**gr and **Y**es related **n**ovel kinase, and **B** **L**ymphocyte **k**inase respectively).

2.1 Src family expression

Table 2 shows the expression pattern of the Src family kinases in various cells and tissues (2). The list is not comprehensive, but rather describes the sites of highest expression of a given kinase, using data derived from both RNA and protein analysis from several higher eukaryotic species. The majority of the Src family kinases have a restricted tissue distribution, and are expressed predominantly in haematopoietic cells. Some, in particular Src, Fyn and Yes, are more widely expressed, such that normal cells usually express at least two of these three kinases. Several of the Src family kinases exist in more than one form, either as a result of alternative splicing or the use of alternative initiation codons, adding to the complexity. While in some cases, e.g. Lyn, both forms of the protein appear to be expressed in all cells (15), in other cases expression of one isoform is restricted to certain cell types, e.g. a form of Fyn (Fyn^T) is expressed only in lymphocytes (16), and one form of Src (neuronal Src or nSrc) is detected, as the name implies, only in neuronal cells (17, 18). Several members of the Src family are expressed to the highest levels in postmitotic cells, for example in neurones, platelets, and macrophages (2).

The Src family kinases probably arose from one primordial gene, since the positions of intron/exon boundaries are frequently conserved (19–22). In addition, in mammals *lck* and *fgr* lie close to one another on one chromosome, and *hck* and *src* on another, suggesting recent gene duplications. Src family members have been found in several lower eukaryotic species, including *Xenopus* (23–25), *Drosophila* (26), *Hydra* (27) and a freshwater sponge (28). In most of these cases, more than one *src*-related gene is present, suggesting that the primitive organism in which *src* first arose has not been identified. The observation that even primitive organisms express multiple Src family kinases suggests specialization of function of the individual kinases. No Src-like kinases have been discovered in the yeasts, and it is therefore thought that they arose coincidentally with multicellularization.

Src family kinases, although synthesized in the cytoplasm (29), are found associated with membranes. The plasma membrane is a particularly prominent location for vSrc (reviewed in 30), but recently it was realized, at least for cSrc, that appreciable quantities also exist in several intracellular membrane compartments. For example, Src has been reported to associate with perinuclear membranes (31),

Table 2 The Src family of protein tyrosine kinases

Gene	Product	Alternative forms	Expression pattern[a]	Viral form	Homology to Src[b]	
					Similarity	Identity
src	Src, 60 kDa	3 (alternative splicing)	Ubiquitous, highest in brain and platelets (not lymphocytes)	Yes, in RSV		
yes	Yes, 62 kDa	—	Ubiquitous, highest in brain, placenta, kidney tubules	Yes, in Y73 and ESV	83%	75%
fyn	Fyn, 59 kDa	2 (alternative splicing)	Ubiquitous, highest in brain, T and B lymphocytes	No	81%	71%
yrk	Yrk, 60 kDa	—	Highest in brain, spleen	No	81%	71%
fgr	Fgr, 55 kDa	—	Granulocytes, macrophages	Yes, in GR-FSV	79%	67%
hck	Hck, 59 kDa	2 (upstream initiation)	Granulocytes, monocytes, B lymphocytes, platelets	No	72%	59%
lck	Lck, 56 kDa	—	T lymphocytes	No	71%	55%
lyn	Lyn, 56 kDa	2 (alternative splicing)	B lymphocytes, placenta, platelets	No	71%	55%
blk	Blk, 57 kDa	—	B lymphocytes	No	69%	56%

[a] Expression data is taken from several species, except for *yrk* (to date only identified in chickens) and *blk* (to date only described in mouse).
[b] Sequences of the human genes are compared, except for *yrk* (chicken), and *blk* (mouse).

endocytic vesicles (32, 33), secretory organelles (34, 35), and growth cones in developing neurones (36). The exact distribution of Src family kinases among different membranes may depend on the cell type and/or the growth state of the cell. In some cells, the plasma membrane form of many Src family proteins appears to be concentrated in areas containing junctions (37), or focal adhesions (38). Some Src family kinases (e.g. Lck) (39), and certain forms of Src kinases (e.g. activated cSrc and vSrc), also associate with the cortical cytoskeleton that underlies the plasma membrane (40, 41).

2.2 Structural topography

Figure 1 shows a schematic of the chicken cSrc protein. It comprises 533 amino acids, of which the initiator methionine is cleaved shortly after synthesis. The second amino acid, glycine, becomes covalently modified with myristic acid, which is necessary but not alone sufficient for the membrane attachment of Src (42). The next 80 or so amino acids form the so-called 'unique domain', which has the least similarity among the Src family kinases. This region of cSrc contains several phosphorylation sites, including a cAMP-dependent protein kinase site (43, 44), and sites that can be phosphorylated by protein kinase C (45) and cdc2 (the serine/ threonine kinase whose activation initiates mitosis) (46). Following on from the unique domain are the SH (Src homology)3 and SH2 domains. These domains, of approximately 60 (SH3) and 110 (SH2) amino acids, are found in a large number of proteins, both kinases and non-kinases, either separately or in combination (for reviews see 47–50). Recent research has shown that both domains form compact, independently-folding structures that are able to interact with other proteins (51–56). The SH2 domain recognizes a motif comprising a short stretch of amino acids containing a phosphorylated tyrosine residue; the specificity of binding is in part provided by the sequences surrounding the phosphotyrosine, such that each SH2 domain is expected to have its own specific profile of associating proteins (reviewed in 50). The recognition sequence for SH3 domain binding is not yet well defined, but may involve proline residues in target proteins (57). The ability of SH2 and SH3 domains to interact with other proteins is of great importance in the functioning of Src family kinases (see below). The majority of the carboxy-terminal amino acids (260 to 516) of cSrc comprise the catalytic domain, which has similarity to the catalytic domain of all protein kinases (58). Contained within these sequences are the ATP binding site (in particular a lysine at position 295; ref. 59), and an autophosphorylation site (Tyr416; ref. 60). The last 'domain' of cSrc is known as the tail. These carboxy-terminal sequences contain a phosphorylation site (Tyr527) whose presence is critical to the normal regulation of cSrc (see below).

To summarize, all members of the Src family share the basic domain structure described above, and contain the same motifs for myristylation, ATP binding, autophosphorylation and tail phosphorylation. The region of greatest difference is the unique domain (which varies in length from 45 to 85 amino acids). The similarity of each Src family kinase to Src itself is shown in Table 2.

2.3 Modes of regulation

The Src family of tyrosine kinases are normally tightly regulated *in vivo*. Genetic analysis has shown that mutations that relieve this regulation convert the proto-oncogene into an oncogene, and furthermore, that such mutations can occur in all the domains of the protein (with the possible exception of the unique domain) (reviewed in 61). Most analysis has been done on the Src protein, and so it will be described here, with reference to other Src family kinases where appropriate. The regulation of Src predominantly occurs via two mechanisms: post-translational modifications and protein:protein interactions.

2.3.1 Ser/Thr phosphorylation

Src is multiply phosphorylated *in vivo*, on serine, threonine and tyrosine residues. The molecule is constitutively phosphorylated on Ser17 by the cAMP-dependent protein kinase (43, 44), and in addition, activation of protein kinase C *in vivo* leads to the phosphorylation of Ser12 (and Ser48 in chicken, but not mammalian Src; ref. 45). The function of either of these phosphorylations in the regulation of Src is not known, and their substitution with alanine residues affects neither kinase activity nor transformation ability. However, mutation of Ser17 site results in constitutive phosphorylation of Src on Ser12 (62). The serine/threonine phosphorylations that arise in the unique domain at mitosis (Thr34, Thr46 and Ser72), probably as a result of cdc2 kinase activation (46, 63), occur coincident with the activation of Src, although they probably do not directly lead to activation (see below).

2.3.2 Tyrosine phosphorylation

There are two major tyrosine phosphorylation sites on cSrc: the autophosphorylation site at Tyr416 (60), and the site in the tail, at Tyr527 (64). *In vivo*, the latter site is usually constitutively phosphorylated, and the enzyme is inactive (65). Tyr527 is only a poor substrate at best for autophosphorylation (66, 67), and furthermore, kinase-inactive forms of Src are phosphorylated *in vivo* at this site (68, 69), suggesting the presence of a Tyr527 kinase other than Src. One such enzyme has recently been identified (70, 72). Called Csk, it is a cytoplasmic enzyme distantly related to Src family kinases, and *in vitro* it phosphorylates not just Src itself, but also several other Src family kinases (73). It is not yet clear whether Csk is the only Tyr527 kinase. While dephosphorylation of Src at Tyr527 has been shown to activate its intrinsic kinase activity *in vitro* (65, 66) less is known about phosphotyrosine phosphatases able to act on this site. In T lymphocytes, CD45, a transmembrane tyrosine phosphatase, is involved in the dephosphorylation of Lck (74, 75). PTPα can activate Src when overexpressed in fibroblasts *in vivo* (76), but it is not clear whether this enzyme is a physiological regulator of Src. Transforming versions of Src, but not normal cSrc, are frequently phosphorylated at Tyr416 *in vivo* (60). In circumstances where Src is activated in normal cells (at mitosis, and after PDGF stimulation — see below), autophosphorylation does not occur (77, 78). Substitution of Phe for Tyr at 416 does not abrogate kinase activity, although it reduces its basal level of activity 2–3 fold, demonstrating that this tyrosine is not strictly required for catalysis (79–81). Perhaps when phosphorylated the tyrosine acts as a binding site for other proteins, particularly those containing SH2 domains.

2.3.3 SH2 and SH3 protein interaction domains

The other major control of Src kinases is believed to occur at the level of protein: protein interaction, involving particularly the SH3 and SH2 domains. For example, a cSrc molecule lacking an SH3 domain is transforming (82), showing that overall this domain normally negatively controls the activity of Src. However, in some circumstances, the SH3 domain is required for cSrc activated by mutation of Tyr527 to Phe to be transforming, suggesting that it may have more than one function (83, 84). Current interest focuses on identifying SH3 binding proteins, and directly testing their effect on Src activity. One such protein, 3BP1, that can bind both the Abl and Src SH3 domains, has been described recently (85), but its effect on Src activity has yet to be elucidated. 3BP1 has homology to the GTPase activating protein of the small GTP-binding protein rho, and its association with SH3 domains requires a proline-rich region in its carboxy-terminal half (85). Interestingly, the neuronal form of Src described earlier, which has a six amino acid insert in its SH3 domain (17, 18), has elevated kinase activity (85). The nSrc SH3 domain interacts only weakly with 3BP1, suggesting that it may negatively control Src activity (85).

The SH2 domain is also thought to have a predominantly negative effect on cSrc activity, because its deletion generates a transforming protein (82). However since Src molecules with other, more subtle SH2 domain mutants can have the opposite effect, i.e. reduction of the transforming potential of activated alleles of Src, or transformation of cells from one species but not another ('host-range mutants'), this domain likely has both positive and negative functions (83, 84; see 61 and 87 for reviews of other mutants). A model to explain the overall negative effects of the SH2 domain on Src activity has been proposed (88), and is shown in Fig 2. Inactive Src molecules are folded in such a way that the SH2 domain is involved in an intramolecular association with its phosphorylated tail; the catalytic domain is

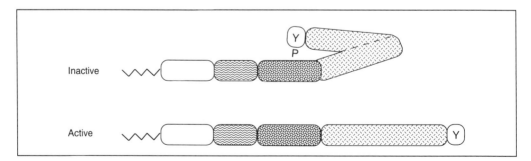

Fig. 2 Model for the control of Src activity. In an inactive configuration, the phosphorylated tail of Src is proposed to interact intramolecularly with the SH2 domain. In the active conformation, dephosphorylation of the tail releases this interaction, leaving both the SH2 domain and the catalytic domain free to interact with other proteins.

inaccessible to substrates, and the SH2 domain unable to interact with other proteins. Dephosphorylation of the tail (or a mutation of Tyr527 to Phe) dramatically lowers the affinity of this interaction, leaving both the catalytic domain and the SH2 domain free to interact with other cellular proteins. There is much circumstantial evidence in favour of this model. For example, an antibody that recognizes the tail of Src activates the kinase, perhaps by sterically hindering the interaction of the tail with the SH2 domain (66). In addition, activated forms of Src bind in an SH2-dependent fashion to a phosphopeptide modelled on the tail sequences, whereas inactive forms do not, presumably because in this case the SH2 domain is already involved in an intramolecular association that restricts the ability of the SH2 domain to interact with other sequences (89). However, the *in vivo* situation is likely to be more complex: for example, the SH3 domain can also influence the ability of the tail negatively to regulate Src activity (Superti-Furga, Fumagalli, Koegl, S.A.C. and Draetta, (1993) *EMBO J.*, **12**, 2625).

A comparison of vSrc and cSrc underscores the importance of the tail in regulating Src activity. The most notable difference between the two proteins is that the 19 carboxy-terminal amino acids of cSrc (including Tyr527) are replaced by an unrelated 12 amino acids, not containing a tyrosine, in vSrc (19). This substitution generates a constitutively active form of the Src protein, which in itself is sufficient to lead to transformation. However, vSrc is more transforming than a cSrc molecule lacking Tyr 527, due to the presence of several point mutations that further increase its transforming potential. For a fully transformed phenotype to be elaborated *in vivo* the autophosphorylation site (Tyr416) must be intact (90). However, *in vitro*, the autophosphorylation site is required for transformation by derepressed forms of cSrc (79, 80), but not by vSrc (91, 92).

2.3.4 Regulation by DNA tumour viruses

The Src family of tyrosine kinases has also been implicated in transformation by some DNA tumour viruses, the polyomaviruses. The mouse polyomavirus transforming protein, middle T antigen, associates with a number of cellular enzymes, including tyrosine kinases of the Src family (principally Src and Yes, but also weakly with Fyn) (93–96). The association of middle T antigen with cSrc increases the intrinsic kinase activity of Src (97–99). This association requires sequences in the tail of cSrc (100), and activation occurs because phosphorylation of Tyr527 is prevented (101). More recently, a distantly-related middle T antigen, from a hamster polyomavirus that causes principally lymphoma, has been characterized. This middle T antigen binds to and activates the Fyn protein, presumably by a similar mechanism (102). The specificity of these interactions, with the mouse middle T antigen binding only weakly to Fyn, and hamster protein associating exclusively with Fyn (and not associating with Src) suggests that there are fundamental differences between Src and Fyn that we have yet to appreciate.

Few other Src family kinases have been subjected to intense mutational analysis. However, in all cases where tested, mutagenesis of the tail (either point mutation or deletion) results in the generation of a transforming protein (103–106). Furthermore, mutation of the Lck protein has also revealed the importance of the SH2 and SH3 domains in regulating activity (107, 108). As is the case with Src, removal of the SH3 domain results in a transforming phenotype. However, the SH2 domain, while it can have a negative role, is also required for Lck activated by substitution of the negative regulatory tyrosine with phenylalanine to transform.

2.4 Functions

It is assumed that the Src family kinases normally function *in vivo* to regulate growth, since first, all members tested override normal growth controls and transform cells when appropriately mutated, and second, vSrc can act as a mitogen (i.e. it drives quiescent cells to enter the cell cycle in the absence of added growth factors) (109). It is therefore particularly intriguing that several members of the Src family are expressed to the highest levels in post-mitotic cells, for example in neurones, platelets and macrophages. This has been interpreted by some to mean that the real function of Src family kinases is not directly in growth control, but instead in regulating processes such as cell shape and adhesion which are dynamic in both dividing and non-dividing cells (110). However, it is also possible that these kinases have different functions in dividing and postmitotic cells, and that cells that can no longer divide are able to tolerate high levels of active kinases (that would transform dividing cells) without detriment. To get an idea in which processes the Src family kinases participate, we should look for situations in which they become activated *in vivo*. In fact, a range of extracellular signals originating in small ligands, cell : cell, and cell : matrix interactions all involve Src family kinases, a few of which will be described below.

2.4.1 Non-kinase receptor association

In haematopoietic cells, the Src family kinases act as catalytic subunits for a large number of transmembrane receptor molecules that themselves lack intrinsic kinase activity. The best example involves the Lck kinase. Lck, which is predominantly expressed in T lymphocytes, exists in stable complexes with two transmembrane molecules called CD4 and CD8, which are proteins required for accessory functions in T cell activation (111, 112). Complex formation requires a Cys–X–Cys sequence in the tail of the CD4 and CD8 molecules, and a Cys–X–X–Cys sequence in the unique domain of Lck, and is thought to be stabilized by bimolecular binding of a metal ion (113, 114). Cross-linking of CD4 and/or CD8 results in significant activation of the Lck kinase activity, probably as the result of dephosphorylation of the tail of Lck (by the lymphocyte specific transmembrane tyrosine phosphatase, CD45), and is accompanied by autophosphorylation (115). This example is unusual in that the complex exists at very high stoichiometry, involving at least 50% of Lck molecules, whereas all other examples of association of Src family kinases with receptors involve at most a few per cent of total molecules. Another Src family kinase, Fyn, also appears to be involved in T cell signalling: in this case, approximately 1% of Fyn has been shown to be associated with the T cell receptor (116).

Many other examples of association of Src family kinases with receptors in haematopoietic cells exist, and readers are referred to recent reviews for a comprehensive list (117, 118). One other cell type, the platelet, is discussed here. Platelets become activated when, after damage to blood vessels, they are brought into contact with both the extracellular matrix and soluble proteins, both of which contain stimulatory molecules, for example collagen and thrombin respectively. In addition, such activated platelets subsequently release yet more stimulatory molecules, including thrombospondin. Stimulation of platelets with such agonists results in waves of increases in tyrosine phosphorylation of several proteins, suggesting activation of several tyrosine kinases (for review see 119). Platelets contain high levels of several Src family kinases, the most highly expressed being Src itself (0.2–0.4% of total cell protein). Thrombin stimulation of platelets appears to stimulate Src activity somewhat (120), as does cross-linking of the FcγRII receptor (120). Furthermore, activation of CD36 (also known as GPIIIB—possibly a collagen receptor) leads to its association with Fyn, Yes and Lyn (121). The molecular basis for these associations is unknown, as are the consequences. Nevertheless, these data suggest that Src family kinases play important roles in signalling in haematopoietic cells. Future work should aim to elucidate the structural bases for the associations and activations detected.

2.4.2 Receptor tyrosine kinase association

The Src family kinases are also involved when cells are stimulated with ligands whose receptors themselves are protein tyrosine kinases. For example, three members of the Src family, Src, Fyn and Yes, are activated when quiescent fibroblasts are stimulated with the growth factor PDGF (77, 123, 124). When activated, the PDGF receptor binds a number of cellular proteins, among them Src family kinases (see Chapter 6). The activation of Src, Fyn, and Yes is coincident with their association with the PDGF receptor, which occurs via their SH2 domains (125). The stoichiometry of association is such that a few per cent of the Src family kinases are associated with the activated PDGF receptor within minutes of ligand stimulation. It is not known whether once bound, the activated Src family kinases are released, or whether they are down-regulated with the receptor. Association results in the phosphorylation of Src on tyrosine and serine residues, but there is no increase in autophosphorylation at Tyr416 (77). The increase in the intrinsic kinase activity of Src may be as a result of the novel amino-terminal phosphorylations, but may also occur because the SH2 domain, once associated with the PDGF receptor, is no longer able to bind to its own tail, releasing this constraint on activity. Why the Src family kinases are activated in response to PDGF is not known, but one reasonable hypothesis is that they phosphorylate critical substrate proteins that are not substrates for the PDGF receptor itself. Indeed, many of the proteins tyrosine phosphorylated in response to PDGF stimulation are also targets for activated Src in transformed cells (see below). A similar activation and association of Src family kinases with cognate tyrosine kinase receptor has also been described in the response to colony stimulating factor 1 (126). Furthermore, the response of cells to

EGF (whose receptor is also a tyrosine kinase) is augmented by the overexpression of Src, although in this case, no association has been detected between Src and the EGF receptor (127).

The above examples suggest that even though Src family kinases do not have extracellular ligand binding domains, they nevertheless get activated by, and participate in the response to, extracellular factors. However, the last example given here of activation of Src does not involve extracellular ligands, and does not occur as cells exit the quiescent state, but rather when they enter mitosis (78). The 2–5 fold mitotic activation of Src is coincident with novel phosphorylation of Src in the unique domain, but there is no autophosphorylation at Tyr416 (46, 63). The amino-terminal phosphorylations can be catalysed by the cdc2 kinase *in vitro* (and this is probably the responsible enzyme *in vivo*), yet phosphorylation of Src by cdc2 *in vitro* does not increase its kinase activity (46, 63). Src molecules lacking Tyr at 527 cannot be further activated in mitosis, suggesting that activation occurs via de-phosphorylation of this site (128, 129). The amino-terminal phosphorylation sites may thus help render Src a better substrate for a Tyr527 phosphatase, since their mutation reduces the mitotic activation of Src (130). The function of Src molecules activated in mitosis is not clear. High resolution immunofluorescence studies suggest that some cSrc is present in the microtubule organizing centre, where its activity could influence microtubule formation (32). Or phosphorylation of membrane and/or cytoskeletal proteins may result in the cell rounding and decreased cell adhesion characteristic of both mitotic and transformed cells.

2.5 Identification of substrates

For a protein to be considered as a physiologically important substrate for an enzyme, at least four criteria should be fulfilled:

- Enzyme and substrate should be in the same subcellular compartment.
- The protein should be phosphorylated on the same sites *in vivo* as are phosphorylated by purified kinase *in vitro*.
- Stoichiometry of phosphorylation should be high.
- Phosphorylation should change the activity (or some other property) of the substrate.

Over the years, much effort has been expended in identifying substrates of Src family kinases. The approach has been twofold: measuring the phosphorylation of candidate proteins *in vitro*, and examining cells transformed by Src family kinases for proteins whose phosphotyrosine content is increased. Many proteins have been identified in these ways, but few fulfil all criteria listed above. For more comprehensive reviews on protein tyrosine kinase substrates see (131, 132).

2.5.1 Use of two-dimensional gels

One of the first approaches taken to identify substrates was to compare the tyrosine phosphorylated proteins of normal and Rous sarcoma virus transformed cells by two-dimensional gel electrophoresis. This led to the identification of a number of rather abundant proteins that were substrates of Src, including some glycolytic enzymes (133). However, in all cases, the stoichiometry of phosphorylation is low, and phosphorylation does not appear to change the activity of the protein. These phosphorylations are therefore thought to be gratuitous, and to occur because of the overexpression of a highly active enzyme. A second approach directed the search towards proteins whose activity is thought to change when cells become transformed. Because changes in cell shape, attachment and communication accompany transformation, proteins involved in these processes were examined. Indeed several components of the cortical cytoskeleton (that underlies the plasma membrane) and focal contacts were found to be tyrosine phosphorylated in Src transformed cells, including calpactin, vinculin, ezrin, talin, paxillin, and the fibronectin receptor (134–140). The effects of phosphorylation on these proteins has not been extensively addressed, although there is some evidence that tyrosine phosphorylation of the fibronectin receptor decreases its affinity for talin (141). A gap junction protein, connexin 43, is also a potential Src substrate, and its tyrosine phosphorylation appears to inhibit gap junctional communication (142–144).

2.5.2 Use of anti-phosphotyrosine antibodies

The development of antibodies specific for phosphotyrosine has enabled the search for substrates to be broadened, and particularly has allowed the identification of proteins that are of low abundance, as well as those that do not resolve on two-dimensional gels. In one approach, such antibodies were used to purify phosphotyrosine-containing proteins from cells transformed by activated Src. Monoclonal antibodies were subsequently raised to these proteins to facilitate their characterization and the eventual cloning of the genes encoding them (145). While the function of most of the proteins identified in this way is still unknown, some have the potential to be physiologically relevant substrates. For example, p80/85 is a cytoskeletal protein whose location changes upon transformation of the cell by Src from the ends of actin stress fibres into podosomes (a term used to described the aberrant focal adhesions present in many transformed cells) (146). Another protein identified in this way, termed Fak, is itself a tyrosine kinase, which is present in areas of focal contact, and may be involved in controlling adhesion of cells (see below) (147). The fact that another tyrosine kinase is a substrate for, and activated by, Src raises the question of whether it is Src or Fak that phosphorylates the focal adhesion and cytoskeletal proteins mentioned earlier.

2.5.3 Signal transducing proteins

Several proteins that have implied roles in signal transduction pathways are also substrates for Src family kinases. These include the GTPase activating protein of Ras (GAP) (148) and phospholipase Cγ (149, 150). The phosphorylation of phospholipase Cγ by the EGF receptor has been shown to increase its enzyme activity (151). However, no changes in GAP activity upon tyrosine phosphorylation have been reported to occur. And finally, very recently a number of proteins have been described that are composed of SH2 or SH2 and SH3 domains, but that lack recognizable catalytic domains of any type. Several of these proteins, e.g. Nck and Shc are phosphorylated on tyrosine in cells containing activated Src, but neither the stoichiometry of phosphorylation nor the activity of these proteins is yet known (152, 153). These proteins, since they consist solely of domains involved in protein:protein interactions, have been called adaptor proteins, and are believed to be involved in signal transduction, linking tyrosine phosphorylation to downstream elements of the signalling pathway, for example Ras activation.

A comparison of cells transformed by activated versions of several Src family kinases has shown that the overall pattern of tyrosine phosphorylated proteins is very similar (154, 155). However, this may only be the case when cells are expressing high levels of the activated forms of these kinases. Other data suggest subtle differences in substrate specificity among members of the Src family. For example cells transformed by the mouse and hamster polyomavirus middle T antigens (where a few per cent of Src and Fyn molecules have been activated respectively) do not have the same profile of tyrosine phosphorylated proteins, with phospholipase Cγ phosphorylated only in the latter case, and GAP only in the former (S.A.C. unpublished observations). Whether this reflects differences in affinity for certain substrates, or different subcellular locations (and therefore different access to substrates) is not yet known.

It is noticeable that several of the substrates of the Src family kinases are also phosphorylated on tyrosine when cells are stimulated with growth factors such as PDGF and EGF (146, 147, 153, 156–158). Either the receptor and non-receptor tyrosine kinases have similar substrate specificities (which is not the case for model substrates *in vitro*), or the phosphorylation of some of these proteins after ligand stimulation of cells is catalysed by Src family kinases that became activated by the receptor in question.

2.6 Why are there so many Src family kinases?

It is interesting to reflect on why there are so many Src family kinases. Each cell type expresses at least two members of the family, and often many more. Yet, in many of the examples given above, more than one member of the family is involved in the same response (for example Src, Fyn, and Yes are all activated by PDGF, and Fyn, Lyn, and Yes all associate with CD36 in platelets). That there is functional redundancy is clearly demonstrated by the use of homologous recombination to generate null alleles of some members of the Src family. In no case are such mutations lethal to the mice; however, in most cases some defect has been measured, for example osteopetrosis in the case of Src (159), lymphocyte abnormalities in the case of Lck and Fyn (160–162), and also changes in long-term potentiation in the case of Fyn (163). The fact that there is loss of function in certain cell types indicates that the functional overlap is not complete. The specific defect detected in each case may reflect a function of a given Src family kinase in a particular cell type that cannot be provided by another member of the family. These may be functions that are mediated by unique domain sequences.

3. cSrc kinase, Csk

Csk was first described as a tyrosine kinase, purified from brain, that had a very restricted substrate specificity, phosphorylating none of the commonly used model substrates except a synthetic polymer of glutamic acid and tyrosine (164). It was subsequently shown to phosphorylate Src at Tyr527 *in vitro*, and decrease its kinase activity (70). Cloning of the gene encoding Csk (72) (which was also cloned independently using a PCR based strategy; 165) revealed a protein of 450 amino acids, with an unexpected similarity to the Src family kinases (overall approximately 59% similarity). Thus, it has both SH3 and SH2 domains followed by a catalytic domain (see Fig. 1). However, it lacks both myristylation sequences and a unique domain and the carboxy-terminal tail sequences contain no tyrosine (so the enzyme cannot be controlled by tyrosine phosphorylation in the same way as the Src family). The catalytic domain of Csk, while highly related to other tyrosine kinases, is unusual in that it lacks a tyrosine in the region corresponding to the autophosphorylation site. Instead, this domain contains two serines and a threonine residue, but they do not appear to be phosphorylated.

Csk is a ubiquitously expressed, cytoplasmic enzyme, and *in vitro* it can phosphorylate all Src family kinases tested on their tail tyrosines (73). Using low stringency hybridizations, no other genes related to Csk have been identified, suggesting that it is not part of a multigene family. However, different transcripts have been identified in lymphoid tissues (165). It is possible, then, that all members of the Src family are controlled by this one kinase, although it cannot yet be ruled out that other kinases, structurally unrelated to Csk, may also regulate the Src family. Whether the catalytic activity of Csk itself is regulated is not yet known.

3.1 Distant cousins of Csk

Recent cloning has identified three other kinases (Tec, Itk, Atk) that, like Csk, have SH2 and SH3 domains, but no tail or myristylation sequences (166–168). Unlike Csk, they do have an amino-terminal extension. All have restricted tissue expression: Tec is expressed in liver, and Atk and Itk in lymphoid cells. Disruption of Atk appears to be the cause of agammaglobulinaemia, and Itk was identified as an interleukin-inducible gene in T cells, suggesting these kinases have important roles in lymphoid development.

4. Focal adhesion associated kinase, Fak

The Fak kinase, a 125 kDa protein also known as p125^{FAK} was first identified as a protein hyperphosphorylated on tyrosine in cells transformed by activated Src, and is widely expressed in adult tissues. Subsequently, cloning of the *fak* gene showed that the protein product had a tyrosine kinase domain, and was active as a kinase when expressed in bacteria (147). Immunofluorescence staining of cells with antibodies specific for Fak showed that it was localized in areas of cell adhesion, hence its name, focal adhesion kinase. While Fak shares a catalytic domain with other tyrosine kinases, which resides approximately in the middle of the molecule (Fig. 1), it has no other recognizable features such as SH2 and SH3 domains, membrane localization sequences, etc.

In normal cells, Fak is one of the most abundant tyrosine phosphorylated proteins. In cells containing activated Src, it is yet more highly phosphorylated on tyrosine, and its intrinsic kinase activity is increased. However, in cells containing non-transforming versions of Src (for example those lacking myristate), Fak is only basally phosphorylated, suggesting that it may be a physiologically important substrate for Src (169).

The function of Fak in normal cells is not yet known, but recent evidence suggest its importance in cell adhesion. Most cells contain cell surface receptors called integrins, which are involved in cell adhesion to extracellular matrix proteins. Recent evidence suggests that integrins, when coupled with ligand, can also initiate intracellular signals. Examples include the CD36 molecule of platelets described in Section 2.4, and the fibronectin receptor in fibroblasts. In both cases one of the major tyrosine phosphorylated proteins detected after integrin engagement is the Fak protein, and this tyrosine phosphorylation of Fak is accompanied by an increase in its intrinsic kinase activity (169–171). In responsive cells small neuropeptides such as bombesin, bradykinin, endothelin and vasopressin also cause an increase in the tyrosine phosphorylation of Fak (172). Taken together, these results suggest that Fak is a point of convergence for the action of Src, peptide growth factors, integrins and neuropeptides. It is possible that these different stimuli result in the phosphorylation of Fak at different sites. Indeed, activated Src and fibronectin treatment of cells have additive effects on Fak activity, suggesting that the latter possibility is correct (169). It is speculated that the function of Fak is to transmit a signal from integrins and other ligands into the cell, or alternatively it may be responsible for phosphorylating focal adhesion proteins such as tensin, paxillin and talin, and thereby regulating the interaction of focal adhesions with the cortical cytoskeleton.

5. The Syk family

The current members of the Syk family are Syk itself, and ZAP-70. Both of these enzymes were first identified by biochemical techniques. Subsequent cloning of their corresponding cDNAs revealed their relationship to one another, and that they have structural features that distinguish them from other non-receptor tyrosine kinases (173, 174). For example, the catalytic domain is preceded by two SH2 domains, and despite the known membrane localization of both Syk and ZAP-70, neither protein has a transmembrane sequence nor a consensus sequence for myristylation (Fig. 1). Both Syk and ZAP-70 are restricted in their tissue expression, with ZAP-70 expressed predominantly in T lymphocytes, and Syk in B lymphocytes and platelets.

The function of both proteins is believed to be in signal transduction. For example, ZAP-70 associates with the T cell receptor component ζ, and is required, together with Lck or Fyn, for its tyrosine phosphorylation (174). Since in an appropriate context, the ζ chain alone of all of the subunits of the T cell receptor is able to elicit all the signalling normally caused by activation of the T cell receptor, and since tyrosine kinase inhibitors prevent this signalling, both Src family kinases and ZAP-70 must play critical roles in T cell activation (175). However, a dissection of the roles of each individual kinase must await the generation of mice carrying null alleles of each. Syk can be activated by cross-linking of surface proteins by lectins in both platelets and B cells (176, 177). The mechanism(s) by which these kinases are regulated is presently unclear.

6. The Jak family

The Jak family currently comprises three members, Jak1, Jak2, and Tyk2. The first two were identified using a PCR strategy (178–180), and the latter by screening of a lymphoid cDNA library (181). The enzymes are widely expressed in both haematopoietic and non-haematopoietic cells. The Jak family derives its name from the fact that the kinases have two catalytic domains, and hence are known as the **Janus kinases**. The proteins lack SH3 domains, and despite their membrane localization, have neither transmembrane sequences nor acylation consensus sequences (Fig. 1). While a sequence that bears limited homology to SH2 domains has been noted (180), some of the consensus residues are not present, and it remains to be determined whether this domain can bind phosphotyrosine-containing peptides. Despite the fact that the proteins contain two kinase-related domains, there is no evidence to suggest that both are functional. Indeed the first of these domains has diverged in some key sequences, and no intrinsic kinase activity of this isolated domain has been detected (179). Perhaps it serves to modulate the binding of substrates and/or regulators.

6.1 Trades for Jak

The function of the Jak family of kinases was obscure until recently, when work on signal transduction in response to interferon converged with the tyrosine phosphorylation field. Interferons are a class of cytokines that have diverse effects such as inhibition of cell growth and viral infection. Treatment of cells with interferon α leads to the rapid activation of transcription of specific genes. An interferon α-specific transcription factor has been described, termed IGSF3. It is composed of four subunits: a 48 kDa DNA binding protein, and three other related subunits of 84, 91, and 113 kDa. These components normally reside in the cytoplasm. Shortly after interferon treatment, the three larger proteins, each of which has an SH2 domain, become phosphorylated on tyrosine (182, 183). A complex is formed of these three proteins, presumably as a result of SH2 domain: phosphotyrosine interactions, and this complex is transported to the nucleus, where it associates with p48, and the active transcription factor is formed.

The identification of the tyrosine kinase responsible for the phosphorylation of p84, 91, and 113 was facilitated by the availability of mutant cell lines unable to respond to interferon α. A gene able to repair the biochemical defect underlying the unresponsiveness was found to encode Tyk2, strongly suggesting that this kinase is normally involved in the interferon pathway (184). Since it has previously been shown that activation of the interferon α-inducible transcription factor could occur using isolated membranes, it is thought that Tyk2 may physically associate with the interferon receptor (which itself lacks a kinase domain); its membrane localization is compatible with such a role.

Treatment of cells with interferon γ induces tyrosine phosphorylation of the 91 kDa protein phosphorylated in response to interferon α. By itself, this protein exhibits distinct DNA binding properties compared with the entire IGSF3 complex, resulting in the transcriptional induction of distinct genes. A cell line mutant in its responsiveness to interferon γ is able to respond when transfected with Jak2 (185) implicating this tyrosine kinase in interferon γ signalling. Moreover, a further mutant line unable to respond to interferon α, β, or γ is complemented by transfection of Jak1 (186).

Tyk2 and Jak1 activities are interdependent in the interferon α pathway; likewise Jak1 and Jak2 in the interferon γ pathway. Thus, the 'active' receptors require a ligand binding protein (the appropriate interferon receptor), either Tyk2 or Jak2 (for interferon α or γ, respectively) and Jak1.

Other cytokines and hormones appear to trigger the Jak kinase pathway leading to phosphorylation of at least the 91 kDa component of IGSF3. These include interleukin (IL)-3, IL-5, IL-6, IL-10, GM-CSF, ciliary neurotrophic factor, EGF, and LIF (187–190). Many of these factors had not previously been implicated in tyrosine kinase signalling. In contrast, nuclear signalling by the EGF receptor had been thought to act primarily through the Ras pathway (see Chapters 4 and 6).

7. The Abl family

The Abl family currently comprises two members, Abl and Arg. Abl was first described as the transforming protein of the Abelson murine leukaemia virus (192), and has also been implicated in chronic myelogenous leukaemia, in which a portion of cAbl is fused to sequences from another protein, Bcr. This fusion protein has high levels of kinase activity, and is able to transform cells *in vitro* (reviewed in 193, 194). Arg was first detected in a cDNA library screening (195). Abl and Arg are highly related to one another, and distantly related to the Src family of tyrosine kinases, having SH3, SH2, and catalytic domains. However they are distinguished by having a unique domain of approximately 70 kDa carboxy-terminal to the catalytic domain (Fig. 1). Contained with this domain is a nuclear localization sequence, and indeed at least 50% of Abl proteins have a nuclear location in normal cells (196).

Both the c-*abl* and the *arg* genes are transcribed into two transcripts that differ in their 5' exons, giving rise to two proteins (in the case of Abl called Abl (I) and Abl (IV)) that differ only in their aminotermini (197). Abl (IV) is myristylated, but Abl (I) appears not to be (198). Only the Abl (IV) protein has been studied in detail. Both Arg and Abl are widely expressed.

The Abl (IV) protein can be rendered oncogenic by deletion of its SH3 domain, suggesting that this domain contains negative regulatory sequences (198). Curiously, one effect of this deletion is to render the nuclear localization signal in the carboxy-terminal domain of the protein non-functional, so that the protein becomes wholly cytoplasmic (196). This cytoplasmic localization appears to be required for transformation to occur. No nuclear substrates for the protooncogenic form of the protein have yet been described, although nuclear proteins phosphorylated on tyrosine do exist.

While the functions of Abl and Arg are not yet known, recent evidence has shown that Abl is able to bind to DNA in a cell cycle-specific fashion (199). The DNA binding domain is localized in the carboxy-terminal domain of Abl, and this region becomes highly phosphorylated on multiple serine and threonine residues in mitotic cells (200). *In vitro* phosphorylation of Abl by cdc2 causes no change in its intrinsic kinase activity, but abolishes its DNA binding (199). Cells that over-express the normal cAbl (IV) protein grow more slowly than wild type cells (196), and transforming versions of Abl are not located in the nucleus, suggesting that its normal function may be in the negative regulation of growth.

Mice homozygous for a null mutation in Abl show poor viability (201, 202). They have profound reductions in B cell and to a lesser extent T cell progenitors. However, abnormalities are not restricted to lymphocytes, but also detected in head and eye development. The lack of a more severe phenotype in mice lacking *abl* might suggest that the closely related *arg* gene product has similar functions.

8. The Fps family

This family comprises the Fps protein itself (also known as Fes) and the Fer protein. Fps sequences were first identified in two transforming retroviruses, one from chickens and one from cats, and the cellular homologue was later described (203). Fer was identified by screening of a cDNA library (204). While Fps expression is restricted to haematopoietic cells, especially macrophages, Fer is more widely expressed, with an alternatively spliced product (FerT) expressed only in testis (205, 206). These proteins are similar to the Src family kinases, but lack an SH3 domain, and instead have a large amino-terminal domain of unknown function (Fig. 1). The Fps protein is located in the cytoplasm, but the Fer protein is present in both the nucleus and the cytoplasm (206).

Little is known about the control of these proteins, but the fact that overexpression of cFps is not sufficient to transform cells suggests that some form of regulation exists. In the viral form of the protein, the 25 amino-terminal amino acids of cFps are replaced with a portion of the viral structural protein, gag. This modification of the amino-terminus appears to be required to uncover the transformation potential of Fps, and suggest that this domain may negatively regulate its activity (207). Transforming versions of Fps require the integrity of the SH2 domain (207). The normal signals that activate Fps and Fer are not known. However, the virally-derived, activated forms of Fps phosphorylate many of the same substrates as activated Src (154).

9. Conclusions

This chapter has attempted to give an overview of the non-receptor protein-tyrosine kinases. Many of these are related to Src; however several kinases with distinct structural motifs have been identified recently, and it seems likely that more members of these families will be identified. While many non-receptor kinases share important features, such as SH2 and SH3 domains, these are not universally found. In addition, it would appear that there is great diversity in the regulation of these enzymes. They also appear to have a broad range of functions, ranging from mitogenesis (for the Src family kinases) to negative growth control (for Csk, Jak2, Tyk2 and cAbl), emphasizing the broad diversity of functions that tyrosine phosphorylation helps to control. And yet, despite the low abundance of phosphotyrosine, and the multitude of protein tyrosine kinases identified to date, it is also striking how overlapping their activities may be, illustrated by the apparent involvement of Fak, Src family kinases, and Syk in platelet aggregation, and the cooperativity between ZAP-70, Lck, and Fyn in T cell activation.

References

1. Jove, R. and Hanafusa, H. (1987) Cell transformation by the viral *src* oncogene. *Ann. Rev. Cell. Biol.*, **3**, 31.
2. Cooper, J. A. (1990) The src family of protein-tyrosine kinases. In *Peptides and protein phosphorylation* (ed. B. E. Kemp). CRC Press, Florida.
3. Sukegawa, J., Semba, K., Yamanashi, Y., Nishizawa, M., Miyajima, N., Yamamoto, Y., and Toyoshima, K. (1987) Characterisation of cDNA clones for the human c-*yes* gene *Mol. Cell. Biol.*, **7**, 41.
4. Semba, K., Nishizawa, M., Miyajima, N., Yoshida, M. C., Sukegawa, J., Yamanashi, Y. *et al.*(1986) *yes*-related protooncogene, *syn*, belongs to the protein-tyrosine kinase family. *Proc. Natl. Acad. Sci. USA*, **83**, 5459.
5. Kawakami, T., Pennington, C. Y., and Robbins, K. C. (1986) Isolation and oncogenic potential of a novel human src-like gene. *Mol. Cell. Biol.*, **6**, 4195.
6. Sudol, M., Greulich, H., Newman, L., Sarkar, A., Sukegawa, J., and Yamamoto, T. (1992) Novel yes-related kinase, Yrk, is expressed at elevated levels in neural and hematopoietic tissues. *Oncogene*, **8**, 823.

7. Nishizawa, M., Semba, K., Yoshida, M. C., Yamamoto, T., Sasaki, M., and Toyoshima, K. (1986) Structure, expression, and chromosomal location of the human c-fgr gene. *Mol. Cell. Biol.*, **6**, 511.

8. Inoue, K., Yamamoto, T., and Toyoshima, K. (1990) Specific expression of human c-*fgr* in natural immunity effector cells. *Mol. Cell. Biol.*, **10**, 1789.

9. Quintrell, N., Lebo, R., Varmus, H. E., Bishop, J. M., Pettenati, M. J., Le Beau, M. M. *et al.* (1987) Identification of a human gene (hck) that encodes a protein tyrosine kinase and is expressed in hematopoietic cells. *Mol. Cell. Biol.*, **7**, 2267.

10. Ziegler, S. F., Marth, J. D., Lewis, D. B., and Perlmutter, R. M. (1987) A novel protein tyrosine kinase gene (*hck*) preferentially expressed in cells of hematopoietic origin. *Mol. Cell. Biol.*, **7**, 2276.

11. Voronova, A. F. and Sefton, B. M. (1986) Expression of a new tyrosine protein kinase is stimulated by retrovirus promoter insertion. *Nature*, **319**, 682.

12. Marth, J. D., Peet, R., Krebs, E. G., and Perlmutter, R. M. (1986) A lymphocyte-specific protein-tyrosine kinase gene is rearranged and overexpressed in the murine T cell lymphoma LSTRA. *Cell*, **43**, 393.

13. Yamanashi, Y., Fukushige, S. I., Semba, K., Sukegawa, J., Miyajima, N., Matsubara, K. I. *et al.* (1986) The *yes*-related cellular gene *lyn* encodes a possible tyrosine kinase similar to p56^{lck}. *Mol. Cell. Biol.*, **7**, 237.

14. Dymecki, S. M., Niederhuber, J. E., and Desiderio, S. V. (1990) Specific expression of a tyrosine kinase gene, *blk*, in B lymphoid cells. *Science*, **247**, 332.

15. Yi, T., Bolen, J. B., and Ihle, J. N. (1991) Hematopoietic cells express two forms of *lyn* kinase differing by 21 amino acids in the amino terminus. *Mol. Cell. Biol.*, **11**, 2391.

16. Cooke, M. P. and Perlmutter, R. M. (1989) Expression of a novel form of the fyn proto-oncogene in hematopoietic cells. *New Biol.*, **1**, 66.

17. Levy, J. B., Dorai, T. D., Wang, L.-H., and Brugge, J. S. (1987) The structurally distinct form of pp60^{c-src} detected in neuronal cells is encoded by a unique c-*src* mRNA. *Mol. Cell. Biol.*, **7**, 4142.

18. Martinez, R., Mathey-Prevot, B., Bernards, A., and Baltimore, D. (1987) Neuronal pp60^{c-src} contains a six-amino acid insertion relative to its non-neuronal counterpart. *Science*, **237**, 411.

19. Takeya, T. and Hanafusa, H. (1983) Structure and sequence of the cellular gene homologous to the RSV src gene and the mechanism for generating the transforming virus. *Cell*, **32**, 881.

20. Anderson, S. K., Gibbs, C. P., Tanaka, A., Kung, H., and Fujita, D. J. (1985) Human cellular *src* gene: nucleotide sequence and derived amino acid sequence of the region encoding coding for the carboxy terminal two-thirds of pp60^{c-src}. *Mol. Cell. Biol.*, **5**, 1122.

21. Katamine, S., Notario, V., Rao, C. D., Miki, T., Cheah, M. S. C., Tronick, S. R., and Robbins, K. C. (1988) Primary structure of the human fgr proto-oncogene product p55^{c-fgr}. *Mol. Cell. Biol.*, **8**, 259.

22. Tronick, S. R., Popescu, N. C., Cheah, M. S. C., Swan, D. C., Amsbaugh, S. C., Lengel, C. R. *et al.* (1985) Isolation and chromosomal localization of the human fgr protooncogene, a distinct member of the tyrosine kinase gene family. *Proc. Natl. Acad. Sci. USA*, **82**, 6595.

23. Steele, R. E., Irwin, M. Y., Knudsen, C. L., Collett, J. W., and Fero, J. B. (1989) The *yes* proto-oncogene is present in amphibians and contributes to the maternal RNA pool in the oocyte. *Oncogene Res.*, **1**, 223.

24. Steele, R. E., Unger, T. F., Mardis, M. J., and Fero, J. B. (1989) The two *Xenopus laevis* SRC genes are co-expressed and each produces functional pp60src. *J. Biol. Chem.*, **264**, 10649.

25. Steele, R. E., Deng, J. C., Ghosn, C. R., and Fero, J. B. (1990) Structure and expression of *fyn* genes in *Xenopus laevis*. *Oncogene*, **5**, 369.

26. Simon, M. A., Drees, B., Kornberg, T., and Bishop, J. M. (1985) The nucleotide sequence and the tissue-specific expression of *Drosophila* c-src. *Cell*, **42**, 831.

27. Bosch, T. C. G., Unger, T. F., Fisher, D. A., and Steele, R. E. (1989) Structure and expression of STK, a *src*-related gene in the simple metazoan *Hydra attenuata*. *Mol. Cell. Biol.*, **9**, 4141.

28. Ottilie, S., Raulf, F., Barnekow, A., Hannig, G., and Schartl, M. (1992) Multiple *src*-related kinase genes, *srk*1–4, in the fresh water sponge *Spongilla lacustris*. *Oncogene*, **7**, 1625.

29. Levinson, A., Courtneidge, S. A., and Bishop, J. M. (1981) Structural and functional domains of RSV transforming protein pp60^{src}. *Proc. Natl. Acad. Sci. USA*, **78**, 1624.

30. Krueger, J. G., Garber, E. A., and Goldberg, A. R. (1983) Subcellular localisation of pp60src in Rous sarcoma transformed cells. *Curr. Top. Microbiol. Immunol.*, **107**, 52.

31. Resh, M. D. and Erikson, R. L. (1985) Highly specific antibody to Rous sarcoma virus src gene product recognizes a novel population of pp60^{v-src} and pp60^{c-src} molecules. *J. Cell Biol.*, **100**, 409.

32. David-Pfeuty, T. and Nouvian-Dooghe, Y. (1990) Immunolocalization of the cellular *src* protein in interphase and mitotic NIH c-src overexpresser cells. *J. Cell Biol.*, **111**, 3097.

33. Kaplan, K. B., Swedlow, J. R., Varmus, H. E., and Morgan, D. O. (1992) Association of p60^{c-src} with endosomal membranes in mammalian fibroblasts. *J. Cell Biol.*, **118**, 321.

34. Parsons, S. J. and Creutz, C. E. (1986) pp60c-src activity detected in the chromaffin granule membrane. *Biochem. Biophys. Res. Comm.*, **134**, 736.

35. Grandori, C. and Hanafusa, H. (1988) P60^{c-src} is complexed with a cellular protein in subcellular compartments involved in exocytosis. *J. Cell Biol.*, **107**, 2125.

36. Barnekow, A., Jahn, R., and Schartl, M. (1990) Synaptophysin: a substrate for the protein kinase pp60^{c-src} in intact synaptic vesicles. *Oncogene*, **5**, 1019.

37. Tsukita, S., Oishi, K., Akiyama, T., Yamanashi, Y., Yamamoto, T., and Tsukita, S. (1991) Specific proto-oncogenic tyrosine kinases of src family are enriched in cell-to-cell adherens junctions where the level of tyrosine phosphorylation is elevated. *J. Cell Biol.*, **113**, 867.

38. Rohrschneider, L. R. (1980) Adhesion plaques of Rous sarcoma transformed cells contain the src gene product. *Proc. Natl. Acad. Sci. USA*, **81**, 3514.

39. Louie, R., King, C. S., Macauley, A., Marth, J. D., Perlmutter, R. M., Eckhart, W., and Cooper, J. A. (1988) p56lck protein-tyrosine kinase is cytoskeletal and does not bind to polyomavirus middle T antigen. *J. Virol.*, **62**, 4673.

40. Hamaguchi, M. and Hanafusa, H. (1987) Association of p60src with Triton X-100-resistant cellular structure correlates with morphological transformation. *Proc. Natl. Acad. Sci. USA*, **84**, 2312.

41. Loeb, D. M., Woolford, J., and Beemon, K. (1987) pp60^{c-src} has less affinity for the detergent-insoluble cellular matrix than do pp60^{v-src} and other viral protein-tyrosine kinases. *J. Virol.*, **61**, 2420.

42. Buss, J. E. and Sefton, B. M. (1985) Myristic acid, a rare fatty acid, is the lipid attached to the transforming protein of Rous sarcoma virus and its cellular homolog. *J. Virol.*, **53**, 7.

43. Collett, M. S., Erikson, E., and Erikson, R. L. (1979) Structural analysis of the avian sarcoma virus transforming protein: sites of phosphorylation. *J. Virol.*, **29**, 770.

44. Roth, C. W., Richert, N. D., Pastan, I., and Gottesman, M. M. (1983) Cyclic AMP treatment of Rous sarcoma virus-transformed Chinese hamster ovary cells increases phosphorylation of pp60src and increases pp60src kinase activity. *J. Biol. Chem.*, **258**, 10768.

45. Gould, K. L., Woodgett, J. R., Cooper, J. A., Buss, J. E., Shalloway, D., and Hunter, T. (1985) Protein kinase C phosphorylates pp60src at a novel side. *Cell*, **42**, 849.

46. Shenoy, S., Choi, J.-K., Bagrodia, S., Copeland, T. D., Maller, J. L., and Shalloway, D. (1989) Purified maturation promoting factor phosphorylates pp60c-src at the sites phosphorylated during fibroblast mitosis. *Cell*, **57**, 763.

47. Mayer, B. J. and Baltimore, D. (1993) Signalling through SH2 and SH3 domains. *Trends Cell Biol.*, **3**, 8.

48. Musacchio, A., Gibson, T., Lehto, V.-P., and Saraste, M. (1992) SH3—an abundant protein domain in search of a function. *FEBS Letts*, **307**, 55.

49. Koch, C. A., Anderson, D., Moran, M. J., Ellis, C., and Pawson, T. (1991) SH2 and SH3 domains: elements that control interactions of cytoplasmic signaling proteins. *Science*, **252**, 668.

50. Pawson, T. and Gish, G. D. (1992) SH2 and SH3 domains: from structure to function. *Cell*, **71**, 359.

51. Waksman, G., Kominos, D., Robertson, S. C., Pant, N., Baltimore, D., Birge, R. B. *et al.* (1992) Crystal structure of the phosphotyrosine recognition domain SH2 of v-*src* complexed with tyrosine-phosphorylated peptides. *Nature*, **358**, 646.

52. Musacchio, A., Noble, M., Pauptit, R., Wierenga, R., and Saraste, M. (1992) Crystal structure of a Src-homology 3 (SH3) domain. *Nature*, **359**, 851.

53. Yu, H., Rosen, M. K., Shin, T. B., Seidel-Dugan, C., Brugge, J. S., and Schreiber, S. L. (1992) Solution structure of the SH3 domain of Src and identification of its ligand-binding site. *Science*, **258**, 1665.

54. Booker, G. W., Breeze, A. L., Downing, A. K., Panayotou, G., Gout, I., Waterfield, M. D., and Campbell, I. D. (1992) Structure of an SH2 domain of the p85α subunit of phosphatidylinositol-3-OH kinase. *Nature*, **358**, 684.

55. Overduin, M., Rios, C. B., Mayer, B. J., Baltimore, D., and Cowburn, D. (1992) Three dimensional solution structure of the src homology 2 domain of c-*abl*. *Cell*, **70**, 697.

56. Eck, M. J., Shoelson, S. E., and Harrison, S. C. (1993) Recognition of a high affinity phosphotyrosyl peptide by the Src homology-2 domain of p56[lck]. *Nature*, **362**, 87.

57. Ren, R., Mayer, B. J., Cicchetti, P., and Baltimore, D. (1993) Identification of a 10-amino acid proline-rich SH3 binding site. *Science*, **259**, 1157.

58. Hanks, S. J., Quinn, A. M., and Hunter, T. (1988) The protein kinase family: conserved features and deduced phylogeny of the catalytic domains. *Science*, **241**, 42.

59. Kamps, M. P., Taylor, S. S., and Sefton, B. M. (1984) Direct evidence that oncogenic tyrosine kinases and cyclic AMP-dependent protein kinase have homologous ATP-binding sites. *Nature*, **310**, 589.

60. Smart, J. E., Oppermann, H., Czernilofsky, A. P., Purchio, A. F., Erikson, R. L., and Bishop, J. M. (1981) Characterization of sites for tyrosine phosphorylation in the transforming protein of Rous sarcoma virus and its cellular homologue. *Proc. Natl. Acad. Sci. USA*, **78**, 6013.

61. Koegl, M. and Courtneidge, S. A. (1992) The regulation of Src activity. *Seminars in Virol.*, **2**, 375.

62. Yaciuk, P., Choi, J.-K., and Shalloway, D. (1989) Mutation of amino acids in pp60^{c-src} that are phosphorylated by protein kinases C and A. *Mol. Cell. Biol.*, **9**, 2453.

63. Morgan, D. O., Kaplan, J. M., Bishop, J. M., and Varmus, H. E. (1989) Mitosis-specific phosphorylation of p60^{c-src} by p34cdc2-associated protein kinase. *Cell*, **57**, 775.

64. Cooper, J. A., Gould, K. L., Cartwright, C. A., and Hunter, T. (1986) Tyr527 is phosphorylated in pp60^{c-src}: implications for regulation. *Science*, **231**, 1431.

65. Courtneidge, S. A. (1985) Activation of pp60^{c-src} kinase by middle-T antigen binding or by dephosphorylation. *EMBO J.*, **4**, 1471.

66. Cooper, J. A. and King, C. S. (1986) Dephosphorylation or antibody binding to the carboxy terminus stimulates pp60^{c-src}. *Mol. Cell. Biol.*, **6**, 4467.

67. Cooper, J. A. and Runge, K. (1987) Avian pp60c-src is more active when expressed in yeast than in vertebrate fibroblasts. *Oncogene Res.*, **1**, 297.

68. Jove, R., Kornbluth, S., and Hanafusa, H. (1987) Enzymatically inactive pp60c-src mutant with altered ATP-binding site is fully phosphorylated in its carboxy-terminal regulatory domain. *Cell*, **40**, 937.

69. Thomas, J. E., Soriano, P., and Brugge, J. S. (1991) Phosphorylation of c-Src on tyrosine 527 by another protein tyrosine kinase. *Science*, **254**, 568.

70. Okada, M. and Nakagawa, H. (1988) Identification of a novel protein tyrosine kinase that phosphorylates pp60^{c-src} and regulates its activity in neonatal rat brain. *Biochem. Biophys. Res. Comm.*, **2**, 796.

71. Okada, M. and Nakagawa, H. (1989) A protein tyrosine kinase involved in regulation of pp60^{c-src} function. *J. Biol. Chem.*, **264**, 20886.

72. Nada, S., Okada, M., MacAuley, A., Cooper, J. A., and Nakagawa, H. (1991) Cloning of a complementary DNA for a protein-tyrosine kinase that specifically phosphorylates a negative regulatory site of p60^{c-src}. *Nature*, **351**, 69.

73. Okada, M., Nada, S., Yamanashi, Y., Yamamoto, T., and Nakagawa, H. (1991) CSK: a protein-tyrosine kinase involved in regulation of src family kinases. *J. Biol. Chem.*, **266**, 24249.

74. Mustelin, T. and Altman, A. (1990) Dephosphorylation and activation of the T cell tyrosine kinase pp56^{lck} by the leukocyte common antigen (CD45). *Oncogene*, **5**, 809.

75. Ostergaard, H. L., Shackleford, D. A., Hurley, T. R., Johnson, P., Hyman, R., Sefton, B. M., and Trowbridge, I. S. (1989) Expression of CD45 alters phosphorylation of the *lck*-encoded tyrosine protein kinase in murine T-cell lymphoma cell lines. *Proc. Natl. Acad. Sci. USA*, **86**, 8959.

76. Zheng, X. M., Wang, Y., and Pallen, C. J. (1992) Cell transformation and activation of pp60^{c-src} by overexpression of a protein tyrosine phosphatase. *Nature*, **359**, 336.

77. Gould, K. and Hunter, T. (1988) Platelet-derived growth factor induces multisite phosphorylation of pp60^{c-src} and increases its protein-tyrosine kinase activity. *Mol. Cell. Biol.*, **8**, 3345.

78. Chackalaparampil, I. and Shalloway, D. (1988) Altered phosphorylation and activation of pp60c-src during fibroblast mitosis. *Cell*, **52**, 801.

79. Kmiecik, T. E. and Shalloway, D. (1987) Activation and suppression of pp60^{c-src} transforming ability by mutation of its primary sites of tyrosine phosphorylation. *Cell*, **49**, 65.

80. Piwnica-Worms, H., Saunders, K. B., Roberts, T. M., Smith, A. E., and Cheng, S. H.

(1987) Tyrosine phosphorylation regulates the biochemical and biological properties of pp60$^{c\text{-}src}$. *Cell*, **49**, 75.

81. Ferracini, R. and Brugge, J. (1990) Analysis of mutant forms of the c-*src* gene product containing a phenylalanine substitution of tyrosine 416. *Oncogene Res.*, **5**, 205.

82. Seidel-Dugan, C., Meyer, B. E., Thomas, S. M., and Brugge, J. S. (1992) Effects of SH2 and SH3 deletions on the functional activities of wild-type and transforming variants of c-Src. *Mol. Cell. Biol.*, **12**, 1835.

83. Hirai, H. and Varmus, H. E. (1990) Site-directed mutagenesis of the SH-2 and SH3-coding domains of c-*src* produces various phenotypes including oncogenic activation of p60$^{c\text{-}src}$. *Mol. Cell. Biol.*, **10**, 1307.

84. Hirai, H. and Varmus, H. E. (1990) Mutations in *src* homology regions 2 and 3 of activated chicken c-*src* that result in preferential transformation of mouse or chicken cells. *Proc. Natl. Acad. Sci. USA*, **87**, 8592.

85. Cicchetti, P., Mayer, B. J., Thiel, G., and Baltimore, D. (1992) Identification of a protein that binds to the SH3 region of Abl and is similar to Bcr and GAP-rho. *Science*, **257**, 803.

86. Levy, J. B. and Brugge, J. S. (1989) Biological and biochemical properties of the c-src+ gene product overexpressed in chicken embryo fibroblasts. *Mol. Cell. Biol.*, **9**, 3332.

87. Parsons, J. T. and Weber, M. J. (1989) Genetics of src: structure and functional organization of a protein tyrosine kinase. *Curr. Top. Microbiol. Immunol.*, **147**, 79.

88. Matsuda, M., Mayer, B. J., Fukui, Y., and Hanafusa, H. (1990) Binding of transforming protein P47gag-crk, to a broad range of phosphotyrosine-containing proteins. *Science*, **248**, 1537.

89. Roussel, R. R., Brodeur, S. R., Shalloway, D., and Laudano, A. P. (1991) Selective binding of activated pp60$^{c\text{-}src}$ by an immobilized synthetic phosphopeptide modeled on the carboxyl terminus of pp60$^{c\text{-}src}$. *Proc. Natl. Acad. Sci. USA*, **88**, 10696.

90. Snyder, M. A. and Bishop, J. M. (1984) A mutation at the major phosphotyrosine in pp60v-src alters oncogenic potential. *Virol.*, **136**, 375.

91. Snyder, M. A., Bishop, J. M., Colby, W. W. and Levinson, A. D. (1983) Phosphorylation of tyrosine-416 is not required for the transforming properties and kinase activity of pp60v-src. *Cell*, **32**, 891.

92. Cross, F. R. and Hanafusa, H. (1983) Local mutagenesis of Rous sarcoma virus: the major sites of tyrosine and serine phosphorylation of p60$^{c\text{-}src}$ are dispensable for transformation. *Cell*, **34**, 597.

93. Courtneidge, S. A. and Smith, A. E. (1983) Polyoma virus transforming protein associates with the product of the c-src gene. *Nature*, **303**, 435.

94. Kypta, R. M., Hemming, A., and Courtneidge, S. A. (1988) Identification and characterization of p59^{fyn} (a *src*-like protein tyrosine kinase) in normal and polyomas virus transformed cells. *EMBO J.*, **7**, 3837.

95. Cheng, S. H., Harvey, R., Espiro, P. C., Semba, K., Yamamoto, T., Toyoshima, K., and Smith, A. E. (1988) Peptide antibodies to the human c-*fyn* product demonstrates pp59$^{c\text{-}fyn}$ is capable of complex formation with the middle-T antigen of polyomavirus. *EMBO J.*, **7**, 3845.

96. Kornbluth, S., Sudol, M., and Hanafusa, H. (1987) Association of the polyomavirus middle T antigen with c-yes protein. *Nature*, **325**, 171.

97. Bolen, J. B. and Israel, M. A. (1985) Middle tumor antigen of polyomavirus transformation-defective mutant Ng59 is associated with pp60$^{c\text{-}src}$. *J. Virol.*, **53**, 114.

98. Courtneidge, S. A., Oostra, B., and Smith, A. E. (1984) Tyrosine phosphorylation and polyoma virus middle T protein. *Cancer Cells*, **2**, 123.

99. Cartwright, C. A., Hutchinson, M. A., and Eckhart, W. (1985) Structural and functional modification of pp60$^{c\text{-}src}$ associated with polyoma middle tumor antigen from infected or transformed cells. *Mol. Cell. Biol.*, **5**, 2647.

100. Cheng, S. H., Piwnica-Worms, H., Harvey, R. W., Roberts, T. M., and Smith, A. E. (1988) The carboxy terminus of pp60$^{c\text{-}src}$ is a regulatory domain and is involved in complex formation with the middle-T antigen of polyomavirus. *Mol. Cell. Biol.*, **8**, 1736.

101. Cartwright, C. A., Kaplan, P. L., Cooper, J. A., Hunter, T., and Eckhart, W. (1986) Altered sites of tyrosine phosphorylation in pp60$^{c\text{-}src}$ associated with polyoma virus middle tumor antigen. *Mol. Cell. Biol.*, **6**, 1562.

102. Courtneidge, S. A., Goutebroze, L., Cartwright, A., Heber, A., Scherneck, S., and Feunteun, J. (1991) Identification and characterisation of the hamster polyomavirus middle T antigen. *J. Virol.*, **65**, 3301.

103. Amrein, K. E. and Sefton, B. M. (1988) Mutation of a site of tyrosine phosphorylation in the lymphocyte-specific tyrosine protein kinase, p56^{lck}, reveals its oncogenic potential in fibroblasts. *Proc. Natl. Acad. Sci. USA*, **85**, 4247.

104. Marth, J. D., Cooper, J. A., King, C. S., Ziegler, S. F., Tinker, D. A., Overell, R. A. *et al.* (1988) Neoplastic transformation induced by an activated lymphocyte-specific protein tyrosine kinase (pp56^{lck}). *Mol. Cell. Biol.*, **8**, 540.

105. Ziegler, S., Levin, S., and Perlmutter, R. M. (1989) Transformation of NIH 3T3 fibroblasts by an activated form of p59^{hck}. *Mol. Cell. Biol.*, **9**, 2724.

106. Kawakami, T., Kawakami, Y., Aaronson, S. A., and Robbins, K. C. (1988) Acquisition of transforming properties of FYN, a normal SRC-related human gene. *Proc. Natl. Acad. Sci. USA*, **85**, 3870.

107. Reynolds, P. J., Hurley, T. R., and Sefton, B. M. (1992) Functional analysis of the SH2 and SH3 domains of the *lck* tyrosine protein kinase. *Oncogene*, **7**, 1949.

108. Veillette, A., Caron, L., Fournel, M., and Pawson, T. (1992) Regulation of the enzymatic function of the lymphocyte-specific tyrosine protein kinase p56^{lck} by the non-catalytic SH2 and SH3 domains. *Oncogene*, **7**, 971.

109. Bell, J. G., Wyke, J. A., and Macpherson, I. A. (1975) Transformation by a temperature sensitive mutant of Rous sarcoma virus in the absence of serum. *J. Gen. Virol.*, **40**, 127.

110. Shalloway, D. and Shenoy, S. (1990) Oncoprotein kinases in mitosis. *Adv. Cancer Res.*, **57**, 185.

111. Rudd, C. E., Trevillyan, J. M., Dasgupta, J. D., Wong, L. L., and Schlossmann, S. F. (1988) The CD4 receptor is complexed in detergent lysates to a protein-tyrosine kinase (pp58) from human T lymphocytes. *Proc. Natl. Acad. Sci. USA*, **85**, 3277.

112. Veillette, A., Bookman, M. A., Horak, E. M., and Bolen, J. B. (1988) The CD4 and CD8 cell surface antigens are associated with the internal membrane tyrosine-protein kinase p56^{lck}. *Cell*, **55**, 301.

113. Shaw, A. S., Whytney, C. J. A., Hammond, C., Amerin, E., Kavathas, P., Sefton, B. M., and Rose, J. K. (1990) Short related sequences in the cytoplasmic domains of CD4 and CD8 mediate binding to the amino-terminal domain of the p56^{lck} tyrosine protein kinase. *Mol. Cell. Biol.*, **10**, 1853.

114. Turner, J. M., Brodsky, M. H., Irving, B. A., Levin, S. D., Perlmutter, R. M., and Littman, D. R. (1990) Interaction of the unique N-terminal region of tyrosine kinase p56^{lck} with cytoplasmic domains of CD4 and CD8 is mediated by cysteine motifs. *Cell*, **60**, 755.

115. Veillette, A., Bookman, M. A., Horak, E. M., Samelson, L. E., and Bolen, J. B. (1989)

Signal transduction through the CD4 receptor involves the activation of the internal membrane tyrosine kinase p56^{lck}. *Nature*, **338**, 257.

116. Samelson, L. E., Phillips, A. F., Luong, E. T., and Klausner, R. D. (1990) Association of the fyn protein-tyrosine kinase with the T-cell antigen receptor. *Proc. Natl. Acad. Sci. USA*, **87**, 4358.

117. Bolen, J. B. (1991) Signal transduction by the SRC family of tyrosine protein kinases in hemopoietic cells. *Cell Growth & Differentiation*, **2**, 409.

118. Bolen, J. B., Rowley, R. B., Spana, C., and Tsygankov, A. Y. (1992) The src family of tyrosine protein kinases in hemopoietic signal transduction. *FASEB J.*, **6**, 3403.

119. Shattil, S. J. and Brugge, J. S. (1991) Protein tyrosine phosphorylation and the adhesive functions of platelets. *Curr. Op. Cell Biol.*, **3**, 869.

120. Clark, E. A. and Brugge, J. S. (1993) Redistribution of activated pp60c-src to integrin-dependent cytoskeletal complexes in thrombin stimulated platelets. *Mol. Cell. Biol.*, **13**, 1863.

121. Huang, M. M., Indik, Z., Brass, L. F., Hoxie, J. A., Schreiber, A. D., and Brugge, J. S. (1992) Activation of FcgRII induces tyrosine phosphorylation of multiple proteins including FcgRII. *J. Biol. Chem.*, **267**, 5467.

122. Huang, M., Bolen, J. B., Barnwell, J. W., Shattil, S. J., and Brugge, J. S. (1991) Membrane glycoprotein IV (CD36) is physically associated with the Fyn, Lyn, and Yes protein-tyrosine kinases in human platelets. *Proc. Natl. Acad. Sci. USA*, **88**, 7844.

123. Ralston, R. and Bishop, J. M. (1985) The product of the proto-oncogene c-*src* is modified during the cellular response to platelet-derived growth factor. *Proc. Natl. Acad. Sci. USA*, **82**, 7845.

124. Kypta, R. M., Goldberg, Y., Ulug, E. T., and Courtneidge, S. A. (1990) Association between the PDGF receptor and members of the *src* family of tyrosine kinases. *Cell*, **62**, 481.

125. Twamley, G., Hall, B., Kypta, R., and Courtneidge, S. A. (1992) Association of Fyn with the activated PDGF receptor: requirements for binding and phosphorylation. *Oncogene*, **7**, 1893.

126. Courtneidge, S. A., Dhand, R., Pilat, D., Twamley, G. M., Waterfield, M. D., and Roussel, M. (1993) Activation of Src family kinases by colony stimulating factor-1, and their association with its receptor. *EMBO J.*, **12**, 943.

127. Luttrell, D. K., Luttrell, L. M., and Parsons, S. J. (1988) Augmented mitogenic responsiveness to epidermal growth factor in murine fibroblasts that overexpress pp60c-src. *Mol. Cell. Biol.*, **8**, 497.

128. Bagrodia, S., Chackalaparampil, I., Kmiecik, T. E., and Shalloway, D. (1991) Altered tyrosine 527 phosphorylation and mitotic activation of p60^{c-src} *Nature*, **349**, 172.

129. Kaech, S., Covic, L., Wyss, A., and Ballmer-Hofer, K. (1991) Association of p60^{c-src} with polyoma virus middle-T antigen abrogating mitosis-specific activation. *Nature*, **350**, 431.

130. Shenoy, S., Chackalaparampil, I., Bagrodia, S., Lin, P.-H., and Shalloway, D. (1992) Role of p34^{cdc2}-mediated phosphorylations in two-step activation of pp60^{c-src} during mitosis. *Proc. Natl. Acad. Sci. USA*, **89**, 7237.

131. Hunter, T., Angel, P., Boyle, W. J., Chiu, R., Freed, E., Gould, K. L. *et al.* (1988) Targets for signal-transducing protein kinases. *Cold Spring Harbor Symp. Quant. Biol.*, **53**, 131.

132. Glenney, J. R. J. (1992) Tyrosine-phosphorylated proteins: mediators of signal transduction from the tyrosine kinases. *Biochim. Biophys. Acta*, **1134**, 113.

133. Cooper, J. A., Reiss, N. A., Schwartz, R. J., and Hunter, T. (1983) Three glycolytic

enzymes are phosphorylated at tyrosine in cells transformed by Rous sarcoma virus. *Nature*, **302**, 218.

134. Radke, K., Gilmore, T., and Martin, G. S. (1980) Transformation by Rous sarcoma virus: a cellular substrate for transformation-specific protein phosphorylation contains phosphotyrosine. *Cell*, **21**, 821.

135. Sefton, B. M., Hunter, T., Ball, E. H., and Singer, S. J. (1981) Vinculin: a cytoskeletal target of the transforming protein of Rous sarcoma virus. *Cell*, **24**, 165.

136. Gould, K. L., Cooper, J. A., Bretscher, A., and Hunter, T. (1986) The protein tyrosine kinase substrate, p81, is homologous to a chicken microvillar core protein. *J. Cell Biol.*, **102**, 660.

137. Pasquale, E. B., Maher, P. A., and Singer, S. J. (1986) Talin is phosphorylated on tyrosine in chicken embryo fibroblasts transformed by Rous sarcoma virus. *Proc. Natl. Acad. Sci., USA*, **83**, 5507.

138. DeClue, J. E. and Martin, G. S. (1987) Phosphorylation of talin at tyrosine in Rous sarcoma virus-transformed cells. *Mol. Cell. Biol.*, **7**, 371.

139. Glenney, J. R. J. and Zokas, L. (1989) Novel tyrosine kinase substrates from Rous sarcoma virus transformed cells are present in the membrane skeleton. *J. Cell Biol.*, **108**, 2401.

140. Hirst, R., Horvitz, A., Buck, C., and Rohrschneider, L. (1986) Phosphorylation of the fibronectin receptor complex in cells transformed by oncogenes that encode tyrosine kinases. *Proc. Natl. Acad. Sci. USA*, **83**, 6470.

141. Tapley, P., Horowitz, A., Buck, C., Duggan, K., and Rohrschneider, L. (1989) Integrins isolated from Rous sarcoma virus-transformed chicken embryo fibroblast. *Oncogene*, **4**, 325.

142. Filson, A. J., Azarnia, R., Beyer, E. C., Loewenstein, W. R., and Brugge, J. S. (1990) Tyrosine phosphorylation of a gap junction protein correlates with inhibition of cell-to-cell communication. *Cell Growth and Differentiation*, **1**, 661.

143. Crow, D. S., Beyer, E. C., Paul, D. L., Kobe, S. S., and Lau, A. F. (1990) Phosphorylation of connexin 43 gap junction protein in uninfected and Rous sarcoma virus-transformed mammalian fibroblasts. *Mol. Cell. Biol.*, **10**, 1754.

144. Swenson, K. I., Piwnica-Worms, H., McNamee, H., and Paul, D. L. (1990) Tyrosine phosphorylation of the gap junction protein connexin 43 is required for the pp60^{v-src} induced inhibition of communication. *Cell Regul.*, **1**, 989.

145. Kanner, S. B., Reynolds, A. B., Vines, R. R., and Parsons, T. T. (1990) Monoclonal antibodies to individual tyrosine-phosphorylated protein substrates of oncogene-encoded tyrosine kinases. *Proc. Natl. Acad. Sci. USA*, **87**, 3328.

146. Wu, H., Reynolds, A. B., Kanner, S. B., Vines, R. R., and Parsons, J. T. (1991) Identification and characterization of a novel cytoskeleton-associated pp60^{src} substrate. *Mol. Cell. Biol.*, **11**, 5113.

147. Schaller, M. D., Borgman, C. A., Cobb, B. S., Vines, R. R., Reynolds, A. B., and Parsons, J. T. (1992) pp125^{FAK}, a structurally distinctive protein-tyrosine kinase associated with focal adhesions. *Proc. Natl. Acad. Sci. USA*, **89**, 5192.

148. Ellis, C., Moran, M., McCormick, F., and Pawson, T. (1990) Phosphorylation of GAP and GAP-associated proteins by transforming and mitogenic tyrosine kinases. *Nature*, **343**, 377.

149. Park, D. J., Rho, H. W., and Rhee, S. G. (1991) CD3 stimulation causes phosphorylation of phospholipase C-γ on serine and tyrosine residues in a human T-cell line. *Proc. Natl. Acad. Sci. USA*, **88**, 5453.

150. Weiss, A., Koretzky, G., Schatzman, R. C., and Kadlecek, T. (1991) Functional activation of the T-cell antigen receptor induces tyrosine phosphorylation of phospholipase C-γ. *Proc. Natl. Acad. Sci. USA*, **88**, 5484.

151. Nishibe, S., Wahl, M. I., Sotomayor-Hernández, S. M. T., Tonks, N. K., Rhee, S. G., and Carpenter, G. (1990) Increase of the catalytic activity of phospholipase C-γ by tyrosine phosphorylation. *Science*, **250**, 1253.

152. Meisenhelder, J. and Hunter, T. (1992) The SH2/SH3 domain-containing protein Nck is recognized by certain anti-phospholipase C-γ monoclonal antibodies, and its phosphorylation on tyrosine is stimulated by platelet-derived growth factor and epidermal growth factor treatment. *Mol. Cell. Biol.*, **12**, 5843.

153. McGlade, J., Cheng, A., Pelicci, G., Pelicci, P. G., and Pawson, T. (1992) Shc proteins are phosphorylated and regulated by the v-Src and v-Fps protein-tyrosine kinases. *Proc. Natl. Acad. Sci. USA*, **89**, 8869.

154. Kamps, M. P. and Sefton, B. M. (1988) Identification of multiple novel polypeptide substrates of the v-*src*, v-*yes*, v-*fps*, v-*ros* and v-*erb*-B oncogenic tyrosine kinase utilizing antisera against phosphotyrosine. *Oncogene*, **2**, 305.

155. Kamps, M. P. and Sefton, B. M. (1988) Most of the substrates of oncogenic viral tyrosine protein kinases can be phosphorylated by cellular tyrosine kinases in normal cells. *Oncogene Res.*, **3**, 105.

156. Wilson, L. K. and Parsons, S. J. (1990) Enhanced EGF mitogenic response is associated with enhanced tyrosine phosphorylation of specific cellular proteins in fibroblasts overexpressing c-src. *Oncogene*, **5**, 1471.

157. Park, D. and Rhee, S. G. (1992) Phosphorylation of Nck in response to a variety of receptors, phorbol myristate acetate, and cyclic AMP. *Mol. Cell. Biol.*, **12**, 5816.

158. Pelicci, G., Lanfrancone, L., Grignani, F., McGlade, J., Cavallo, F., Forni, G. *et al.* (1992) A novel transforming protein (SHC) with an SH2 domain is implicated in mitogenic signal transduction. *Cell*, **70**, 93.

159. Soriano, P., Montgomery, C., Geske, R., and Bradley, A. (1991) Targeted disruption of the c-src proto-oncogene leads to osteopetrosis in mice. *Cell*, **64**, 693.

160. Appleby, M. W., Gross, J. A., Cooke, M. P., Levin, S. D., Qian, X., and Perlmutter, R. M. (1992) Defective T cell receptor signaling in mice lacking the thymic isoform of p59[fyn]. *Cell*, **70**, 751.

161. Stein, P. L., Lee, H.-M., Rich, S., and Soriano, P. (1992) pp59[fyn] mutant mice display differential signaling in thymocytes and peripheral T cells. *Cell*, **70**, 741.

162. Molina, T. J., Kishihara, K., Siderovski, D. P., van Ewijk, W., Narendran, A., Timms, E. *et al.* (1992) Profound block in thymocyte development in mice lacking p56[lck]. *Nature*, **357**, 161.

163. Grant, S. G. N., O'Dell, T. J., Karl, K. A., Stein, P. L., Soriano, P., and Kandel, E. R. (1992) Impaired long-term potentiation, spatial learning, and hippocampal development in *fyn* mutant mice. *Science*, **258**, 1903.

164. Okada, M. and Nakagawa, H. (1988) Protein tyrosine kinases in rat brain: neonatal rat brain expresses two types of pp60[c-src] and a novel protein tyrosine kinase. *J. Biochem.*, **104**, 297.

165. Partanen, J., Armstrong, E., Bergman, M., Mäkelä, T. P., Hirvonen, H., Huebner, K., and Alitalo, K. (1991) Cyl encodes a putative cytoplasmic tyrosine kinase lacking the conserved tyrosine autophosphorylation site (Y416src). *Oncogene*, **6**, 2013.

166. Mano, H., Ishikawa, F., Nishida, J., Hirai, H., and Takaku, F. (1990) A novel protein-tyrosine kinase, *tec*, is preferentially expressed in liver. *Oncogene*, **5**, 1781.

167. Vetrie, D., Vorcchovsky, I., Sideras, P., Holland, J., Davies, A., Flinter, F. *et al.* (1993) The gene involved in X-linked agammaglobulinaemia is a member of the *src* family of protein-tyrosine kinases. *Nature*, **361**, 226.

168. Siliciano, J. D., Morrow, T. A., and Desiderio, S. V. (1992) Itk, a T-cell inducible specific tyrosine kinase gene inducible by interleukin-2. *Proc. Natl. Acad. Sci. USA*, **89**, 11194.

169. Guan, J.-L. and Shalloway, D. (1992) Regulation of focal adhesion-associated protein tyrosine kinase by both cellular adhesion and oncogenic transformation. *Nature*, **358**, 690.

170. Kornberg, L. J., Earp, H. S., Turner, C. E., Prockop, C., and Juliano, R. L. (1991) Signal transduction by integrins: increased protein tyrosine phosphorylation caused by clustering of β1 intergrins. *Proc. Natl. Acad. Sci. USA*, **88**, 8392.

170. Kornberg, L., Earps, H. S., Parsons, J. T., Schaller, M., and Juliano, R. L. (1992) Cell adhesion or integrin clustering increases phosphorylation of a focal adhesion-associated tyrosine kinase. *J. Biol. Chem.*, **267**, 23439.

171. Zachary, I., Sinnett-Smith, J., and Rozengurt, E. (1992) Bombesin, vasopressin and endothelin stimulation of tyrosine phosphorylation in Swiss 3T3 cells. Identification of a novel tyrosine kinase as a major substrate. *J. Biol. Chem.*, **267**, 19031.

173. Taniguchi, T., Kobayashi, T., Kondo, J., Takahashi, K., Nakamura, H., Susuki, J. *et al.* (1991) Molecular cloning of a porcine gene *syk* that encodes a 72-kDa protein-tyrosine kinase showing high susceptibility to proteolysis. *J. Biol. Chem.*, **266**, 15790.

174. Chan, A. C., Iwashima, M., Turck, C. W., and Weiss, A. (1992) ZAP-70: a 70kd protein-tyrosine kinase that associates with the TCR ζ chain. *Cell*, **71**, 649.

175. Irving, B. A. and Weiss, A. (1991) The cytoplasmic domain of the T cell receptor z chain is sufficient to couple to receptor-associated signal transduction pathways. *Cell*, **64**, 891.

176. Ohta, S., Taniguchi, T., Asahi, M., Kato, Y., Nakagawara, G., and Yamamura, H. (1992) Protein-tyrosine p72^{syk} is activated by wheat germ agglutinin in platelets. *Biochem. Biophys. Res. Comm.*, **185**, 1128.

177. Yamada, T., Taniguchi, T., Nagai, K., Saitoh, H., and Yamamura, H. (1991) The lectin wheat germ agglutinin stimulates a protein-tyrosine kinase activity of p72^{syk} in porcine splenocytes. *Biochem. Biophys. Res. Comm.*, **180**, 1325.

178. Wilks, A. (1989) Two putative protein-tyrosine kinases identified by application of the polymerase chain reaction. *Proc. Natl. Acad. Sci. USA*, **86**, 1603.

179. Wilks, A. F., Harpur, A. G., Kurban, R. R., Ralph, S. J., Zürcher, G., and Ziemiecki, A. (1991) Two novel protein-tyrosine kinases, each with a second phosphotransferase-related catalytic domain define a new class of protein kinase. *Mol. Cell. Biol.*, **11**, 2057.

180. Harpur, A. G., Andres, A. C., Ziemiecki, A., Aston, R. R., and Wilks, A. F. (1992) JAK2, a third member of the JAK family of protein tyrosine kinases. *Oncogene*, **7**, 1347.

181. Firmbach-Kraft, I., Byers, M., Shows, T., Dalla-Favera, R., and Krolewski, J. J. (1990) *tyk2*, prototype of a novel class of non-receptor tyrosine kinase genes. *Oncogene*, **5**, 1329.

182. Fu, X.-Y. (1992) A transcription factor with SH2 and SH3 domains is directly activated by an interferon α-induced cytoplasmic protein tyrosine kinase(s). *Cell*, **70**, 323.

183. Schindler, C., Shuai, K., Prezioso, V. R., and Darnell, J. E. Jr. (1992) Interferon-dependent tyrosine phosphorylation of a latent cytoplasmic transcription factor. *Science*, **257**, 809.

184. Velazquez, L., Fellous, M., Stark, G. R., and Pellegrini, S. (1992) A protein tyrosine kinase in the interferon α/β signaling pathway. *Cell*, **70**, 313.

185. Watling, D., Guschin, D., Müller, M., Silvennoinen, O., Whitthuhn, B. A., Quelle, F. W. *et al.* (1993) Complementation by the protein tyrosine kinase JAK2 of a mutant cell line defective in the interferon-γ signal transduction pathway. *Nature*, **366**, 166.

186. Müller, M., Briscoe, J., Laxton, C., Guschin, D., Ziemiecki, A., Silvennoinen, O. *et al.* (1993) The protein tyrosine kinase JAK1 complements defects in interferon-α/β and -γ signal transduction. *Nature*, **366**, 129.

187. Larner, A. C., David, M., Feldman, G. M., Igarashi, K., Hackett, R. H., Webb, D.S.A. *et al.* (1993) Tyrosine phosphorylation of DNA binding proteins by multiple cytokines. *Science*, **261**, 1730.

188. Ruff-Jamison, S., Chen, C., and Cohen, S. (1993) Induction by EGF and interferon-γ of tyrosine phosphorylated DNA binding proteins in mouse liver nuclei. *Science*, **261**, 1733.

189. Silvennoinen, O., Schindler, C., Schlessinger, J., and Levy, D. E. (1993) Ras-independent growth factor signaling by transcription factor tyrosine phosphorylation. *Science*, **261**, 1736.

190. Sadowski, H. B., Shuai, K., Darnell, J. E. Jr, and Gilman, M. Z. (1993) A common nuclear signal transduction pathway activated by growth factor and cytokine receptors. *Science*, **261**, 1739.

191. Bonni, A., Frank, D. A., Schindler, C., and Greenberg, M. E. (1993) Characterization of a pathway for ciliary neurotrophic factor signaling to the nucleus. *Science*, **262**, 1575.

192. Goff, S. P., Gilboa, E., Witte, O. N., and Baltimore, D. (1980) Structure of the Abelson murine leukemia virus genome and the homologous cellular gene: studies with cloned viral DNA. *Cell*, **22**, 777.

193. Wang, J.-J. (1992) Modelling chronic myelogenous leukemia. *Curr. Biol.*, **2**, 70.

194. Baltimore, D. (1992) The *abl* gene and cellular transformation. *Curr. Opinion Oncology*, **4**, 32.

195. Perego, R., Ron, D., and Kruh, G. D. (1991) *Arg* encodes a widely expressed 145kDa protein-tyrosine kinase. *Oncogene*, **6**, 1899.

196. Van Etten, R. A., Jackson, P., and Baltimore, D. (1989) The mouse type IV c-*abl* gene product is a nuclear protein, and activation of transforming ability is associated with cytoplasmic localization. *Cell*, **58**, 669.

197. Ben-Neriah, Y., Bernards, A., Paskind, M., Daley, G. Q., and Baltimore, D. (1986) Alternative 5' exons in c-*abl* mRNA. *Cell*, **44**, 577.

198. Jackson, P. and Baltimore, D. (1989) N-terminal mutations activate the leukemogenic potential of the myristylated form of c-*abl*. *EMBO J.*, **8**, 449.

199. Kipreos, E. T. and Wang, J. Y. J. (1992) Cell cycle-regulated binding of c-Abl tyrosine kinase to DNA. *Science*, **256**, 382.

200. Kipreos, E. T. and Wang, J. Y. J. (1990) Differential phosphorylation of c-Abl in cell cycle determined by cdc2 kinase and phosphatase activity. *Science*, **248**, 217.

201. Schwartzberg, P. L., Stall, A. M., Hardin, J. D., Bowdish, K. S., Humaran, T., Boast, S. *et al.* (1991) Mice homozygous for the *abl*^m1 mutation show poor viability and depletion of selected B and T cell populations. *Cell*, **65**, 1165.

202. Tybulewicz, V. L. J., Crawford, C. E., Jackson, P. K., Bronson, R. T., and Mulligan, R. C. (1991) Neonatal lethality and lymphopenia in mice with a homozygous disruption of the c-*abl* proto-oncogene. *Cell*, **65**, 1153.

203. Mathey-Prevot, B., Hanafusa, H., and Kawai, S. (1982) A cellular protein is immunologically crossreactive with and functionally homologous to the Fujinami sarcoma virus transforming protein. *Cell*, **28**, 897.

204. Pawson, T., Letwin, K., Lee, T., Hao, Q.-L., Heisterkamp, N., and Groffen, J. (1989) The FER gene is evolutionarily conserved and encodes a widely expressed member of the FPS/FES protein-tyrosine kinase family. *Mol. Cell. Biol.*, **9**, 5722.

205. MacDonald, I., Levy, J., and Pawson, T. (1985) Expression of the mammalian c-*fes* protein in hematopoietic cells and identification of a distinct *fes*-related protein. *Mol. Cell. Biol.*, **5**, 2543.

206. Hao, Q.-L., Ferris, D. K., White, G., Heisterkamp, N., and Groffen, J. (1991) Nuclear and cytoplasmic location of the FER tyrosine kinase. *Mol. Cell. Biol.*, **11**, 1180.

207. Foster, D. A., Shibuya, M., and Hanafusa, H. (1985) Activation of the transformation potential of the cellular *fps* gene. *Cell*, **42**, 105.

208. Johnson, K. A. and Stone, J. C. (1990) Delineation of functional determinants in the transforming protein of Fujinami sarcoma virus. *J. Virol.*, **64**, 3337.

8 | Genetic approaches to protein kinase function in lower eukaryotes

SIMON E. PLYTE

1. Introduction

The study of protein kinase function in higher eukaryotes such as mammalia has been generally restricted to biochemical analysis of purified proteins due to limitations of genetic approaches in these species. Most of these enzymes were first identified biochemically and their genes isolated subsequent to peptide sequence analysis. However, the promiscuous nature of protein kinases in phosphorylating proteins *in vitro* has complicated study of their function in cells. In contrast, the tractability of lower eukaryotes to genetic analysis has resulted in a distinct approach in which genes involved in a given process are initially identified via phenotypic effects. These genetic strategies, of course, do not specifically target protein kinase genes but this family of proteins has surfaced at a surprising frequency during analysis of regulatory genes, which is a testimony to their adaptability and ubiquity as biological tools.

Once identified, the relative ease of genetic manipulation in lower eukaryotes enables the investigation of kinase function *in vivo*. Thus, mutational analysis can be used to determine whether a gene product is required in a particular pathway. Additionally, genetic epistatic experiments, between two genes in a particular pathway, may be applied to determine the order of action of those gene products. During the last decade, functional homologues of many mammalian protein kinases have been identified in lower eukaryotes, allowing a degree of extrapolation in the workings of the proteins in different species. For example, the Raf serine/threonine kinase is required in the Sevenless, Ellipse, and Torso signalling pathways of *Drosophila* and in the Ras-mediated stimulation of transcription factors in mammalian tumourigenesis (1–3). Similarly, the Fus3 and Kss1 kinases of the yeast mating response are related to the mitogen-activated protein kinases that function in mitogen signalling (4,5). So by analysing kinase function in genetically tractable organisms, an indication of their function in higher eukaryotes can often be inferred. To this end a vast amount of information regarding the role of protein kinases in cell cycle control has been obtained from studies in yeast (discussed in Chapter 5). Here I will outline several examples of signal transduction pathways that have been determined by genetic means:

1. Signal transduction during R7 photoreceptor development in *Drosophila melanogaster* and during vulval induction in *Caenorhabditis elegans* as examples of receptor tyrosine kinase signalling pathways.

2. The role of serine/threonine kinases in intercell signalling and cell fate determination during *Drosophila* embryogenesis.

3. The role of cAMP-dependent protein kinase during the life-cycle of the slime mold, *Dictyostelium discoideum*.

4. The pheromone signalling pathway in *Saccharomyces cerevisiae* as parallels of the mitogen-activated protein kinase cascades in mammals.

2. Genetic manipulation in *Drosophila*

2.1 Signal transduction from the *sevenless*-encoded receptor

Genetic dissection of the pathway responsible for the development of the *Drosophila* UV photoreceptor (R7 cell) has revealed a wealth of information regarding the mechanism of action of receptor tyrosine kinases. The fruit fly compound eye is comprised of approximately 800 ommatidia. Each ommatidium is composed of a precise array of 20 cells: eight neuronal photoreceptor cells (R1–8), four lens secreting cells and eight other accessory cells. During development, the R7 cell is the last of the eight photoreceptors to be recruited from the accessory cells and is responsible for UV detection (reviewed in 6 and 7). Its differentiation is thought to be controlled by the spatial and temporal expression of specific proteins on the R7 and R8 photoreceptors (reviewed in 8 and 9). The *sevenless* mutation (the gene encodes a receptor tyrosine kinase; 10) results in complete loss of the R7 photoreceptor from each ommatidium and adult flies are unable to respond to UV stimulation (11, 12). Figure 1 illustrates the genes which have thus far been implicated in Sevenless signalling. Interaction of the transmembrane protein encoded by *bride of sevenless (boss)* with the Sevenless receptor tyrosine kinase results in stimulation of the pathway and ligand–receptor internalization (13, 14). Sevenless receptor phosphorylation is thought to activate the product of the *son of sevenless (sos)* gene via the Drk protein and inhibit the product of *gap1* which is a Ras inhibitor (15). Drk is an adapter protein that tethers the Sevenless receptor to Sos via its SH2 and SH3 domains (16). The Sos protein is a putative activator of guanine nucleotide exchange related to the fission yeast *CDC25* gene (17) and this factor is thought to promote GTP binding and activation of Ras (18). The downstream effects of activated Ras appear to be mediated by the *D-raf* protein (a Ser/Thr kinase encoded by the *polehole* allele) and other proteins encoded by *enhancers of sevenless (E(sev))* (1, 18). One of these genes has recently been shown to be the fruit fly homologue of MAP kinase (L. Zipursky, personal communication). This pathway leads to the

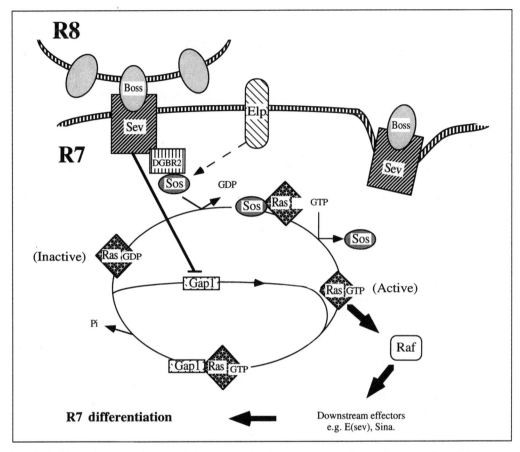

Fig. 1 The Sevenless signalling pathway. Interaction between Bride of sevenless (Boss) and Sevenless receptor (Sev) results in Ras activation and ultimately to photoreceptor development (for details see text). In this figure an interaction between the Ellipse (Elp) receptor and Sos is also shown to demonstrate signalling pathway crosstalk. After an initial interaction, the Sev–Boss complex is internalized. Sos, son of sevenless; Sina, seven in absentia; DGBR2, *Drosophila* growth factor receptor-bound protein 2.

regulation of the *seven in absentia (sina)* gene product which is a transcription factor and ultimately to R7 photoreceptor development (1, 13–15, 18–20).

2.1.1 Loss-of-function analysis

A variety of genetic techniques have been used to study the mechanism of receptor tyrosine kinase signalling in the Sevenless pathway. Screening for recessive, viable, loss-of-function mutations which result in the *sevenless* phenotype is a direct method for identifying genes in the pathway. Flies are mutagenized, either chemically or by UV irradiation, and progeny screened for the phenotype. The *boss, sina*, and *gap*1 genes were initially identified in this way (13, 19, 21, 22). This approach only reveals genes dispensable for viability, i.e. that do not perform essential functions in other processes and pathways. Once a gene has been identified, mosaic analysis can be used to determine whether the gene acts in a cell autonomous or non-autonomous manner. Firstly, a cell autonomous phenotypic marker must be identified which is proximal to the gene of interest. In studies with *boss*, the *chaoptic (chp)* gene, which effects rhombdomere morphology, was used as the phenotypic marker (22). In flies heterozygous for *chp* and *boss*, homozygous patches of cells were produced by X-ray induced mitotic recombination. Cells mutant for *boss* were then identified by their *chaoptic* phenotype. The R7 photoreceptor failed to develop in ommatidia where the R8 cell expressed the *boss* mutation, suggesting that the Boss transmembrane protein, expressed exclusively on the R8 cell, specifies the fate of the R7 photoreceptor cell. Mosaic analysis has shown that the *sevenless, sos, gap1, sina*, and other *E(sev)* genes are required in a cell autonomous manner for R7 photoreceptor development (12, 18, 19).

2.1.2 P-element insertional mutagenesis

'Enhancer trap' techniques can be used to identify genes which give rise to a particular phenotype. Using transposable P-element vectors, the β-galactosidase gene fused to an enhancerless promoter can be randomly integrated into the *Drosophila* genome. The resultant expression of β-galactosidase may then reflect the expression pattern of a nearby gene (23–25). If the β-galactosidase construct also disrupts the host gene, a phenotype may be scored. Genomic sequences flanking the site of P-element insertion can be recovered by plasmid rescue (26) and used to screen genomic libraries to reveal the disrupted gene. The *sos* and *seven-up* genes were identified in this way (15, 27). Seven-up is a member of the steroid receptor family and is required in the R1, R3, R4, and R6 photoreceptors for specification of their fate (27). These four photoreceptors develop R7 characteristics in *seven-up* mutant ommatidia (27).

2.1.3 Phenotypic rescue

Another approach to identify members of a particular signalling pathway is to look for second-site dominant suppressors of the phenotype. Flies already mutant for the phenotype of interest are fed a chemical mutagen and progeny screened for rescue of that phenotype. For example, an activated Ras protein was able to rescue R7 photoreceptor development in *boss* and *sevenless* null flies (28). However, it failed to rescue the phenotype in flies carrying a mutation in *sina* which suggested that Ras acts downstream of Boss and Sevenless and upstream of Sina. The *sos* gene was also identified as a second site suppressor of *sevenless* in flies expressing a mutant Sevenless receptor (29). However, rescue was only observed when Sevenless protein was produced, suggesting a Sos–Sevenless interaction. Similar experiments have demonstrated that Drk is required downstream of Sevenless and upstream of Ras (16). Additionally, this protein has been demonstrated to interact physically with the Sos and Sevenless proteins, consistent with genetic data linking these latter two gene products (16).

Such genetic epistatic experiments have been widely used to study signal transduction pathways. However, they are limited to ordering the action of genes and cannot provide information about the number of steps that exist between identified genes. The dominant *sos* allele was able to enhance the 'rough eye' phenotype resulting from constitutive activation of the *ellipse* encoded receptor tyrosine kinase (the fly EGF receptor homologue) (30), suggesting that the *sos* protein is common to both the Sevenless and EGF signalling pathways. A constitutively activated *raf* gene was able to rescue R7 development in *sevenless* mutant flies (1). In contrast, reduced levels of D-raf suppressed the effects of a constitutively activated Ras protein, suggesting that D-raf acts downstream of Sevenless and Ras in this pathway (1). Ras and D-raf are also required in the Torso receptor signalling pathway (2, 31). Like Sevenless and Ellipse, Torso is a receptor tyrosine kinase (30) and is required for the determination of terminal embryonic structures (32, 33). In oocytes containing a mutant Torso mRNA, correct terminal differentiation can be achieved by injection of Ras-1 protein (31). However, this rescue is dependent on the presence of maternally derived Raf. In addition, a loss of function mutation in *raf* can suppress a gain of function mutation in *torso* which together with the Ras1 injection studies suggests that Ras1 and Raf act downstream of Torso. A recent genetic screen for dominant suppressors of a gain of function *torso* allele identified 40 separate mutations (in seven complementation groups). Ras 1 and Sos were identified together with several other mutations that were shown to be common to other tyrosine kinase activated pathways (34).

2.1.4 Pathway-restricted genetic screening

One of the problems with the techniques described above is the potential for the mutated protein to interfere with various other pathways, as demonstrated by mutations in *sos*, *ras*, and *D-raf*. When proteins common to more than one signal transduction pathway are mutated, the phenotype of interest may be obscured by secondary effects resulting from disruption of several pathways. Simon *et al.* (18) devised a sensitive genetic screen to circumvent the problems of signalling pathway crosstalk. They attempted to uncover genes that were required for signalling through a weakened Sevenless pathway by screening for second gene mutations that were functionally inefficient but not inactive, which therefore selectively exacerbated the *sevenless* phenotype. This technique allows for the identification of genes common to more than one pathway. Firstly a temperature-sensitive mutant of *sevenless* was derived to enable manipulation of the efficiency of this signalling pathway. Flies were then mutagenized and progeny grown at a temperature where the Sevenless pathway was barely functional. Under these conditions a subtle mutation in another component of the pathway, resulting in as little as a two-fold reduction in its efficiency, would be sufficient to block Sevenless signalling and prevent R7 development but not affect the role of the secondary gene in other pathways. Seven independent gene mutations (termed *enhancers of sevenless*, *E(sev)*), including the *ras* and *sos* genes, were identified as having a role in Sevenless signalling (18). In addition, *ras*, *sos*, and two other *E(sev)* genes were shown to have a role in the Ellipse pathway. Activation of Ras appears to be the primary response to Sevenless and Ellipse receptor signalling and may be a common mechanism for the activation of some tyrosine kinase receptor pathways (see Section 2.1).

2.2 Maternal effect genes and embryogenesis

Certain genes exhibit a maternal and zygotic requirement during embryogenesis. That is, products of a maternally-derived transcript deposited in the oocyte, and zygotic transcripts of the same gene may perform discrete tasks essential for correct development during embryogenesis. Analysis of the role of maternally-derived mRNA has benefited from the dominant female-sterile technique and X-ray induced mitotic recombination, which together allow investigation of the maternal requirement of such genes without disruption of their zygotic function (35, 36). In heterozygous flies, some oocyte-producing germ cells are made homozygous by X-ray induced mitotic recombination. These germ-line cells give rise to one oocyte and 15 nurse cells. The nurse cells are responsible for synthesizing and depositing maternal mRNAs into the oocyte and hence deposit a mutated maternal mRNA. Upon fertilization, the mutated zygotic gene from the oocyte is rescued by a native paternal gene. Any disruption to embryogenesis is then a direct result of the mutation in the maternal gene product. Many genes involved in the establishment of embryonic polarity have been identified by this technique (reviewed in 37).

2.2.1 Genetics of the Shaggy protein kinase

The *Drosophila zeste-white3/shaggy* gene encodes a serine/threonine kinase homologous to mammalian glycogen synthase kinase-3 (GSK-3) (38–41). This gene exhibits both maternal and zygotic effects (42–46). Zygotically-derived proteins have pleiotropic roles during embryogenesis, as mutation of the gene results in an abnormal syncytial blastoderm and larval death (43). Differential splicing generates several distinct *sgg* proteins sharing a common catalytic core. Generation of mosaic patches of *shaggy⁻* cells in adult flies by X-ray induced mitotic recombination results in homeotic transformations (42–45). Here, cells destined to become epidermal develop neuronal characteristics. This homeotic transformation is thought to arise from disruption of a signalling pathway involving the *notch* gene product which acts to specify cell fate in several small groups of cells called proneural clusters (45, 46). These pronueral clusters comprise 7–8 cells that all have equal potential to become neuronal. During development one cell becomes dominant and differentiates neuronally, producing a sensory bristle. At the same time this cell induces the surrounding cells to adopt an epidermal cell fate via a mechanism of lateral inhibition (46, 47). The Notch transmembrane protein is thought to be the receptor for this inhibitory signal because in clusters mutant for *notch*, all cells produce a sensory bristle (47). Likewise, several neuronal cells per cluster are observed in *shaggy* mutant flies suggesting that this kinase is involved in transduction of the inhibitory signal (45, 47). An epistatic relationship between *shaggy* and loss and gain of function alleles of *notch* indicate that *shaggy* does indeed lie downstream of *notch* in this signalling pathway (45); *shaggy* may also be involved in the formation of these proneural clusters as mutants lead to ectopic proneural regions on the anterior wing blade (44).

In contrast, mutations in the maternal gene product result in the complete absence of the denticle belt from the anterior part of each segment (44). The maternal gene product functions in a signalling pathway that helps specify the position of the parasegmental boundary during embryogenesis. Intrasegmental patterning during *Drosophila* development is regulated by cell–cell communication (48–52). The boundary between the segment polarity genes *wingless* (the *Drosophila* homologue of Wnt 1 (53)) and *engrailed* (a homeodomain transcription factor; 54) -expressing cells in the developing embryo specifies the position of the so-called parasegmental groove (55–57). Mutations which effect *engrailed* and *wingless* expression produce observable morphological changes to the cuticle of developing flies. Several genes have been identified that affect segmental morphology and regulate *engrailed* and *wingless* expression (57). Maternally-derived *shaggy* protein apparently plays a role here in the negative regulation of *engrailed* (57, 58). A series of pair-rule and gap genes lay down the restricted domains of *engrailed* and *wingless* expression whereby regions of cells only express one or other gene (57). Later, after pair-rule gene expression has ceased, *engrailed* and *wingless* expression become interdependent on a mutual cell–cell signalling process (56, 57). This inductive pathway has been extensively studied and is illustrated in Fig. 2. In the anterior of the *wg*-competent domain, *wingless* expression is negatively regulated by the *patched* gene product (a transmembrane protein, restricted to this region) whilst *engrailed* expression is suppressed posteriorly by *shaggy* and *naked* (43, 55, 59–65). In cells adjacent to the *engrailed*-expressing region, the inhibitory effect of *patched* on *wingless* is negated by *engrailed*-induced expression of the *hedgehog* gene (60, 65, 66). Hedgehog is thought to be a transmembrane protein that likely interacts with the Patched receptor on the surface of the neighbouring cell (65–67). This blockade of Patched action maintains *wingless* expression. Wingless is a secreted factor which diffuses to the *en*-expressing cells countering the negative effect of Shaggy and Naked on *engrailed* expression (57, 58, 68). Epistatic experiments have shown Naked and Shaggy to operate in distinct pathways as *wingless/shaggy* and *wingless/naked* double mutants have opposite phenotypes (58). The *fused* gene product (a serine/threonine kinase) also acts in the *engrailed*-expressing cells to positively regulate the expression of *engrailed* (69). Fused possibly works in direct competition with the Naked and Shaggy pathways as *fused/naked* double mutants have a normal phenotype. Disruption of *fused* results in premature extinction of *engrailed* expression due to unopposed suppression by Shaggy and Naked (67, 70).

Thus, genetic approaches have revealed a complex reciprocal interaction between cells. In this case the mutual induction of *engrailed* and *wingless* in juxtaposed cells maintains the relative position of the parasegmental groove during all the morphogenetic movements of embryogenesis. The 'interactions' are primarily based on genetic evidence and have yet to be demonstrated biochemically. However, it is difficult to imagine how far a biochemical approach would have progressed in the absence of the genetics. In contrast, the Sevenless pathway comprises a

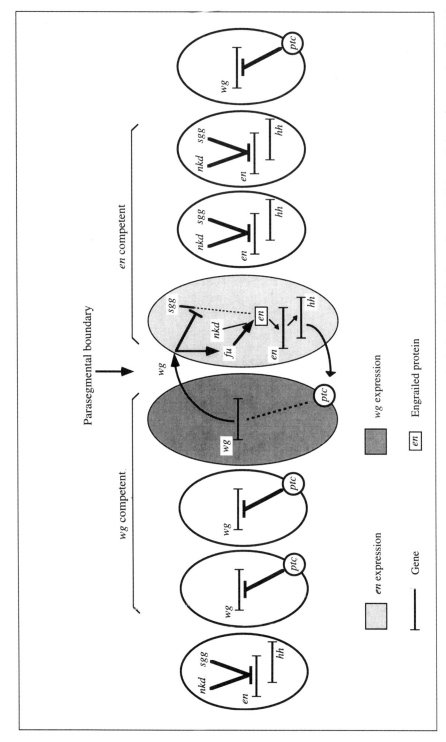

Fig. 2 Possible interactions between gene products in the *engrailed* (*en*) and *wingless* (*wg*) inductive pathway. Half the cells in each parasegment are *en* competent (capable of expressing *en*) and half are *wg* competent, but the *en* and *wg* expression domains are restricted to two adjacent stripes dependent on a mutual cell-signalling pathway. The expression of *wingless* and *engrailed* in competent cells is prevented by the actions of *patched* (*ptc*) and *shaggy* (*sgg*)/*naked* (*nkd*) proteins, respectively. The *wingless* protein diffuses to the *engrailed*-expressing cell and negates *sgg* repression of *engrailed*. In addition it stimulates *fused* (*fu*) product activity which counters the negative effects of *naked*. *Engrailed* activity then leads to expression of *hedgehog* (*hh*). The *hedgehog* protein then leads to the inhibition of *patched* function in the neighbouring cell which allows the continual expression of *wingless*.

more familiar organization of proteins, several of which were first identified in mammals via their biochemical properties and associations, and provides independent evidence for their involvement in intracellular signal transduction.

3. Vulval development in *Caenorhabditis* elegans
3.1 Vulval induction

The nematode *C. elegans* is a genetically-manipulable multicellular organism which has yielded an abundance of signalling pathway information. Cells within the developing nematode are visible by microscopy and fate maps of every cellular division have been compiled. With this information, the consequences of mutations on any of the processes occurring throughout development can be recognized at the single cell level. Development of the nematode vulva has provided insights into the mechanism of action of receptor tyrosine kinases (reviewed in 71–73). The adult vulva develops from three of six embryonic cells termed vulval precursor cells (VPCs). These cells have equal potential to develop into one of three cell types denoted 1°, 2° and 3° as shown by cell ablation studies (74). The 1° and 2° cell types give rise to the vulva proper. The mechanism of induction and inhibition of the VPCs in vulval development has been used as a model system for investigating cell–cell signalling and is cartooned in *Fig. 3*. An inductive signal from the gonadal anchor cell stimulates an EGF receptor-like tyrosine kinase (Let-23) on the surface of the VPC. The *lin-3* gene product is most likely the inductive molecule and is similar to epidermal growth factor and transforming growth factor-α (75). Stimulation of the Let-23 receptor is thought to override an inhibitory signal emanating from the syncytial hypoderm (acting via the *lin-15* protein) and activate a Ras-like protein (Let-60). Let-60 then activates other downstream components, including Lin-45 (the homologue of Raf) and Lin-1 leading to 1° and 2° cell fates (78). The cell closest to the anchor cell usually adopts the 1° fate. This cell is then thought to help induce the two flanking cells to adopt the 2° fate via a mechanism of lateral inhibition (mediated by an EGF-like protein encoded by *lin-12*) (72, 75–81). The remaining three cells adopt the 3° fate and later fuse with the hypodermis.

Over 30 separate genes have been identified by mutation analysis that play a role in this signalling mechanism and await biochemical characterization (71). Mutations which effect the cell fates of these six VPCs can be classed into two types: vulvaless and multivulva. In vulvaless mutants, all cells adopt the 3° cell fate as a result of disruption of the inductive signalling pathway. In multivulva mutants, all of the cells develop as either 1° or 2° due to continual stimulation of the inductive signal. In some multivulva mutants, termed hypervulval, two 1° cells are found juxtaposed and are thought to result from disruption of the lateral inhibitory signal (73). Therefore, scoring of these phenotypes in mutational experiments has enabled several genes to be assigned to particular pathways.

Gonad ablation in loss-of-function *lin-15* mutants does not effect the resulting multivulval phenotype which suggests that this gene does not act in the gonads or

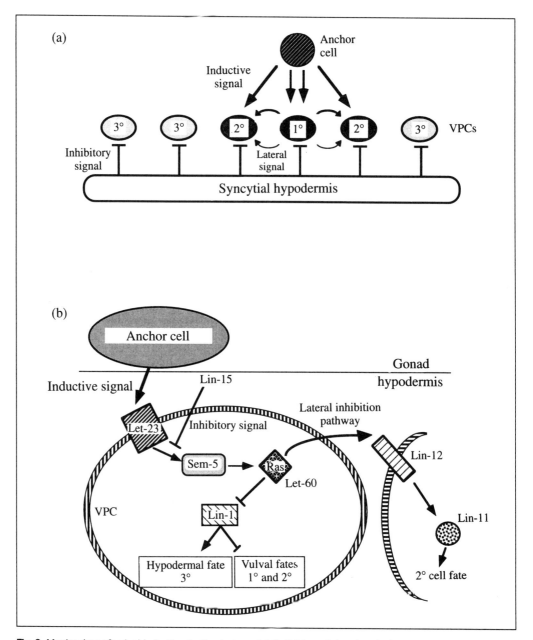

Fig. 3 Mechanism of vulval induction in *C. elegans*. (a) Cell fates of the six vulval precursor cells (VPCs). An inductive signal from the anchor cell overcomes an inhibitory signal from the syncytial hypodermis and stimulates a 1° or 2° cell fate in three of the VPCs. A signal from the 1° cell helps to induce the two neighbouring cells to adopt the 2° cell fate. The three remaining VPCs adopt the 3° cell fate. (b) The putative order of gene products thus far identified in the vulval inductive pathway. The Let-23 receptor tyrosine kinase transduces an inductive signal from the anchor cell by activation of Let-60 Ras. Let-60 Ras activation inhibits Lin-1 and lifts the repression to 1° and 2° cell fate determination. In addition, Let-60 Ras activation stimulates the lateral inhibitory pathway which results in juxtaposed cells adopting the 2° cell fate (acting via Lin-12).

exert its effect by inhibition of the inducing factor (71). Genetic mosaic and ablation analysis has shown that Lin-15 does not emanate from the gonads or VPCs and it is now thought to arise from the surrounding syncytial hypodermis (82). The *let-23*, *let-60*, *let-45*, and *sem-5* mutations were identified as suppressors of the multivulva phenotype resulting from a *lin-15* mutation and act downstream of this signal (76, 81–84). Genetic analysis of various *let-60* mutants has demonstrated that this Ras protein acts as the switch in determining the developmental fate of the VPCs (79). Mutations that decrease the effectiveness of Let-60 produce a vulvaless phenotype (3° cell fate), whereas those which increase its activity produce the multivulval phenotype (1° and 2° cell fates) (79, 80, 83). These findings have been supported by gene dosage experiments. During vulval development in *C. elegans*, receptor-induced Ras activity appears to be mediated by the Raf serine/threonine kinase in a manner analagous to mammalian signalling pathways. Undoubtedly several other serine/threonine kinases, such as MAP kinase homologues, will surface as downstream mediators of Ras and Raf in this signalling pathway.

The *sem-5* gene was identified in a screen for suppressors of the multivulva phenotype in *lin-5* mutants. The protein was sequenced and shown to be composed of an SH2 domain flanked by two SH3 domains (81). Proteins containing SH2 domains bind to peptides containing a phosphorylated tyrosine residue and many are associated with receptor tyrosine kinases (85, see Chapters 6 and 7). It is likely that *sem-5* interacts with Let-23 or its substrates and is involved in the transduction of the receptor signal to Let-60. Additionally, there is growing evidence that SH3 domains interact with guanine nucleotide exchange factors and GTPases which are required for the regulation of Ras proteins (85). It is tempting to speculate that, after activation of the Let-23 receptor, Sem-5 binds either to the receptor or one of its substrates and causes activation of Ras by an interaction with Ras regulatory proteins. Genetic and biochemical studies with the *Drosophila* homologue of Sem-5 (Drk) support this idea (16, see Section 2.1). The identification of Sem-5 in *C. elegans* prompted a search for a homologous protein in mammals which resulted in the isolation of Grb2. Recent biochemical evidence has shown a direct interaction between Grb2, the EGF receptor and the mammalian homologue of Sos (86–89). Thus, genetic identification of Sem-5 together with biochemical characterization of the mammalian homologue has furthered our understanding of how receptor tyrosine kinases activate Ras.

3.2 Lateral inhibition

A hypervulval phenotype was observed in a screen for genes which changed the phenotype in *lin-15* mutants (78). In these animals 1° cells were found juxtaposed suggesting disruption of the lateral inhibitory signal. The new mutation was identified as *lin-12* (encoding another EGF receptor-like transmembrane tyrosine kinase) and has been suggested to act in reception of the lateral inhibitory signal (90, 91). Further, overexpression of *lin-12* results in all cells adopting the 2° cell fate (92). This mechanism resembles the Notch-mediated lateral inhibition of sensory bristle formation in fruit flies and a *C. elegans* homologue of *shaggy* was recently identified (45, 46, Section 2.2). However, 2° cells can form in the absence of any other VPCs which suggests that the lateral signal is not essential for 2° cell fate. Mutations in *lin-11* (which encodes a homeodomain transcription factor) result in only one of the three possible progeny cell types found in 2° cell siblings (71, 77, 93). This suggests that Lin-11 acts in the Lin-12 signalling pathway to specify formation of the correct daughter cells.

Together, the fruit fly and nematode studies demonstrate the central importance of signal transducing proteins to processes such as development and differentiation. Several common principles recur, such as the role of receptor tyrosine kinases in signal reception, SH2 motif-containing proteins in recruiting transducing proteins, and protein-serine kinases acting to disseminate the signals throughout the cells. Equally important to these feed-forward pathways are those that antagonize or desensitize. Thus, inductive cycles can be established involving mutually dependent signals locking cells into complex relationships. By definition, disruption of these components yields phenotypic changes, suggesting there is very limited redundancy. In view of the large number of protein kinase genes expressed per cell (probably >200), these enzymes must be far more specific in their *in vivo* functions than is often apparent *n vitro*.

4. Signal transduction in slime mould

Cells of the slime mould *Dictyostelium discoideum* persist as single-celled amoeba until environmental conditions trigger aggregation and multicellular development (94). Multicellularity arises from aggregation of individual ameoboid cells and not from a process of cell division. Cell growth and development are completely separate and can be controlled precisely in the laboratory (95). Under conditions of starvation vast quantities of *Dictyostelium* will synchronously initiate a developmental programme which is reversible until the last stages of maturation (95). Mutation and gene disruption experiments are relatively straightforward in this slime mould as most of the life cycle is spent in the haploid state (for a guide to *Dictyostelium* biology and manipulation see ref. 95). cAMP is responsible for the coordination and regulation of many cellular processes during the life cycle of *Dictyostelium* (95–101). It is involved in the initial aggregation of amoeba by acting as an extracellular chemoattractant, as well as inducing chemotaxis and early gene expression (96). Additionally, during the multicellular phase of the life cycle, cAMP is required intracellularly for coordination of morphogenic movements and differential gene expression (101).

Upon starvation, a few cells start emitting pulses of cAMP and the surrounding amoebae respond by moving chemotactically towards the source. The cAMP pulses are received by G-protein-coupled adrenergic receptors which then activate three distinct pathways leading to chemotaxis, gene expression, and activation of cAMP-dependent protein kinase (PKA; reviewed in 98 and 102). Genetic experiments have shown that receptor activation of phospholipase C and guanyl cyclase leads to gene expression and chemotaxis, respectively (103, 104). Receptor adaptation occurs within a few seconds which facilitates reception of the next flux of cAMP. This adaptation is thought to be mediated by specific kinases as phosphorylated receptors are observed *in vivo* several seconds after cAMP stimulation and analogous protein kinases desensitize mammalian adrenergic receptors (97–110). After aggregation the cells undergo a slug phase and are composed of several distinct cell types (see Fig. 4(a)). Under conditions of high humidity and low light levels, the slug will migrate until it reaches a more favourable environment where culmination is initiated. Here, cells differentiate further and redistribute to produce a mature fruiting body. Several Ser/Thr and Tyr kinases have now been isolated and many play a role in *Dictyostelium* development.

4.1 Identification of protein kinases

Slime mould casein kinase II has been purified and cloned but as yet a specific function for this kinase remains unknown (106). This kinase has been implicated in a variety of mammalian cellular processes, including regulation of glycogen metabolism and transcription factors (107, 108). Disruption of the slime mould gene appears lethal (106). A slime mould homologue of GSK-3 has recently been isolated and appears to play a role in PKA-regulated stalk cell differentiation (A. Harwood and S. Plyte, unpublished observations). In support of such a role, this kinase has also been implicated in the regulation of transcription factors and in fate determination in fruit flies (see Section 2.1.3; 39, 58, 109). Several protein kinase genes have been isolated by homology cloning techniques and await genetic and biochemical characterization (110–113). A multigene family comprising at least five putative Ser/Thr kinases was identified by a PCR-based screen (112, 113). Two putative Ser/Thr kinases (termed Dd PK1 and Dd PK2) were identified by screening a cDNA library with oligonucleotides directed towards conserved regions of eukaryotic protein kinases (110). *Dd PK1* RNA levels were seen to decrease during aggregation and reaccumulate later in development, suggesting a role in culmination (110). Overexpression of *Dd PK2* caused rapid development and affected the intracellular level of cAMP in *Dictyostelium* (114). Tan and Spudich (111) used antiphosphotyrosine antibodies to screen an expression library and isolated two fusion proteins that exhibited protein tyrosine kinase activity (termed *Dd PYK1* and *Dd PYK2*). These tyrosine kinases were the first to be identified in *Dictyostelium* and share homology with both mammalian Ser/Thr and Tyr kinases. This is consistent with the idea that the catalytic domains of Tyr kinases were derived from an archetypal protein Ser/Thr kinase although it is presently unclear whether the Dd PYKs exclusively phosphorylate tyrosine (115).

4.2 The role of cAMP-dependent protein kinase

The structure and function of PKA is discussed in Chapter 1. This protein kinase plays a special role in *Dictyostelium* biology as cAMP appears to coordinate several developmental events. Firtel and Chapman (116) have demonstrated that over-expression of the regulatory subunit of PKA (Rm), mutated to prevent cAMP binding, results in a block to aggregation (99). These cells do not express genes which are normally induced in the multicellular stages of development. Additionally, cells overexpressing the catalytic subunit (*Dd PK3*) undergo accelerated growth whilst those having a disrupted gene fail to aggregate (117). Overexpression studies of *Dd PK2* and *Dd PK3* have indicated a role for PKA in the formation of spores (101, 114, 117). More elaborate experiments, using stage specific promoters, have provided further insights into the function of this kinase during development (101). During the slugging and culmination stages of development cells destined to form the stalk tube (pst cells) can be divided into several cell types. Two of these cell types can be distinguished by their differential gene expression: pstB cells express two genes encoding extracellular matrix proteins (termed *ecmA* and *ecmB*) whilst pstA cells only express one, *ecmA* (118–120). Both genes are induced by the stalk cell morphogen, DIF (119, 120). During culmination the pstA cells migrate to the mouth of the stalk tube and are transformed into pstB cells by induction of the *ecmB* gene (Fig. 4(a)). The pstB cells then migrate into and form the developing stalk tube, in a process likened to a 'reverse fountain'.

A mutant regulatory subunit of PKA (acting as a dominant inhibitor) was fused to the stage-specific *ecmA* promoter to study the effect of blocking the action of this kinase during multicellular differentiation. *Dictyostelium* transformed with this vector (*ecmA-Rm*) exhibited prolonged migration and failed to culminate into mature fruiting bodies (101). This implied that PKA is required for the choice between continued migration and culmination. A more detailed analysis demonstrated that no pstB cells were produced as the cells failed to express the *ecmB* gene. Heterotypic aggregates, produced by mixing wild type and mutant amoebae, produced slugs that were able to culminate normally. However, none of the cells containing the *ecmA-Rm* construct was seen to migrate to the tip of the slug and was not found in the stalk tube (101). *Dictyostelium* transformed with either an *ecmB* promoter fused to the *Rm* gene or with an *ecmA* or *ecmB* promoter fused to a constitutively active PKA gene undergo normal development. These data suggested that protein kinase activity was required for *ecmB* gene induction and pstA cell migration. Harwood *et al.* have proposed that a repressor in the pstA cells, which blocks DIF induction, prevents precocious expression of the *ecmB* gene (101) (Fig. 4(b)). When pstA cells migrate to the mouth of the stalk tube this repression is lifted, by the action of cAMP-dependent protein kinase, resulting in *ecmB* gene expression.

5. Serine/threonine kinases in yeast

Many protein kinases have been identified in both budding (*Saccharomyces cerevisiae*) and fission (*Schizosaccharomyces pombe*) yeasts (121–123). The kinases involved in cell cycle progression have been extensively studied in fission yeast

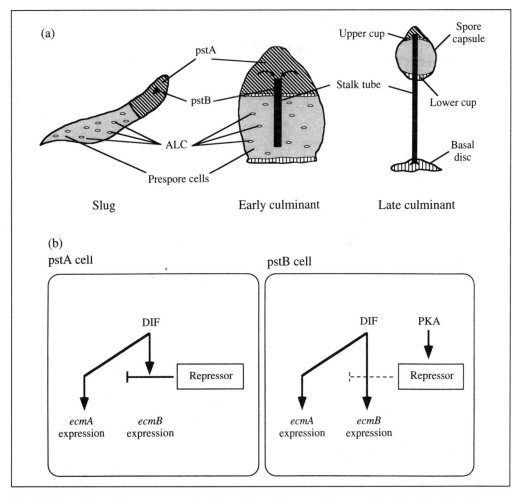

Fig. 4 (a) Distribution of various cell types during different stages of *Dictyostelium discoideum* development. During the slugging phase, prestalk A (pstA) cells and a small node of prestalk B (pstB) cells are situated in the anterior region of the slug. The anterior-like cells (ALC) are found distributed amongst the prespore cells. During culmination the pstA cells migrate to the region where the pstB cells are located. Upon arrival they are transformed into pstB cells and migrate into and form the stalk tube which lifts the spore-containing capsule into the air. (b) One of the putative roles of cAMP-dependent protein kinase (PKA) during *Dictyostelium* development. The inductive action of DIF to promote pstB cell differentiation is inhibited in pstA cells by a repressor protein. Upon migration of the pstA cells to the mouth of the stalk tube this repression is lifted by the induced action of PKA, allowing pstB cell differentiation. This prevents ectopic differentiation of prestalk cells in other regions of the organism.

and are discussed in Chapter 5. A gamut of data regarding kinase function in a diverse number of signalling pathways, including transcription, translation, and meiosis has been obtained from studies in yeast. Here, two pathways exemplifying the advantages of yeast in analysis of function are described.

5.1 The yeast mating response

In budding yeast, diploidy results from fusion of two different haploid cell types termed a and α. Each cell secretes a mating factor (either a or α pheromone) which interacts with the corresponding receptor on the opposite cell type to promote mating. Receptor binding initiates a signal transduction pathway leading to cell cycle arrest, conjugation and ultimately meiosis (reviewed in 123, 124). Mutational analysis produces two phenotypes: responsive and non-responsive (125). Here, the mating response is either constitutively activated or completely non-responsive to pheromone. Much of this pathway has been determined by combining these types of mutation in genetic epistatic experiments. Upon pheromone binding, both receptors elicit the same response by stimulation of a common pathway.

Eight genes have thus far been identified that act downstream of the receptor/G-protein complex and are shown in Fig. 5. Six of these are serine/threonine kinases: *FUS3, KSS1, STE20, STE11, STE7, STE20* (126–131). Fus3 and Kss1 are related to Spk1 in fission yeast and to mammalian mitogen-activated protein kinases (see Chapter 4 and Fig. 6). Moreover, Ste7 is related to mammalian MAP kinase kinase (MEK) and to Byrl in fission yeast; Ste11 is related to the fission yeast kinase Byr2

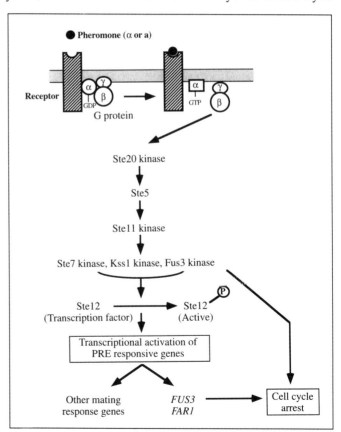

Fig. 5 Involvement of protein kinases in the mating response of *S. cerevisiae*. Pheromone binding stimulates guanine nucleotide exchange on the α subunit of the G-protein–receptor complex which promotes dissociation of the β and γ subunits and activates the mating response kinase cascade. The relative ordering of the components has been established by a combination of genetic and biochemical experiments. Some of the relationships remain ambiguous. PRE, pheromone responsive elements.

and to MEKK in mammals. A dominant allele of *STE11* was able to rescue mating in *STE4* mutants but not in cells carrying mutations in *STE7*, *FUS3*, and *KSS1* (132, 133). This demonstrated that Ste11 acted before these other genes in the mating response pathway. In response to pheromone stimulation the Ste7, Kss1, and Fus3 kinases become hyperphosphorylated (134, 135). In cells harbouring an overactive *STE11* allele, and in the absence of pheromone binding, the Ste7 and Fus3 are found in a hyperphosphorylated state (133–135). Furthermore, this phosphorylation is not autocatalytic as it is still observed in a kinase-dead mutant of Fus3 (134). Pheromone-induced hyperphosphorylation of Ste7 is dependent on functional Ste11 (135) and Fus3 is phosphorylated on tyrosine and threonine by Ste7 *in vitro* (136) placing this before Ste7 in the pathway. These observations are consistent with the topography of the mammalian MAP kinase cascade (see Fig. 6). The downstream target of this pathway appears to be a transcription factor encoded by *STE12* (137, 138). This factor is a phosphoprotein and binds to pheromone responsive elements (PRE) found upstream of target genes (137, 139, 140). Ste12 is rapidly phosphorylated in response to pheromone binding and this results in increased transcription of pheromone-responsive genes (139). (In fission yeast a

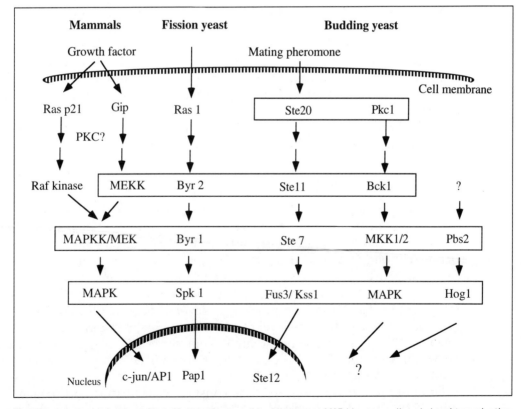

Fig. 6 Comparison of mammalian, fission yeast, and budding yeast MAP kinase mediated signal transduction pathways. The boxed proteins share homology.

downstream target of Spk1$^+$ is a transcription factor encoded by *pap1*$^+$ (141).) *FUS3* is one of these pheromone-responsive genes and also has a role in cell cycle arrest (128, 142). G$_1$ arrest is achieved in part by the phosphorylation-mediated regulation of cyclins which then inhibit CDC28 activity (reviewed in 124, 143–146). Far1 is a pheromone-induced protein associated with the CDC28/cyclin complex and promotes cell cycle arrest (147). The induced interaction between Far1 and the cyclin complex has recently been shown to be dependent on Fus3 activity and thus a direct link between pheromone signalling and cell cycle arrest has been established (147).

The remarkable degree of conservation of intermediate elements of the pheromone pathway in budding yeast and pathways in fission yeast and mammals, including preservation of the order of action, suggests these interactions evolved before the divergence of the ancestors of these organisms (148; Fig. 6). There are other pathways in budding yeast including a protein kinase C cascade that mediates cell wall synthesis. Here the Bck1 kinase (related to Ste11, Byr2 and MEKK) acts upstream of two closely related proteins encoded by *MKK1* and *MKK2* which are similar to Ste7/Byr1/MEK (149, 150, reviewed in 151). In turn, the *MKK1/2* gene products act on Mpk1 (also termed Slt2) which is related to Fus3, Kss1, Spk1$^+$ and MAP kinases (152, 153). A further cascade mediating responses to osmolarity involves the MAP kinase-relative Hog1 and the MEK-relative Pbs2 (154, 155). Whereas the intermediate proteins, the protein kinases, are highly conserved, their input signals and output targets are quite different. This is a common theme of signal transducing systems. Many intermediary protein kinases are evolutionary dinosaurs, remnants from a bygone age perhaps trapped by their own efficiency in performing a transductory function that has been adapted as needs demand. A second example is mammalian GSK-3 which can functionally substitute for its fruit fly Shaggy counterpart and has close relatives in yeast and slime mould (39, 45, 58).

5.2 Entry into meiosis

Once mating has occurred other signalling pathways are initiated leading to meiosis. Again, several Ser/Thr kinases play a central role in this process (reviewed in 122). In budding yeast, entry into meiosis appears to be regulated by two pathways; one is activated by the *RAS* product and the other by the mating response pathway. The *RAS*-activated pathway, which is initiated upon starvation, is mediated by PKA and results in a switch from vegetative growth to meiosis (reviewed in 156). Mutational analysis of several components of the *RAS* pathway have demonstrated that the level of cAMP is the major determinant for entry into meiosis (122). Loss-of-function mutations in either *CDC25* (an activator of *RAS*), *RAS2* (an activator of adenyl cyclase, *CYR1*) or *CYR1* itself result in low levels of cAMP and lead to sporulation (156–159). Conversely, mutations in the regulatory subunit of PKA (*BCY1* gene) result in constitutive activation and prevent sporulation (122). Thus, high levels of cAMP prevent entry into meiosis and low levels induce sporulation (122, 159). Activation of this pathway leads to derepression of an inducer of meiosis (encoded by *IME1*) and promotes sporulation (160–162). The mating response pathway also leads to activation of Ime1. Some genes from the mating type locus repress the activity of a negative regulator of meiosis (encoded by *RME1*) which normally antagonizes Ime1 (163). Ime1 induces transcription of a putative Ser/Thr kinase encoded by *IME2* which presumably phosphorylates proteins involved in meiotic progression (160, 164). Two further genes have been identified which positively regulate Ime1 activity, *MCK1* (a Ser/Thr kinase related to GSK-3/Shaggy) (165, 166) and *IME4* (167) which may act to ensure coordination of events within the cells, as the former has roles in sporulation and centromere formation. Unlike the pheromone kinase cascade, most of the genes identified in the meiotic programme do not appear homologous to proteins involved in meiosis in mammals, which may reflect the restriction of the meiotic process to specialized germ cells in multicellular organisms.

6. Concluding remarks

Reversible protein phosphorylation is a recurrent theme in biology and it is evident that many signalling pathways present in lower eukaryotes persist in higher organisms. Figure 6 compares the mammalian MAP kinase pathway with other yeast pathways involving homologous kinases. The MAP kinase pathway was mainly determined by biochemical analysis whilst the pathways in yeast were elucidated by genetic means. Work in *Drosophila* and *C. elegans* has accelerated our understanding of the mechanism of action of receptor tyrosine kinases. Counterparts for some of the components in these pathways have been identified in higher eukaryotes, for example the identification of Sem-5 in receptor tyrosine kinase signalling in *C. elegans* prompted the search for an homologous protein in higher eukaryotes. To this end, the mammalian homologue of Sem-5, GRB2, was identified (168) as well as its counterpart in *Drosophila, Drk* (16). Combination of biochemical data from mammalian systems with genetic data from lower eukaryotic studies allows a tentatative ordering of many of the components of a Ras-mediated receptor tyrosine kinase signalling cascade. Genetic approaches have contributed significantly to the construction of the cascade and these techniques should rapidly provide clues as to the missing components that lead to transcriptional activation.

Over the past decade, the realization that the fundamental processes of growth control, cell cycle regulation and signal transduction are similar in all eukaryotes has revolutionized these fields. The increased tractability of simpler eukaryotes to genetic analysis offers gene discovery based on function rather than purely structure. Genes identified in one organism can be rapidly isolated in others, stimulating discovery of further genes. Conservation of function allows examination of the properties of one component in a heterologous system. Out of this enormous convergence of interests around the 'basic processes' of transduction, protein kinases have emerged as common denominators. They are implicated in every aspect of cellular regulation and represent the most varied class of proteins known.

Application of both genetic and biochemical approaches to their study will undoubtedly reveal more of their manifold talents.

References

1. Dickson, B., Sprenger, F., Morrison, D., and Hafen, E. (1992) Raf functions downstream of Ras 1 in the Sevenless signal transduction pathway. *Nature*, **360**, 600.
2. Ambrosio, L., Mahowald, A. P., and Perrimon, N. (1989) Requirement of the *Drosophila raf* homologue for *torso* function. *Nature*, **342**, 288.
3. Kyriakis, J. M., App, H., Zhang, X., Banerjee, P., Brautigan, D. L., Rapp, U. R., and Avruch, J. (1992) Activation of MAP kinase kinase by *Raf-1*. *Nature*, **358**, 417.
4. Boulton, T. G., Nyle, S. H., Robbins, D. J., Ip, N. Y., Radziejewska, E., Morgenbesser, S. D. *et al.* (1990) ERKS: a family of protein serine/threonine kinases that are activated and tyrosine phosphorylated in response to insulin and NGF. *Cell*, **48**, 389.
5. Thomas, G. (1992) MAP kinase by any other name smells just as sweet. *Cell*, **68**, 3.
6. Tomlinson, A. and Ready, D. F. (1987) Cell fate in the *Drosophila* ommatidium. *Dev. Biol.*, **123**, 264.
7. Tomlinson, A. (1988) Cellular interactions in the developing *Drosophila* eye. *Development*, **104**, 183.
8. Rubin, G. M. (1991) Signal transduction and the fate of the R7 photoreceptor in *Drosophila*. *TIG*, **7**, 372.
9. Rubin, G. M. (1989) Development of the *Drosophila* retina: induction events studied at the single cell resolution. *Cell*, **57**, 519.
10. Hafen, E., Basler, K., Edstroem, J. E., and Rubin, G. M. (1987) *sevenless*, a cell specific homeotic gene of *Drosophila*, encodes a putative transmembrane receptor with a tyrosine kinase domain. *Science*, **236**, 55.
11. Campos-Ortega, J. A., Jurgens, G., and Hofbauer, A. (1979) Cell clones and pattern-formation: studies on *sevenless*, a mutant of *Drosophila melanogaster*. *Roux's Arch. Dev. Biol.*, **186**, 27.
12. Harris, W. A., Stark, W. S., and Walker, J. A. (1976) Genetic dissection of the photoreceptor system in the compound eye of *Drosophila melanogaster*. *J. Physiol.*, **256**, 415.
13. Kramer, H., Cagan, R. L., and Zipursky, S. L. (1991) Interaction of the *bride of sevenless* membrane-bound ligand and the *sevenless* tyrosine kinase receptor. *Nature*, **352**, 207.
14. Cagan, R. L., Kramer, H., Hart, A. C., and Zipursky, S. L. (1992) The *bride of sevenless* and *sevenless* interaction: internalisation of a transmembrane ligand. *Cell*, **69**, 393.
15. Gaul, U., Mardon, G., and Rubin, G. M. (1992) A putative Ras GTPase activating protein acts as a negative regulator of signalling by the *sevenless* receptor tyrosine kinase. *Cell*, **68**, 1007.
16. Olivier, J. P., Raabe, T., Henkenmeyer, M., Dickson, B., Mbamalu, G., Margolis, B. *et al.* (1993) A *Drosophila* SH2/SH3 protein couples the *sevenless* receptor tyrosine kinase to *Sos*, a putative guanine nucleotide exchange factor. *Cell*, **73**, 179.
17. Jones, S., Vignais, M. L., and Broach, J. R. (1991) The CDC25 protein of *S. cerevisiae* promotes exchange of guanine nucelotides bound to Ras. *Mol. Cell Biol.*, **11**, 2641.
18. Simon, M. A., Bowtell, D. D. L., Dodson, G. S., Laverty, T. R., and Rubin, G. M. (1991) Ras1 and a putative guanine nucleotide exchange factor perform crucial steps in signalling by the *sevenless* protein tyrosine kinase. *Cell*, **67**, 701.

19. Cathew, R. W. and Rubin, G. M. (1990) *seven in absentia*, a gene required for specification of the R7 cell fate in the *Drosophila* eye. *Cell*, **63**, 561.

20. Van Vactor, D. L., Cagan, R. L., Kramer, H., and Zipursky, S. L. (1991) Induction in the compound eye of *Drosophila*: multiple mechanisms restrict R7 induction to a single retinal precursor cell. *Cell*, **67**, 1145.

21. Hart, A. C., Kramer, H., Van Vactor, D. L., Paidhungat, M., and Zipursky, S. L. (1990) Induction of cell identity in the *Drosophila* retina: the *bride of sevenless* protein is predicted to contain a large extracellular domain and seven transmembrane segments. *Genes Dev.*, **4**, 1835.

22. Reinke, R. and Zipursky, S. L. (1988) Cell–cell interaction in the *Drosophila* retina: the *bride of sevenless* gene is required in photoreceptor cell R8 for R7 cell development. *Cell*, **55**, 321.

23. Beir, E., Vaessin, H., Sheperd, S., Lee, K., McCall, K., Barbel, S. *et al.* (1989) Searching for pattern and mutations in the *Drosophila* genome with a p-*lacZ* vector. *Genes Dev.*, **3**, 1273.

24. Bellen, H., O'Kane, C. J., Wilson, C., Grossniklaus, U., Kurth-Pearson, R., and Gehring, W. J. (1989) P-element mediated enhancer detection: a versatile method to study development in *Drosophila*. *Genes Dev.*, **3**, 1288.

25. Wilson, C., Kurth-Pearson, R., Bellen, H. J., O'Kane, C. J., Grossniklaus, U., and Gehring, W. J. (1989) P-element mediated enhancer detection: an efficient method for isolating and characterising developmentally regulated genes in *Drosophila*. *Genes Dev.*, **3**, 1301.

26. Hanahan, D., Lane, D., Lipsich, L., Wigler, M., and Botchan, M. (1980) Characteristics of an SV40-plasmid recombinant and its movement into and out of the genome of a murine cell. *Cell*, **22**, 127.

27. Mlodzik, M., Hiromi, Y., Weber, U., Goodman, C. S., and Rubin, G. M. (1990) The *Drosophila seven-up* gene, a member of the steroid receptor gene superfamily, controls photoreceptor cell fates. *Cell*, **60**, 211.

28. Fortinin, M. E., Simon, M. A. and Rubin, G. M. (1992) Signalling by the *sevenless* protein tyrosine kinase is mimicked by Ras 1 activation. *Nature*, **355**, 559.

29. Rogge, R. D., Kariovich, C. A., and Banerjee, U. (1991) Genetic dissection of a neurodevelopmental pathway: *son of sevenless* functions downstream of the *sevenless* and EGF receptor tyrosine kinases. *Cell*, **64**, 398.

30. Sprenger, F., Stevens, L. M., and Nusslein-Volhard, C. (1989) The *Drosophila* gene *torso* encodes a putative receptor tyrosine kinase. *Nature*, **338**, 478.

31. Lu, Z., Chou, T. Z., Williams, N. G., Roberts, T., and Perrimon, N. (1993) Control of cell fate determination by $p21^{ras}$ Ras1, an essential component of *torso* signalling in *Drosophila*. *Genes Dev.*, **7**, 621.

32. Baker, N. E. and Rubin, G. M. (1989) Effect on eye development of dominant mutations in *Drosophila* homologue of the EGF receptor. *Nature*, **340**, 150.

33. Klingler, M., Erdelyi, M., Szabad, J., and Nusslein-Volhard, C. (1988) Function of *torso* in determining the terminal analgen of the *Drosophila* embryo. *Nature*, **335**, 275.

34. Doyle, H. J. and Bishop, J. M. (1993) Torso, a receptor tyrosine kinase required for embryonic pattern formation, shares substrates with Sevenless and EGF-R pathways in *Drosophila*. *Genes Dev.*, **7**, 633.

35. Perrimon, N. (1984) Clonal analysis of dominant female-sterile, germline-dependent mutations in *Drosophila melanogaster*. *Genetics*, **108**, 927.

36. Garcia-Bellido, A. and Robbins, L. C. (1983) Viability of female germ-line cells homozygous for zygotic lethals in *Drosophila melanogaster*. *Genetics*, **103**, 235.

37. St Johnston, D. and Nusslein-Volhard, C. (1992) The origin of pattern and polarity in the *Drosophila* embryo. *Cell*, **68**, 201.

38. Woodgett, J. R. (1990) Molecular cloning and expression of glycogen synthase kinase-3/factor A. *EMBO J.*, **9**, 2431.

39. Plyte, S. E., Hughes, K., Nikolakaki, E., Pulverer, B. J., and Woodgett, J. R. (1992) Glycogen synthase kinase-3: functions in oncogenesis and development. *Biochim. Biophys. Acta*, **1114**, 147.

40. Siegfried, E., Perkins, L. A., Capaci, T. M., and Perrimon, N. (1990) Putative protein kinase product of the *Drosophila* segment polarity gene, *zeste-white* 3. *Nature*, **354**, 825.

41. Bourouis, M., Moore, P., Ruel, L., Grau, Y., Heitzler, P., and Simpson, P. (1990) An early embryonic product of the gene *shaggy* encodes a serine/threonine protein kinase related to the CDC28/cdc2+ subfamily. *EMBO J.*, **9**, 2877.

42. Simpson, P., El Messal, M., Moscoso del Prado, J., and Ripol, P. (1988) Stripes of positional homologies across the wing blade of *Drosophila melanogaster*. *Development*, **103**, 391.

43. Perrimon, N. and Smouse, D. (1989) Multiple functions of a *Drosophila* homeotic gene, *zeste-white3*, during segmentation and neurogenesis. *Dev. Biol.*, **135**, 287.

44. Blair, S. S. (1992) *shaggy* (*zeste-white3*) and the formation of supernumerary bristle precursors in the developing wing blade of *Drosophila*. *Dev. Biol.*, **152**, 265.

45. Ruel, L., Bourouis, M., Heitzler, P., Pantesco, V., and Simpson, P. (1993) *Drosophila* shaggy kinase and rat glycogen synthase kinase-3 have conserved activities and act downstream of Notch. *Nature*, **362**, 557.

46. Simpson, P. and Carteret, C. (1990) Proneural clusters: equivalence groups in the epithelium of *Drosophila*. *Development*, **110**, 927.

47. Heitzler, P. and Simpson, P. (1991) The choice of cell fate in the epidermis of *Drosophila*. *Cell*, **64**, 1083.

48. Perrimon, N. and Mahowald, A. P. (1988) Maternal contributions to early development in *Drosophila*. In *Primers in developmental biology* (ed. G. Malacinski), p. 305. Macmillan, New York.

49. Konrad, K. D., Engstrom, L., Perrimon, N., and Mahowald, A. P. (1985) Genetic analysis of oogenesis and the role of maternal gene expression in early development. In *Developmental biology: a comprehensive synthesis* (ed. L. Browder), p. 577. Plenum Press, New York.

50. Manseau, L. J. and Scupbach, T. (1988) The egg came first, of course! *Trends Genet.*, **5**, 400.

51. Ingham, P. W. (1988) The molecular genetics of embryonic pattern formation in *Drosophila*. *Nature*, **335**, 25.

52. Ingham, P. W. and Martinez-Arias, A. (1992) Boundaries and fields in early embryos. *Cell*, **68**, 221.

53. Rijsewijk, F., Schuermann, M., Wagenaar, E., Parren, P., Weigel, D., and Nusse, R. (1987) The *Drosophila* homolog of the mouse mammary oncogene *int-1* is identical to the segment polarity gene *wingless*. *Cell*, **50**, 649.

54. Fjose, A., McGinnis, W. J., and Gehring, W. J. (1985) Isolation of a homeobox-containing gene from the *engrailed* region of *Drosophila* and the spatial distribution of its transcripts. *Nature*, **313**, 284.

55. Martinez-Arias, A., Baker, N. E., and Ingham, P. W. (1988) Role of segment polarity genes in the definition and maintenance of cell states in the *Drosophila* embryo. *Development*, **103**, 157.

56. Vincent, J. P. and O'Farrell, P. H. (1992) The state of *engrailed* expression is not clonally transmitted during early *Drosophila* development. *Cell*, **68**, 923.

57. Ingham, P. W. (1991) Segment polarity genes and cell patterning within the *Drosophila* body segment. *Curr. Op. Genet. Dev.*, **1**, 261.

58. Siegfried, E., Chou, T., and Perrimon, N. (1992) *wingless* signalling acts through *zeste-white3*, the *Drosophila* homolog of glycogen synthase kinase-3, to regulate *engrailed* and establish cell fate. *Cell*, **71**, 1167.

59. Nakano, Y., Guerrero, I., Hidalgo, A., Taylor, A., Whittle, J. R. S., and Ingham, P. W. (1989) A protein with several possible membrane-spanning domains encoded by the *Drosophila* segment polarity gene *patched*. *Nature*, **341**, 508.

60. Hidalgo, A. and Ingham, P. (1990) Cell patterning in the *Drosophila* segment: spatial regulation of the segment polarity gene *patched*. *Development*, **110**, 291.

61. Ingham, P. W., Taylor, A. M., and Nakano, Y. (1991) Role of *Drosophila patched* gene in positional signalling. *Nature*, **353**, 184.

62. Hooper, J. E. and Scott, M. P. (1989) The *Drosophila patched* gene encodes a putative membrane protein required for segmental patterning. *Cell*, **59**, 75.

63. Heemskerk, J., DiNardo, S., Kostriken, R., and O'Farrell, P. H. (1991) Multiple modes of *engrailed* regulation in the progression towards cell fate determination. *Nature*, **352**, 404.

64. Sampedro, J. and Guerrero, I. (1991) Unrestricted expression of the *Drosophila* gene *patched* allows a normal segment polarity. *Nature*, **353**, 187.

65. Ingham, P. W. and Hidalgo, A. (1993) Regulation of *wingless* transcription in the *Drosophila* embryo. *Cell*, **117**, 283.

66. Lee, J. L., von Kessler, D. P., Parks, S., and Beachy, P. A. (1992) Secretion and localised transcription suggests a role in positional signalling for products of the segmentation gene *hedgehog*. *Cell*, **71**, 33.

67. Mohler, J. and Vani, K. (1992) Molecular organisation and embryonic expression of the *hedgehog* gene involved in cell–cell communication in segmental patterning of *Drosophila*. *Development*, **115**, 957.

68. Gonzales, F., Swales, L., Bejsovec, A., Skaer, H., and Martinez-Arias, A. (1992) Secretion and movement of the *wingless* protein in the epidermis of the *Drosophila* embryo. *Mech. Dev.*, **35**, 43.

69. Preat, T., Therond, P., Lamour-Isard, C., Limbourg-Bouchon, B., Tricoire, H., Erk, I. et al. (1990) A putative serine/threonine protein kinase encoded by the segment-polarity *fused* gene of *Drosophila*. *Nature*, **347**, 87.

70. Limbourg-Bouchon, B., Busson, D., and Lamour-Isnard, C. (1990) Interactions between *fused*, a segment polarity gene in *Drosophila*, and other segmentation genes. *Development*, **112**, 417.

71. Ferguson, E. L., Sternberg, P. W., and Horvitz, H. R. (1987) A genetic pathway for the specification of the vulval cell lineages of *Caenorhabditis elegans*. *Nature*, **326**, 259.

72. Horvitz, H. R. and Sternberg, P. W. (1991) Multiple intercellular signalling systems control the development of the *Caenorhabditis elegans* vulva. *Nature*, **351**, 535.

73. Sternberg, P. W. and Horvitz, H. R. (1991) Signal transduction during *C. elegans* vulval induction. *TIG*, **7**, 366.

74. Sternberg, P. W. and Horvitz, H. R. (1986) Pattern formation during vulval development in *Caenorhabditis elegans*. *Cell*, **44**, 761.

75. Hill, R. J. and Sternberg, P. W. (1992) The gene *lin-3* encodes an inductive signal for vulval development in *C. elegans*. *Nature*, **358**, 470.

76. Han, M., Golden, A., Han, Y., and Sternberg, P. W. (1993) *C. elegans lin-45 raf* gene participates in *let-60 ras*-stimulated vulval differentiation. *Nature*, **363**, 133.

77. Sternberg, P. W. (1988) Lateral inhibition during vulval induction in *Caenorhabditis elegans*. *Nature*, **335**, 551.

78. Sternberg, P. W. and Horvitz, H. R. (1989) The combined action of two intercellular signalling pathways specifies three cell fates during vulval induction in *C. elegans*. *Cell*, **58**, 679.

79. Beitel, G. J., Clark, S. G., and Horvitz, H. R. (1990) *Caenorhabditis elegans Ras* gene *let-60* acts as a switch in the pathway of vulval induction. *Nature*, **348**, 503.

80. Han, M. and Sternberg, P. W. (1990) *let-60*, a gene that specifies cell fates during *C. elegans* vulval induction, encodes a ras protein. *Cell*, **63**, 921.

81. Clark, S. G., Stern, M. J., and Horvitz, H. R. (1992) *C. elegans* cell-signalling gene *sem-5* encodes a protein with SH2 and SH3 domains. *Nature*, **356**, 340.

82. Herman, R. K. and Hedgecock, E. M. (1990) Limitation of the size of the vulval primordium of *Caenorhabditis elegans* by *lin-15* expression in surrounding hypodermis. *Nature*, **348**, 169.

83. Han, M., Arioan, R. V., and Sternberg, P. W. (1990) The *let-60* locus controls the switch between vulval and nonvulval fates in *Caenorhabditis elegans*. *Genetics*, **126**, 899.

84. Aroian, R. V. and Sternberg, P. W. (1991) Multiple functions of *let-23*, a *Caenorhabditis elegans* receptor tyrosine kinase gene required for vulval induction. *Genetics*, **128**, 251.

85. Pawson, T. and Gish, G. D. (1992) SH2 and SH3 domains: from structure to function. *Cell*, **71**, 356.

86. Rozakis-Adcock, M., Fernley, R., Wade, J., Pawson, T., and Bowtell, D. (1993) The SH2 and SH3 domains of mammalian Grb2 couple the EGF receptor to the Ras activator mSos1. *Nature*, **363**, 83.

87. Li, N., Batzer, A., Daly, R., Yajnik, V., Skolnik, E., Chardin, P., Bar-Sagi, D. *et al.* (1993) Guanine nucelotide-releasing factor hSos1 binds Grb2 and links receptor tyrosine kinases to Ras signalling. *Nature*, **363**, 85.

88. Gale, N. W., Kaplan, S., Lowenstein, E. J., Schlessinger, J., and Bar-Sagi, D. (1993) Grb2 mediates the EGF-dependent activation of guanine nucelotide exchange on Ras. *Nature*, **363**, 88.

89. Egan, S. E., Giddings, B. W., Brooks, M. W., Buday, L., Sizeland, A. M., and Weinberg, R. A. (1993) Association of Sos Ras exchange protein with Grb2 is implicated in tyrosine kinase signal transduction and transformation. *Nature*, **363**, 45.

90. Greenwald, I. S. (1985) *lin-12*, a nematode homeotic gene, is homologous to a set of mamalian proteins that includes epidermal growth factor. *Cell*, **43**, 583.

91. Yochem, J., Weston, K., and Greenwald, I. S. (1988) The *Caenorhabditis elegans lin-12* genc encodes a transmembrane protein with overall homology to *notch*. *Nature*, **335**, 547.

92. Greenwald, I. S., Sternberg, P. W., and Horvitz, H. R. (1983) The *lin-12* locus specifies cell fates in *Caenorhabditis elegans*. *Cell*, **34**, 435.

93. Ferguson, E. and Horvitz, H. R. (1989) The multivulva phenotype of certain *Caenorhabditis elegans* mutants results from defects in two functionally redundant pathways. *Genetics*, **123**, 109.

94. Devreotes, P. N. (1982) Chemotaxis. In *The development of* Dictyostelium discoideum (ed. W. Loomis), p. 117. Academic Press, San Diego.

95. Spudich, J. A. (ed.) (1987) *Dictyostelium discoideum*: molecular approaches to cell biology. In *Methods in cell biology*, Vol. 28. Academic Press, London.

96. Dottin, R. P., Bodduluri, S. R., Doody, J. F., and Haribabu, B. (1991) Signal transduction and gene expression in *Dictyostelium discoideum*. *Dev. Genet.*, **12**, 2.

97. Anschultz, A., Um, H-D., Tao, Y-P., and Klein, C. (1991) Regulation of protein phosphorylation in *Dictyostelium discoideum*. *Dev. Genet.*, **12**, 14.

98. Firtel, R. A. (1991) Signal transduction pathways controlling multicellular development in *Dictyostelium*. *TIG*, **7**, 381.

99. Simon, M., Driscoll, D., Mutzel, R., Part, D., Williams, J., and Veron, M. (1989) Overproduction of the regulatory subunit of the cAMP-dependent protein kinase blocks the differentiation of *Dictyostelium discoideum*. *EMBO J.*, **8**, 2039.

100. Pitt, G. S., Milona, N., Borleis, J., Lin., K. C., Reed, R. R., and Devreotes, P. N. (1992) Structurally distinct and stage specific adenyl cyclase genes play different roles in *Dictyostelium* development. *Cell*, **69**, 305.

101. Harwood, A. J., Hopper, N. A., Simon, M. N., Driscoll, D. M., Veron, M., and Williams, J. G. (1992) Culmination in *Dictyostelium* is regulated by the cAMP-dependent protein kinase. *Cell*, **69**, 615.

102. Firtel, R. A., van Haarstert, P. J. M., Kimmel, A. R., and Devreotes, P. (1989) G protein linked signal transduction pathways in development: *Dictyostelium* as an experimental system. *Cell*, **58**, 235.

103. Kumagi, A., Pupillo, M., Gundersen, R., Miake-Tye, R., Devreotes, P. N., and Firtel, R. A. (1989) Regulation and function of Gα proteins in *Dictyostelium*. *Cell*, **57**, 265.

104. Klein, P. S., Sun, T. J., Saxe, C. L., Kimmel, A. R., Johnson, R. L., and Devreotes, P. N. (1988) A chemoattractant receptor controls development in *Dictyostelium discoideum*. *Science*, **241**, 1467.

105. Johnson, R. L., Grundersen, R., Lilly, P., Pitt, G. S., Pupillo, M., Sun, T. J. *et al.* (1989) G-protein-linked signal transduction systems control development in *Dictyostelium*. *Development*, **107**, 75.

106. Kikkawa, U., Mann, S. K. O., Firtel, R. A., and Hunter, T. (1992) Molecular cloning of casein kinase II α subunit from *Dictyostelium discoideum* and its expression in the life cycle. *Mol. Cell. Biol.*, **12**, 5711.

107. Meisner, H. and Czech, M. P. (1991) Phosphorylation of transcriptional factors and cell cycle dependent proteins by casein kinase II. *Curr. Opin. Cell Biol.*, **3**, 474.

108. Hathaway, G. M. and Traugh, J. A. (1982) Casein kinases—multipotential protein kinases. *Curr. Top. Cell. Regul.*, **21**, 101.

109. Woodgett, J. R. (1991) A common denominator linking glycogen metabolism, nuclear oncogenes and development. *Trends Biochem. Sci.*, **19**, 177.

110. Burki, E., Anjard, C., Scholder, J. C., and Reymond, C. D. (1991) Isolation of two genes encoding putative protein kinases regulated during *Dictyostelium discoideum* development. *Gene*, **102**, 57.

111. Tan, J. L. and Spudich, J. A. (1990) Developmentally regulated protein-tyrosine kinase genes in *Dictyostelium discoideum*. *Mol. Cell. Biol.*, **10**, 3578.

112. Haribabu, B. and Dottin, R. P. (1991) Homology cloning of protein kinase and phosphoprotein phosphatase sequences of *Dictyostelium discoideum*. *Dev. Genet.*, **12**, 45.

113. Haribabu, B. and Dottin, R. P. (1991) Identification of a protein kinase multigene family of *Dictyostelium discoideum*: molecular cloning and expression of a cDNA encoding a developmentally regulated protein kinase. *Proc. Natl. Acad. Sci. USA*, **88**, 1115.

114. Anjard, C., Pinaud, S., Kay, R. R., and Reymond, C. D. (1992) Overexpression of *Dd*

PK2 protein kinase causes rapid development and affects the intracellular cAMP pathway of *Dictyostelium discoideum*. *Development*, **115**, 785.

115. Hunter, T. (1987) A thousand and one protein kinases. *Cell*, **50**, 823.

116. Firtel, R. A. and Chapman, A. L. (1990) A role for cAMP-dependent protein kinase A in early *Dictyostelium* development. *Genes Dev.*, **4**, 18.

117. Mann, S. K. O., Yonemoto, W. M., Taylor, S. S., and Firtel, R. A. (1992) DdPK3, which plays essential roles during *Dictyostelium* development, encodes the catalytic subunit of cAMP-dependent protein kinase. *Proc. Natl. Acad. Sci. USA*, **89**, 10701.

118. Jermyn, K., Berks, M., Kay, R., and Williams, J. (1987) Two distinct classes of prestalk-enriched messenger RNA sequences in *Dictyostelium discoideum*. *Development*, **100**, 745.

119. Williams, J., Ceccarelli, A., McRobbie, S., Mahbubani, H., Kay, R. R., Early, A. *et al.* (1987) Direct induction of *Dictyostelium* prestalk gene expression by DIF provides evidence that DIF is a morphogen. *Cell*, **49**, 185.

120. McRobbie, S. J., Jermyn, K. A., Duffy, K., Blight, K., and Williams, J. G. (1988) Two DIF-inducible, prestalk specific mRNAs of *Dictyostelium* encode extracellular matrix proteins of the slug. *Development*, **104**, 275.

121. Hoekstra, M. F., Demaggio, A. J., and Dhillon, N. (1991) Genetically, identified kinases in yeast. I: transcription, translation, transport and mating. *TIG*, **7**, 256.

122. Hoekstra, M. F., Demaggio, A. J., and Dhillon, N. (1991) Genetically, identified kinases in yeast. II: DNA metabolism and meiosis. *TIG*, **7**, 293.

123. Sprague, G. F. (1991) Signal transduction in yeast mating. *TIG*, **7**, 393.

124. Marsh, L., Neiman, A. M., and Herskowitz, I. (1991) Signal transduction during pheromone response in yeast. *Ann. Rev. Cell. Biol.*, **7**, 699.

125. Hartwell, L. H. (1980) Mutants of *Saccharomyces cerevisiae* unresponsive to cell division control by polypeptide mating hormone. *J. Cell. Biol.*, **85**, 811.

126. Teague, M. A., Chaleff, D. T., and Errede, B. (1986) Nucleotide sequence of the yeast regulatory gene *STE7* predicts a protein homologous to protein kinases. *Proc. Natl. Acad. Sci. USA*, **83**, 7371.

127. Elion, E. A., Grisafi, P. L., and Fink, G. R. (1990) *FUS3* encodes a *cdc2/CDC28*-related kinase required for the transition from mitosis to conjugation. *Cell*, **60**, 649.

128. Fujimura, H. (1990) Molecular cloning of the *DAC2/FUS3* gene essential for pheromone induced G1-arrest of the cell cycle in *Saccharomyces cerevisiae*. *Curr. Genet.*, **18**, 395.

129. Rhodes, N., Conell, L., and Errede, B. (1990) Ste11 is a protein kinase required for cell type specific transcription and signal transduction in yeast. *Genes Dev.*, **4**, 1862.

130. Courchesne, W. E., Kunisawa, R., and Thorner, J. (1989) A putative protein kinase overcomes pheromone-induced arrest of cell cycling in *S. cerevisiae*. *Cell*, **58**, 1107.

131. Leberer, E., Dignard, D., Harcus, D., Thomas, D. Y., and Whiteway, M. (1992) The protein kinase homologue Ste20 is required to link the yeast pheromone response G-protein βγ subunits to downstream signalling components. *EMBO J.*, **11**, 4815.

132. Cairns, B. R., Ramer, S. W., and Kornberg, R. D. (1992) Order of action of components in the yeast pheromone response pathway revealed with a dominant allele of the Ste11 kinase and the multiple phosphorylation of the Ste7 kinase. *Genes Dev.*, **6**, 1305.

133. Stevenson, B. J., Rhodes, N., Errede, B., and Sprague, G. F. (1992) Constitutive mutants of the protein kinase Ste11 activate the yeast pheromone response pathway in the absence of the G protein. *Genes Dev.*, **6**, 1293.

134. Gartner, A., Nasmyth, K., and Ammerer, G. (1992) Signal transduction in *Saccharomyces cerevisiae* requires tyrosine and threonine phosphorylation of Fus3 and Kss1. *Genes Dev.*, **6**, 1280.

135. Zhou, Z., Gartner, A., Cade, R., Ammerer, G., and Errede, B. (1993) Pheromone-induced signal transduction in *Saccharomyces cerevisiae* requires the sequential function of three protein kinases. *Mol. Cell. Biol.*, **13**, 2069.

136. Errede, B., Gartner, A., Zhou, Z., Nasmyth, K., and Ammerer, G. (1993) MAP kinase-related Fus3 from *S. cerevisiae* is activated by Ste7 *in vitro*. *Nature*, **362**, 261.

137. Dolan, J. W., Kirkman, C., and Fields, S. (1989) The yeast Ste12 protein binds to the DNA sequence mediating pheromone induction. *Proc. Natl. Acad. Sci. USA*, **86**, 5703.

138. Dolan, J. W. and Fields, S. (1990) Overproduction of the yeast Ste12 protein leads to constitutive transcriptional activation. *Genes Dev.*, **4**, 492.

139. Errede, B. and Ammerer, G. (1989) Ste12, a protein involved in cell type specific transcription and signal transduction in yeast is part of protein–DNA complexes. *Genes Dev.*, **3**, 1349.

140. Song, O. K., Dolan, J. W., Yuan, Y. I. O., and Fields, S. (1991) Pheromone-dependant phosphorylation of the yeast Ste12 protein correlates with transcriptional activation. *Genes Dev.*, **5**, 741.

141. Toda, T., Shimanuki, T., and Yanagida, M. (1991) Fission yeast genes that confer resistance to staurosporine encode an AP-1-like transcription factor and a protein kinase related to the mammalian ERK1/MAP2 and budding yeast Fus3 and Kss1 kinases. *Genes Dev.*, **5**, 60.

142. Elion, E. A., Brill, J. A., and Fink, G. R. (1991) *FUS3* represses *CLN1* and *CLN2* and in concert with *KSS1* promotes signal transduction. *Proc. Natl. Acad. Sci. USA*, **88**, 9392.

143. Moll, T., Tebb, G., Surana, U., Robitsch, H., and Naysmith, K. (1991) The role of phosphorylation and the *CDC28* protein kinase in cell cycle regulated nuclear import of the *S. cerevisiae* transcription factor *SWI5*. *Cell*, **66**, 743.

144. Tyres, M., Tokiwa, G., Nash, R., and Futcher, B. (1992) The Cln3–Cdc28 kinase complex of *Saccharomyces cerevisiae* is regulated by proteolysis and phosphorylation. *EMBO J.*, **11**, 1773.

145. Goebl, M. G. and Wirey, M. (1991) The yeast cell cycle. *Curr. Opin. Cell Biol.*, **3**, 242.

146. Norbury, S. L. and Nurse, P. (1991) Cell cycle regulation in the yeasts *Saccharomyces cerevisiae* and *Schizosaccharomyces pombe*. *Ann. Rev. Cell Biol.*, **7**, 227.

147. Tyres, M. and Futher, B. (1993) Far1 and Fus3 link the mating pheromone signal transduction pathway to three G1-phase Cdc28 kinase complexes. *Mol. Cell Biol.*, **13**, 5659.

148. Sprague, G. F. (1992) Kinase cascade conserved. *Current Biology*, **2**, 587.

149. Lee, K. S. and Levin, D. E. (1992) Dominant mutations in a gene encoding a putative protein kinase (*BCK1*) bypass the requirement for a *Saccharomyces cerevisiae* protein kinase C homolog. *Mol. Cell. Biol.*, **12**, 172

150. Irie, K., Takase, M., Lee, K. S., Levin, D. E., Araki, H., Matsumoto, K., and Oshima, Y. (1993) *MKK1* and *MKK2*, which encode *Saccharomyces cerevisiae* mitogen-activated protein kinase-kinase homologs, function in the pathway mediated by protein kinase C. *Mol. Cell. Biol.*, **13**, 3076.

151. Errede, B. and Levin, D. E. (1993) A conserved kinase cascade for MAP kinase activation in yeast. *Curr. Opinions in Cell Biol.*, **5**, 254.

152. Lee, K. S., Irie, K., Gotoh, Y., Watanabe, Y., Araki, H., Nishida, E., Matsumoto, K., and Levin, D. E. (1993) A yeast mitogen-activated protein kinase homolog (Mpk1p) mediates signalling by protein kinase C. *Mol. Cell. Biol.*, **13**, 3067.

153. Torres, L., Martin, H., Garcia-Saez, M. I., Arroyo, J., Molina, M., Sanchez, M., and

Nombela, C. (1991) A protein kinase gene complements the lytic phenotype of *Saccharomyces cerevisiae lyt2* mutants. *Mol. Microbiol.*, **5**, 2845.

154. Brewster, J. L., de Valoir, T., Dwyer, N. D., Winter, E., and Gustin, M. C. (1993) An osmo-sensing signal transduction pathway in yeast. *Science*, **259**, 1760.

155. Boguslawski, G. and Polazzi, J. O. (1987) Complete nucleotide sequence of a gene conferring polymixin B resistance on yeast: similarity of the predicted polypeptide to protein kinases. *Proc. Natl. Acad. Sci. USA*, **84**, 5848.

156. Broach, J. (1991) *Ras* genes in *Saccharomyces cerevisiae*: signal transduction in search of a pathway. *Trends Genet.*, **7**, 28.

157. Camorus, J. H., Halekine, M., Gondre, B., Garreau, H., Boy-Marcotte, E., and Jacquet, M. (1986) Characterisation, cloning and sequence of the *CDC25* gene which controls the cyclic AMP level in *Saccharomyces cerevisiae*. *EMBO J.*, **5**, 375.

158. Mitsuzawa, H., Uno, I., Oshima, T., and Ishikawa, T. (1989) Isolation and characterisation of temperature-sensitive mutations in the *Ras 2* and *CYR1* genes of *Saccharomyces cerevisiae*. *Genetics*, **123**, 739.

159. Toda, T., Cameron, S., Sass, P., Zoller, M., Scott, J. D., McMullen, B. *et al.* (1987) Cloning and characterisation of *BCY1*, a locus encoding a regulatory subunit of the cyclic AMP-dependent protein kinase in *Saccharomyces cerevisiae*. *Mol. Cell. Biol.*, **7**, 1371.

160. Smith H. E. and Mitchell, A. P. (1989) A transcriptional cascade governs entry into meiosis in *Saccharomyces cerevisiae*. *Mol. Cell Biol.*, **9**, 2142.

161. Matsuura, A., Treinin, M., Mitsuzawa, H., Kassir, Y., Uno, I., and Simchen, G. (1990) The adenylate cyclase/protein kinase cascade regulates early entry into meiosis in *Saccharomyces cerevisiae* through the gene *IME1*. *EMBO J.*, **9**, 3225.

162. Simchen, G. and Hassir, Y. (1989) Genetic regulation of differentiation towards meiosis in the yeast *Saccharomyces cerevisiae*. *Genome*, **31**, 95.

163. Mitchell, A. P. and Herkowitz, I. (1986) Activation of meiosis and sporulation by repression of the *RME1* product in yeast. *Nature*, **319**, 738.

164. Mitchell, A. P., Driscoll, S. E., and Smith, H. E. (1990) Positive control of sporulation specific genes by the *IME1* and *IME2* products in *Saccharomyces cerevisiae*. *Mol. Cell Biol.*, **10**, 2104.

165. Shero, J. H. and Hieter, T. (1991) A supressor of a centromere DNA mutation encodes a putative protein kinase (*MCK1*). *Genes Dev.*, **5**, 549.

166. Neigeborn, L. and Mitchell, A. P. (1991) The yeast *MCK1* gene encodes a protein kinase homologue that activates early meiotic gene expression. *Genes Dev.*, **5**, 533.

167. Shar, J. C. and Clancy, M. J. (1992) *IME4*, a gene that modulates MAT and nutritional control of meiosis in *Saccharomyces cerevisiae*. *Mol. Cell Biol.*, **12**, 1078.

168. Lowenstein, E. J., Daly, R. J., Batzer, A. G., Li, W., Margolis, B., Lammers, R. *et al.* (1992) Close similarity between an EGF receptor binding protein GRB2 and the C. *elegans* signal transduction protein *sem-5* suggests a mechanism for growth factor control of *ras* signaling. *Cell*, **70**, 431.

Index